STUDY GUIDE
to accompany

LIFE
AN INTRODUCTION TO BIOLOGY

STUDY GUIDE

to accompany

LIFE

AN INTRODUCTION TO BIOLOGY

Third Edition

BECK • LIEM • SIMPSON

by

KENNETH T. WILKINS
Baylor University

HarperCollins*Publishers*

For permission to use copyrighted material, grateful acknowledgment is made to the copyright holders:

Page 44, Figure 5.1; *page 51*, Figure 5.3; *page 57*, Figure 6.1: Copyright © 1991 by Kristine Rassmussen. Used with permission.

Page 74, Labelling Exercise: From S. M. Dietz, *Laboratory and Field Problems in Botany 1*. Copyright © 1959 Wm. C. Brown, Publishers, Dubuque, Iowa.

Page 117, Discussion Question 6: From *Human Biology: Form, Function, & Adaptation*, by William DeWitt. Copyright © 1989, Scott, Foresman. Reprinted with permission.

Page 184, Discussion Question 2: From *Biology*, 1st ed., by Claude A. Villee and Eldra Solomon, Figure 14-9, page 307. Copyright © 1985, Saunders College Publishing. Reprinted with permission.

Page 257, Figure 23.1: From *Zoology* by Mitchell, Mutchmor, and Dolphin. Copyright © 1988, Benjamin/Cummings Publishing Company, Menlo Park, Calif. Reprinted with permission.

Page 276, Discussion Question 5: From *Principles of Anatomy and Physiology*, 6th ed., by Gerard J. Tortora and Nicholas P. Anagnostakos, Figure 20-2. Copyright © 1990 Biological Sciences Textbooks, Inc., A & P Textbooks, Inc., and Elia-Sparta, Inc. Reprinted with permission.

Page 429, Figure 38.2 (top): David M. Phillips/Visuals Unlimited.

Page 429, Figure 38.2 (bottom): Runk/Schoenberger/Grant Heilman.

Page 437, Figure 39.1: L. E. Roth, Univ. of Tennessee/BPS.

Page 484, Figure 43.2 (bipinnaria): From *Zoology* by Mitchell, Mutchmor, and Dolphin. Copyright © 1988, Benjamin/Cummings Publishing Company, Menlo Park, Calif. Reprinted with permission.

Page 536, Figure 48.2: From *Basic Concepts of Ecology* by C. B. Knight, Figures 2-13 and 2-14, page 667. Copyright © 1965, by Macmillan Publishing Company. Reprinted with permission.

STUDY GUIDE to accompany **LIFE:** An Introduction to Biology, *Third Edition*

Copyright © 1991 by HarperCollins Publishers Inc.

All rights reserved. Printed in the United States of America. No part of this book may be used or reproduced in any manner whatsoever without written permission, except in the case of brief quotations embodied in critical articles and reviews. For information address HarperCollins Publishers Inc., 10 East 53rd Street, New York, NY 10022.

ISBN: 0-06-040606-2

91 92 93 94 9 8 7 6 5 4 3 2 1

CONTENTS

PREFACE v

PART 1 THE SCIENCE OF BIOLOGY   1
    1 THE WORLD OF LIFE 1
    2 THE HISTORY AND SCOPE OF BIOLOGY 9

PART 2 BIOCHEMICAL AND CELLULAR BASIS OF LIFE   16

    SECTION 1 CHEMICAL FUNDAMENTALS 16
    3 ATOMS AND MOLECULES 16
    4 ORGANIC MOLECULES IN LIVING ORGANISMS 29

    SECTION 2 CELLS: UNITS OF LIFE 41
    5 CELL STRUCTURE 41
    6 MEMBRANES AND CELL WALLS 55
    7 CELL FUNCTIONS 66

    SECTION 3 ENERGY AND ENERGETICS 77
    8 METABOLISM AND THE TRANSFER OF ENERGY 77
    9 HOW CELLS EXTRACT ENERGY FROM THE ENVIRONMENT 90
    10 CAPTURING THE SUN'S ENERGY: PHOTOSYNTHESIS 101
    11 BIOSYNTHESIS OF CELL CONSTITUENTS 113

    SECTION 4 HEREDITY AND GENETICS 120
    12 HOW CELLS REPRODUCE 120
    13 GENETICS 133
    14 THE PHYSICAL BASIS OF HEREDITY 142
    15 SEX AND HUMAN HEREDITY 154
    16 THE NATURE OF GENES 165
    17 MECHANISMS OF GENE ACTIVITY 177
    18 REGULATION OF GENE ACTIVITY 189
    19 DEVELOPMENT OF FORM IN THE ANIMAL BODY 202
    20 DEVELOPMENTAL MECHANISMS 215
    21 HOW ORGANISMS REPRODUCE 226

PART 3 HOMEOSTASIS 243

SECTION 1 DEFENDING AGAINST ADVERSITY 243
22 BLOOD 243
23 IMMUNITY 255

SECTION 2 MAINTAINING INTERNAL CONSTANCY 267
24 TRANSPORT 267
25 EXCRETION 280
26 NUTRITION AND METABOLISM 293
27 RESPIRATION 307

SECTION 3 INTEGRATION AND COORDINATION 317
28 CHEMICAL COORDINATION 317
29 NEURAL COORDINATION 333
30 NERVOUS SYSTEMS 344
31 EFFECTORS 357
32 BEHAVIOR 369
33 SOCIAL ASPECTS OF BEHAVIOR 379

PART 4 BIOLOGY OF ORGANISMS AND POPULATIONS 387

SECTION 1 EVOLUTION 387
34 ELEMENTARY PROCESSES OF EVOLUTION 387
35 ADAPTATION 398
36 ORIGIN OF SPECIES 407

SECTION 2 THE DIVERSITY OF LIFE 417
37 PRINCIPLES OF CLASSIFICATION 417
38 KINGDOM OF THE SMALL: MONERA 427
39 PROTISTS 436
40 FUNGI 447
41 PLANTS 457
42 INVERTEBRATE ANIMALS 469
43 CHORDATES AND ANIMAL PHYLOGENY 480
44 HOW DIVERSITY AROSE: THE HISTORY OF LIFE 493

SECTION 3 ECOLOGY 503
45 POPULATIONS 503
46 ECOSYSTEMS 512
47 COMMUNITIES 524
48 THE GEOGRAPHY OF LIFE 534

SECTION 4 HUMAN LIFE 546
49 HUMAN POPULATIONS AND ENVIRONMENTS 546

APPENDIXES 554
ANSWERS TO MATCHING QUESTIONS 554
ANSWERS TO TESTING YOUR UNDERSTANDING 555

PREFACE

Welcome to the wonderful world of biology! Of all the subjects you will study during your college career, biology clearly is the most relevant to your life. Not only can study of biology help prepare you for a career, but it will also help you understand how you as an organism interact with the other living and nonliving components of the world around you.

Be forewarned, however. Biology may not be an easy subject to master. You will need to devote much quality time and intellectual energy to study of the subject. You will need to learn a new vocabulary of biological terminology. But success in this field (as in most others) requires much more than memorization of jargon. Success requires that you learn how to work with the terminology to learn the concepts and principles that underlie the discipline of biology.

It is my hope that this study guide will facilitate your learning and appreciation of this fascinating field. For this study guide to help you, you must use it actively. Although it is not intended to be a substitute for the textbook that it accompanies, you will find it beneficial to read its appropriate sections (Chapter Overview and Topic Summaries) and the text *before* a lecture or class or lab meeting on a particular topic. Such "prereading" will give you an idea of topics and terms that will be addressed in the classroom. During class, you should take general notes while you concentrate on the lecture or discussion. Then, soon after class, you should reread the study guide and spend time with the text to obtain detailed information on the topics discussed in class.

After reading, your active use of the study guide should involve definition of terms (in each chapter, Terms to Understand). Then, be sure that you go beyond mere definitions to learn the significance of the terms to the principles and concepts treated in the chapter. Next, a series of matching exercises and chart exercises provides an opportunity to use the terminology and to summarize information presented in the chapter. Each chapter also includes several discussion questions; some of these can be answered directly by drawing information from the reading, whereas others require synthetic thinking, often drawing on information presented in previous chapters.

A section of multiple-choice/true-false questions (Testing Your Understanding) concludes each chapter. From each set of four statements, you identify the false statement, and then you provide a corrected version of the false statement. Answers to these and to the matching questions are provided in the appendixes.

You must realize that your ability to answer correctly all of the questions provided in the guide does not necessarily mean that you know all of the material, but rather that you have a reasonable understanding of the particular topics that the questions addressed. I encourage you to use the exercises as springboards to develop your own study questions on other topics. Be aware, too, that each professor has his or her own philosophy and approach to the subject. So, be prepared to modify your use of this study guide to match the approach used by your professor. The understanding and knowledge of this subject that you obtain, and the satisfaction that you derive, will be directly related to the effort that you invest. Best wishes!

I would like to acknowledge the many people who figured prominently in the completion of this study guide. I thank Randy Moore for providing the opportunity. I thank students and fellow faculty members at Baylor University for continuing intellectual stimulation. Reviewers and editors at HarperCollins provided many forms of indispensible assistance. Special thanks goes to David Fox for his contribution. Most significantly, my wife, Chris, and my son, Brent, continually offered the combination of encouragement and patience that is required for such a project to be completed successfully

Kenneth T. Wilkins

STUDY GUIDE

to accompany

LIFE

AN INTRODUCTION TO BIOLOGY

CHAPTER 1--THE WORLD OF LIFE

CHAPTER OVERVIEW

Life, ranging from microscopic one-celled organisms to immense multicellular organisms, is all around us. Life exists in many environmental settings: on land, in water, in air. Three such views are seen in Chapter 1; these and other situations share many features.

We humans are but one species among millions of species, and our existence depends on many other species. Indeed, the continued existence of many other species depends upon the lack of carelessness on the part of humans. The diverse array of form, size, behavior, and ability seen among organisms is the result of adaptation to particular environmental conditions. Such diversity has resulted from evolutionary experimentation over long periods of time-- hundreds of millions to billions of years.

TOPIC SUMMARIES

A. <u>"Know Thyself"</u>--The human species is <u>not</u> separable from its physical and biological surroundings. We obtain energy, nutrients, and other materials only by consuming other organisms (or their products). Human welfare depends on our understanding of how we interact with other species and with the non-living environment. The text presents three views of life (forest, coral reef, icecap) to illustrate this dependent and interactive perspective of life.

B. <u>Common Elements in the Sketches</u>.--Despite their differences, forests, reefs, polar icecaps, and all other ecosystems share many important features:

(1) Organization: Living and non-living parts of a system are not randomly thrown together, but instead <u>exist in orderly arrangements</u>. Such organization exists within an organism itself and beyond the level of the organism as well (Fig. 1.1). A cell is the lowest organizational level that can be "alive." Table 1-1 in the text lists five levels below the cellular level: subatomic particles, atoms, molecules, macromolecules, and cellular organelles. Living levels beyond the cell are tissues (muscle tissue made of muscle cells), then organs (a muscle such as the biceps brachii of the upper arm), then systems of organs (muscular system), and finally the entire organism. Groups of organisms are progressively grouped into populations, evolutionarily related (phyletic) lineages, and then communities. The most complex and inclusive level of organization is the entire living world positioned in space and time. Biology is studied at each level.

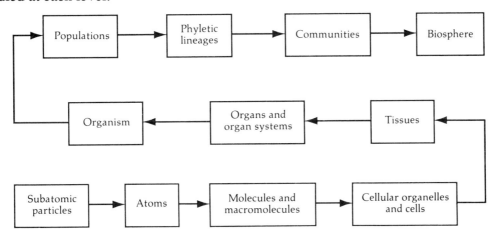

Fig. 1.1: Levels of biological organization

(2) Exchange of Materials and Energy: Every living cell (and, consequently, every organism) requires a continual supply of energy and of materials (vitamins, minerals, etc.). Most plants (and a few other organisms) obtain energy and materials from the physical environment (i.e., sun, soil, air), but others obtain energy and materials by consuming these plants or other organisms positioned lower on the <u>food chain</u>. Each species occupies a more-or-less fixed position in such an energy-and-materials flow scheme. Most energy captured by organisms originates as sunlight, which continuously strikes our planet. Nutrients and minerals of many kinds, however, are available on Earth in limited amounts. They are <u>recycled</u> between the biological and physical environments.

(3) Self-Regulation and Control: The orderly arrangement of life at any level of organization automatically and continuously tends to degenerate into disorder. Living organisms can be thought of as entities which resist this loss of order. <u>Homeostasis</u> is the maintenance of order in an organism despite the vastly changing nature of the environmental factors affecting the organism. For example, an animal might spend energy to maintain a suitable body temperature when environmental temperature is lower (by shivering) or higher (by panting) than is suitable. Self-regulation in animals often involves the nervous system or hormonal (endocrine) systems.

(4) Reproduction: All species can produce new individuals of that species. Reproduction does more than just <u>replace</u> the aging and ill and individuals lost in other ways. Reproduction involves various processes by which <u>variation</u> arises: new individuals may have different combinations of characters than seen in parents (or even new characters), some of which may allow individuals to be <u>better adapted</u> to changing environmental conditions. This is a critical aspect of the process of evolution.

(5) Evolutionary Change: All things, living and non-living, change with the passage of time. Continents move, mountains rise and erode, sea level rises and falls, glaciers advance and shrink. Species change from one generation to the next, and in time adapt to the changing environment. Such changes usually require vast periods of time--thousands or many millions of years. These changes are almost invisible over the relatively brief human life span. Much of the biologist's understanding of <u>evolutionary change</u> results from study of the <u>fossil record</u>. Adaptation through evolutionary change has produced the diversity of species now residing on Earth. All modern species evolved from one (or more) common ancestral species that lived billions of years ago.

(6) Responsiveness and Behavior: All individuals of all species respond to changes in their surroundings. They engage in various activities which allow acquisition of energy, materials, mates, or otherwise enable that organism to continue its own (or its species') existence.

(7) Geography of Life: Biological and physical events occur over space and in three dimensions. A species does not normally exist just anywhere, but can be found in (some of the) places where suitable conditions (temperature, soil type, sunlight, etc.) exist. Hence, the general <u>geographic distribution</u> of penguins is the cold waters of parts of the southern hemisphere. You might realize (and correctly so) that the northern hemisphere also has correspondingly cold waters in and near Arctic regions. Why then do no penguins live in high northern latitudes? Perhaps they could <u>if</u> they had access to the region. Possibly, hot equatorial waters keep the penguins to the south. Or, perhaps the presence of predators makes the Arctic inhospitable. The geography of life is affected by many factors.

(8) History of Life: As indicated above, life exists in time. Fossils document the existence of life several billions of years ago. Fossils also allow us to trace life's history of change and to determine evolutionary relationships (i.e., which species gave rise to which

other species). There is only one history of life on Earth; the ultimate goal of evolutionary biologists is to unravel the clues required to learn that history and to understand the principles and mechanisms of change.

TERMS TO UNDERSTAND: Define and identify the following terms and names. How do they apply to the field of biology as presented in Chapter 1?

biology

biosphere

lithosphere

hydrosphere

atmosphere

geology

solar radiation

climate

herbivorous

carnivorous

food chain

environment

cycling of matter

metabolism

energy

photosynthesis

homeostasis

evolution

adaptation

diversification

behavior

ecology

PEOPLE TO KNOW: Identify each of the following individuals and summarize their major contributions to biology as presented in Chapter 1.

Walter B. Cannon

Charles Darwin

MATCHING EXERCISES

Different fields of biology examine life at different levels of organization. Match the fields of biology with the appropriate levels of organization. More than one field of biology might apply to more than one level of organization.

LEVEL OF ORGANIZATION	FIELD OF BIOLOGY

_____ 1. tissues	A. physiology

_____ 2. population	B. cytology

_____ 3. organism	C. ecology

_____ 4. community	D. anatomy

_____ 5. molecules	E. evolution

_____ 6. cell	F. histology

_____ 7. organelles	G. biochemistry

CHART EXERCISES

Complete the following chart to summarize how the "common elements" are seen in each of three views of life. Draw on information presented in Chapter 1 as well as from your own experience to complete the chart. Two cells of the chart are completed for you.

COMMON ELEMENTS	FOREST	CORAL REEF	POLAR ICECAP
Organization			
Materials and Energy Exchange	Carnivores (mountain lions) feed on herbivores (deer).		
Self-Regulation and Control			
Reproduction			
Evolutionary Change			
Patterns of Behavior			
Geographical Limits			
History		Corals progressively build the reef.	

DISCUSSION QUESTIONS

1. What is the relevance of the phrase "know thyself" to the study of biology?

2. List, in sequence from the least to most inclusive, the levels of organization characteristic of living systems. Give an example of each level.

3. What is the ultimate source of energy for life on Earth? Give an example of how energy flows in an ecosystem.

4. What is a geological <u>plate</u>? What is the importance of plate movement to the study of biology?

TESTING YOUR UNDERSTANDING

For each of the test items below, three of the lettered alternatives are true and the other one is false. Determine which alternative is false and write its letter in the blank to the left of the question number. On the blank line below alternative D, write a corrected version of the false statement.

_____ 1. A. Evolution can be studied at any level of biological organization.
B. The population level is more inclusive than the community level.
C. The cell, tissue and organ levels are levels at which anatomy and physiology can be studied.
D. The organ level is more inclusive than is the tissue level.

Correction: _____

_____ 2. A. Humans can be considered to be occupants primarily of the lithosphere.
B. All habitats in which Earth's living organisms normally occur are subsets of the biosphere.
C. Solar radiation must pass through the atmosphere before it reaches the hydrosphere.
D. Plankton reside in various habitats of the lithosphere.

Correction: _____

_____ 3. A. The Earth's crust is made of a series of semi-rigid plates upon which the continents are positioned.
B. Geological evidence indicates that the continents have held approximately their same geographic locations over the past 200 million years.
C. Collision of slowly moving plates can produce mountain ranges such as the Himalayas.
D. The rate of plate movement is only a few centimeters per year.

Correction: _____

_____ 4. A. In the food chain of a coral reef, jellyfishes feed on corals which feed on plankton.
B. Hawks are carnivores whereas squirrels are herbivores.
C. Planet Earth is essentially a "closed system" with regard to minerals and nutrients because they are present in limited amounts.
D. As pertains to energy, our planet is an "open system" because the sun supplies a constant stream of radiant energy.

Correction: _____

_____ 5. A. Earth is approximately 4.5 billion to 5 billion years old.
B. The first appearance of life on Earth probably occurred over three eons ago.
C. Although the waters of the oceans were originally fresh, they became salty due to erosion of salt from the land.
D. The core of Earth, although molten during its early history, has cooled and hardened.

Correction: _____

_____ 6. A. Energy can be defined as the capacity to do work.
B. Photosynthesis is the process by which solar radiation is incorporated into biological tissues.
C. Photosynthesis is an example of a metabolic process.
D. Although animals require energy to carry out their metabolic processes, plants do not require energy for body metabolism because they can perform photosynthesis.

Correction: _____

_____ 7. A. To maintain homeostasis, an organism relies on many different mechanisms of self-regulation and control.
B. Maintenance of order within an organism's body is accomplished by several different organ systems, each of which operates independently of other organ systems.
C. One of the means by which invasion of the body by foreign matter is combatted is by use of antibodies.
D. Homeostasis can be defined as an organism's efforts to restore its equilibrium when it has been disturbed.

Correction: _____

_____ 8. A. A fundamental role of reproduction is the continuation of a species.
B. A fundamental role of reproduction is the production of genetic variability.
C. The phrase "like begets like" means that offspring have the same genetic make-up that their parents have.
D. Genetic variability within a species is important because it allows adaptation to changing environmental conditions.

Correction: _____

_____ 9. A. Humans are relative newcomers in the course of Earth's history.
B. Deserts which occur on different continents usually are inhabited by the same species of plants on each of the continents.
C. The movements of plates of the Earth's crust probably has affected the geographic distributions of many kinds of organisms.
D. Fossils are useful in piecing together the evolutionary history of life on Earth.

Correction: _____

CHAPTER 2--THE HISTORY AND SCOPE OF BIOLOGY

CHAPTER OVERVIEW

Earliest human efforts toward understanding biological phenomena were shrouded with mystery and superstition. This perspective often entailed viewing Earth's species as being arranged in a "ladder of nature," with humans residing at the pinnacle. Modern biologists see each species as occupying its own summit. Modern biological knowledge is gained through application of the scientific method.

During the early centuries of biological study, most scientific effort was spent in arranging the known organisms into classifications. More recently, however, most biologists have gone beyond this fundamental level of naming and description of species to learn how organisms function and how they interact with other organisms and with their environment. Our present state of biological knowledge is the product of contributions by many individuals.

TOPIC SUMMARIES

A. <u>Science and the Scientific Method</u>.--Our earliest recorded knowledge of biology stems from the studies and writings of ancient philosophers such as Aristotle. The contributions made by the <u>ancients</u> were mainly descriptions of observed natural phenomena; little of what they observed, however, did they understand. Progress towards true understanding of the natural world came with the development of experimental and empirical methods during the 15th through 19th centuries in western European culture.

<u>Science</u> is an activity in which questions about the world are answered not by beliefs or attitudes, but instead by <u>observation and experimentation</u>. Observation of a phenomenon normally leads to a tentative explanation (an <u>hypothesis</u>) for the observed events. Hypotheses must be <u>testable and falsifiable</u>; they must be constructed so that relevant information (data) can be collected by measurement or experimentation that can refute the hypothesis, and, thereby prove it wrong. If an hypothesis is thusly disproven, it is rejected and modified for further evaluation.

Each field of study experiences several stages in its progress towards being a full-fledged science. Hence, the logical first step is to identify the "players" by answering questions such as: What kinds of plants occur in Hawaii? or What are the parts that make up a cell? Listing and then description of these players is the next step. The preceding steps often are referred to as <u>natural history</u>. This is followed by <u>classification</u> of the described items to impart some sort of order to (or to discover the existing relationships among) the array of items. Once these steps have been done, then begins scientific pursuit of the "how" questions to reveal the mechanisms by which biological processes occur.

One of the great values of science is that it can be <u>predictive</u>; understanding a phenomenon on a small scale can allow prediction of events on a large scale. As scientists, we often study events in one species (a model) and then apply those findings more generally. Hence, medical researchers study physiological effects of drugs in lab rodents, rabbits, or other model species in efforts to anticipate effects of drugs in humans.

B. <u>The Span of Biology</u>.--Biology is the study of life. It includes many fields of study. One approach to "cutting the cake" of biology is according to the <u>kinds of organisms</u> (layers of cake) studied (Fig. 2.1). Another approach is to study <u>processes</u> (slices which cross each layer) or other aspects of biology characteristic of all kinds of organisms. Most biologists operate in a middle ground in which they study certain processes in a certain group of organisms.

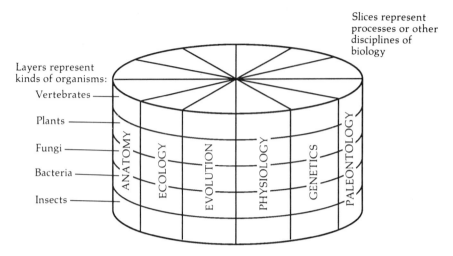

Fig. 2.1: A sliced layer-cake model of fields of biological study

C. Living Organisms: Are They Machines?--Not only does each empirical field of biology have a history, but the philosophical views of biology also have a history. Several incongruent views regarding the essence and explicability of life prevailed in the 19th century: The scientifically-inclined people held the mechanistic view that life can be explained in terms of physical and chemical laws. The opposing vitalists argued that life has a special essence and therefore cannot be reduced to a mass of chemicals. Finalism is a doctrine that the evolution of life progresses towards a predetermined endpoint.

D. Is There a Ladder of Nature?--Organisms are often referred to as having "higher" or "lower" positions in a hierarchy of species with the implication that higher forms are "better" than lower ones. This concept of a "ladder of nature," called scala naturae, has prevailed since the time of Aristotle. Indeed, we generally place our own species (Homo sapiens) at the pinnacle of this ladder. With the arrival of an evolutionary view of nature, however, most scientists discarded scala naturae, realizing instead that each species tends to be adequately (perhaps even admirably) adapted to its particular position in nature.

TERMS TO UNDERSTAND: Define and identify the following terms and names. How do they apply to the field of biology as presented in Chapter 2?

scientific method

hypothesis

natural history

classification

systematics

binomial

science

technology

observation

mechanism

vitalism

finalism

organicism

holism

scala naturae

H. M. S. Beagle

Galapagos

theory

PEOPLE TO KNOW: Identify each of the following individuals and summarize their major contributions to biology as presented in Chapter 2.

Aristotle

Charles Darwin

John Ray

Carolus Linnaeus

Conte de Buffon

Georges Cuvier

Rene Descartes

Henri Bergson

Hans Driesch

William Harvey

Plato

Francis Bacon

Immanuel Kant

Bonnet

Jean Baptiste Lamarck

Goethe

MATCHING EXERCISES

Match the biological field of study with its definition.

FIELD OF BIOLOGY		DEFINITION
____	1. systematics	A. the study of mankind
____	2. protistology	B. the study of birds
____	3. ichthyology	C. the study of protists
____	4. botany	D. the study of cultivated plants and domesticated animals in the context of food production

_____ 5. psychology   E. the study of animals

_____ 6. mycology   F. the study of fungi

_____ 7. ornithology   G. the study of bacteria

_____ 8. anthropology   H. the study of human disease and its treatment

_____ 9. zoology   I. the study of plants

_____ 10. malacology   J. the study of amphibians and reptiles

_____ 11. herpetology   K. the study of thought and behavior

_____ 12. bacteriology   L. the study of mollusks

_____ 13. medicine   M. the study of the diversity of organisms

                               N. the study of fishes

CHART EXERCISES

Fill in the chart to indicate the steps in the scientific method and what activity occurs in each step.

STEP	ACTIVITY
1.	
2.	
3.	
4.	
5.	

DISCUSSION QUESTIONS

1. What is the importance of classification to a discipline such as biology?

2. Write a one-paragraph sketch of Charles Darwin's life. What were the three primary contributions of his book, *The Origin of Species by Means of Natural Selection*, to biology?

3. Explain the concept of vitalism. Compare vitalism with mechanism and with finalism.

TESTING YOUR UNDERSTANDING

For each of the test items below, three of the lettered alternatives are true and the other one is false. Determine which alternative is false and write its letter in the blank to the left of the question number. On the blank line below alternative D, write a corrected version of the false statement.

_____ 1. A. The study of birds can correctly be called ornithology or zoology.
B. Systematics is the study of the diversity of organisms.
C. A scientist who studies snails is a malacologist.
D. A herpetologist studies reptiles and fishes.

Correction: _____

_____ 2. A. Agriculture is a field of study in which a botanist might work.
B. A physician is also a zoologist.
C. Mycology is a field in which human thought and behavior are studied.
D. Protists are the organisms studied by a protistologist.

Correction: _____

_____ 3. A. Well-designed scientific experiments can be conducted only in a laboratory.
B. The scientific method involves testing of hypotheses which can be falsified.
C. An hypothesis is a tentative explanation of available facts.
D. A scientific theory cannot be based on beliefs or superstitions that cannot be tested empirically.

Correction: _____

_____ 4. A. Hypotheses and theories can be proven to be correct explanations from observed phenomena.
  B. A theory is an hypothesis that has been tested often enough to convince scientists that the theory is probably correct.
  C. Teleonomic explanations relate biological structures and functions to their apparent purposes.
  D. An example of a "physiologic" explanation is that circulation of the blood is caused by contractions of muscle fibers in the heart.

  Correction: _____

_____ 5. A. John Ray discovered the circulation of blood in humans.
  B. Aristotle was an early philosopher who compiled descriptions of animal life.
  C. Charles Darwin was a 19th century British naturalist who sailed on the <u>Beagle</u>.
  D. One of Curvier's important scientific contributions was to include extinct species into classifications of living species.

  Correction: _____

_____ 6. A. The binomial system of nomenclature that biologists use even today was devised by Swedish botanist Carolus Linnaeus.
  B. Jean Baptiste Lamarck and Charles Darwin proposed versions of evolution which stated that traits acquired by an organism can be inherited by offspring.
  C. Buffon was an early natural historian.
  D. Natural selection is an essential element of Darwin's theory of evolution.

  Correction: _____

_____ 7. A. Rene Descartes was a mechanist who argued that life could be explained in terms of physical and chemical laws.
  B. Vitalists, such as Hans Driesch, believed that life has a special essence that cannot be explained mechanistically.
  C. Vital forces have been called by a variety of terms including <u>entelechy</u>, <u>scala naturae</u> and <u>elan vital</u>.
  D. Many vitalists were also supporters of finalism, a doctrine stating that the history of life progressed toward a certain goal by a preordained plan.

  Correction: _____

CHAPTER 3--ATOMS AND MOLECULES

CHAPTER OVERVIEW

This chapter begins a two-chapter treatment of fundamental principles that regulate the chemistry of life. We start at subcellular levels with a description of subatomic particles called protons, neutrons, and electrons. Next, we examine how these particles are assembled into atoms, and then how molecules are formed by combining atoms of the same or different elements. Such atoms may be joined by covalent bonds, ionic bonds, and hydrogen bonds, and attracted to each other by van der Waal forces and hydrophobic interactions.

The elements found most abundantly in living organisms are carbon, oxygen, hydrogen, and nitrogen, although virtually all other naturally-occurring elements are found in living tissues as well. Water is a simple compound of crucial biological importance. Most biological reactions occur in aqueous (watery) solutions. The importance of water to life revolves around its many special properties: expansion on freezing, ability to attract other water molecules, capability to hold large amounts of heat, and excellence as a solvent.

TOPIC SUMMARIES

A. <u>General Chemical Definitions</u>.--All materials, whether living or non-living, are made of <u>chemicals</u>. Chemicals are produced by combining one or more of the slightly more than 100 <u>elements</u>. The smallest amount (or smallest particle) of an element that still has the characteristics of that element is an <u>atom</u> of that element. Atoms of one or more elements can join (<u>bond</u>) in specific ways to form <u>compounds</u>. The basic unit of a compound is a <u>molecule</u>.

Let's consider the simple sugar, glucose, as an example: The <u>chemical formula</u> of glucose is $C_6H_{12}O_6$. This molecule is made of 24 atoms of three different elements--six of these are carbon (C) atoms, 12 are hydrogen (H) atoms, and six are oxygen (O) atoms. The chain and ring forms of the chemical formula for glucose are shown in Fig. 3.1. We will return repeatedly to glucose in later chapters.

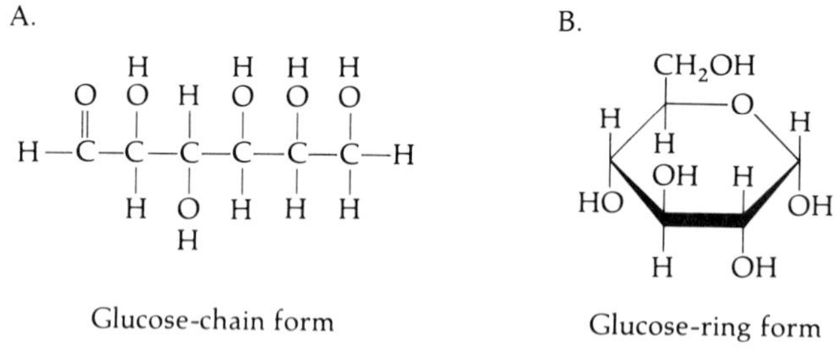

Fig. 3.1: Chain and ring forms of the glucose molecule

B. <u>Structure of Atoms</u>.--Each atom is made of one or more (usually three) kinds of <u>subatomic particles</u> arranged in a particular way. <u>Protons</u> and <u>neutrons</u> cluster in the center of the atom to form the <u>nucleus</u>. <u>Electrons</u> rapidly move around the nucleus at various distances from the nucleus in <u>orbitals</u>. Fig. 3.2 shows the arrangement of subatomic particles in atoms of

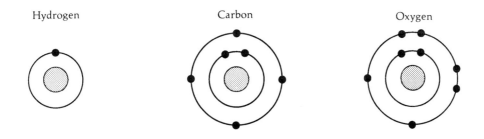

Fig. 3.2: Models of atoms of hydrogen, carbon, and oxygen

Protons and electrons are called <u>charged particles</u> because they carry electrical charges. The charges for these particles (positive for protons; negative for electrons) are opposite, and the charges are of equal magnitude. So, an atom with equal numbers of protons and electrons is balanced electrically. Neutrons are electrically neutral.

Each of these subatomic particles has mass. Protons and neutrons have equal masses, designated as one <u>mass unit</u>. Electrons also have mass, but the mass is so small that it is negligible and can be ignored; hence, we say the mass of an electron is zero.

Atoms of the 100+ elements are made of different combinations of protons, neutrons, and electrons. Therefore, we can refer to each element by a unique number: the <u>atomic number</u> is the number or protons <u>or</u> the number of electrons. (Recall that the electrical balance of an atom results from presence of equal numbers of protons and electrons.) Each atom also has an <u>atomic mass</u>, the sum of the number of particles (protons and neutrons) in the nucleus.

Although each element has a unique atomic number, many elements exist in multiple forms called <u>isotopes</u>. Each isotope has a different number of neutrons and, hence, a different atomic mass. Yet, the chemical bonding and other properties of all isotopes of an element are similar. Some isotopes are <u>unstable</u> or <u>radioactive</u>; they spontaneously give off energy (<u>decay</u>) as the number of neutrons adjusts towards a stable isotope. Refer to Figure 3-3 in your text to see the differences in atomic structure for isotopes of hydrogen and carbon.

C. <u>Chemical Bonding</u>.--The joining of elements into compounds occurs by chemical bonding. Several types of bonding exist, and each type of bond has a different strength. To understand bonding, you must understand the positioning of electrons around an atomic nucleus. We can think of electrons as residing in a series of concentric <u>orbitals</u> or <u>shells</u> around a nucleus. The first shell can contain a maximum of two electrons and the next shell can contain eight electrons. (Refer to a chemistry textbook if you desire more details about numbers of electrons in other shells.) Only the outermost shell can contain fewer than the full complement of electrons. It is the number of electrons in the outermost shell that largely determines the bonding (reaction) characteristics of that element. Basically (and anthropomorphically speaking), atoms "like" to fill all their electron shells. Hence, reactions are largely the results of atoms' efforts to get enough electrons to fill those outer shells.

(1) Covalent Bonding: One means of getting those electrons is to <u>share</u> them with another atom. Figure 3-6 in the text shows two atoms of chlorine. The outer shell of each chlorine atom contains only seven electrons. Each atom "borrows" one electron from the outer shell of the other and each behaves as if its outer shell were full. Chlorine is much more stable as this bonded compound than as individual atoms. This type of bonding where electrons are shared is called <u>covalent bonding</u>.

(2) Ionic Bonding: An alternate means for an atom to achieve an outer shell full of electrons is to take the needed number of electrons from another atom or to give the excess electrons to another atom. By such gain or loss of electrons, however, the electrical balance of the atom is lost so that the remainder of the atom (called an <u>ion</u>) now has a positive or a negative charge. A <u>cation</u> is a positively charged ion having more protons than electrons; an <u>anion</u> is a negatively charged ion having more electrons than protons. The <u>valence</u> number of a cation indicates how many excess protons are present in the nucleus, or to look at it from the opposite perspective, how many electrons are needed to bring the atom back into electrical balance. For example, a magnesium cation (Mg^{++}) has two extra protons, and needs two electrons to regain its electrical balance. Conversely, the valence of an anion indicates the number of protons it needs to gain (or the number of electrons it needs to lose) to become neutral. For example, the fluoride anion (F^-) has one electron to lose, or needs to gain one proton, to achieve electrical balance.

As your text states, such <u>ionic bonding</u> is really just an extreme version of covalent bonding in which the electrons are shared so unequally that the bond is purely the result of attraction of electrical opposites. Such unequal positioning of electrons in a molecule produces <u>polar molecules</u>--molecules with some areas having positive charge and other areas having negative charge. The importance of polarity will become apparent below in section E (Water).

(3) Hydrogen Bonding: Removing the only electron from a hydrogen atom produces the positively charged hydrogen ion (H^+). The hydrogen ion can electrically attract free electrons or ions or molecules that are negatively charged. The electrostatic attraction is a weak version of an ionic bond. Despite the relative weakness of this bond, hydrogen bonding is very important biologically. For example, hydrogen bonding is important in determining the three-dimensional structure of many long-chain molecules, such as proteins. Hydrogen bonding also is the primary means by which water molecules attract each other. We will discuss these examples below in Part E (Water).

(4) Van der Waals Forces: These are a minor type of attraction between molecules located very close to each other due to interactions of electrons of one molecule with the nucleus of another molecule. Van der Waals forces are weaker than hydrogen bonds, yet in combination with hydrogen bonds they can help stabilize the three-dimensional structure of large molecules like proteins.

(5) Hydrophobic Interactons: The last kind of interaction we will consider here is one that repels, rather than attracts, much in the way that like poles of two magnets repel each other. Mixtures of a polar liquid (vinegar) with a non-polar liquid (oil) must be agitated to overcome such forces to keep the components from separating. Hydrophobic interactions are important in the structure and function of cell membranes, as will be seen in Chapter 6.

D. <u>Elements in Living Organisms</u>.--Even though all of the 92 naturally occurring elements are found in tissues of living organisms, the vast majority of body mass (about 96% for humans) consists of only four elements: carbon, oxygen, hydrogen, and nitrogen. Interestingly, of these four, only oxygen is also among the four most commonly available elements in Earth's crust. The special properties of carbon, oxygen, hydrogen, and nitrogen have allowed them to play special biological roles.

(1) Carbon: The fields of organic chemistry and biochemistry center around molecules containing carbon. Much of Earth's reservoir of carbon is in the form of gaseous carbon dioxide (CO_2) in the atmosphere. Carbon also is stored as fossil fuels (petroleum, coal, etc.), dissolved in water (as carbonic acid when reacting in water), and is bound into the tissues of

living organisms. Carbon's ability to react with many elements and compounds is due to the number of electrons (four) in its outer shell.

(2) Hydrogen: Because hydrogen is an excellent source of electrons or protons, it is involved in most biochemical reactions. We will comment below (Part F, Ionic Dissociaiton) on the importance of hydrogen in acid/base reactions.

(3) Oxygen: Both as free oxygen (O_2) and combined into water and carbon dioxide, oxygen is crucial to virtually all organisms in the energy-releasing process of cellular respiration, and to many organisms in the energy-storing process of photosynthesis.

(4) Nitrogen: Nitrogen is found in two important classes of macromolecules--proteins (important as antibodies, enzymes, and structural molecules) and nucleic acids (important in heredity). It enters the bodies of most organisms as ammonia, nitrites, nitrates, or proteins. Waste nitrogen usually leaves bodies of animals as urea, uric acid, or ammonia.

(5) Minerals and Trace Elements: Elements needed by living organisms in lesser amounts than carbon, oxygen, hydrogen, and nitrogen, but still in moderate amounts, are called minerals: sodium (Na), magnesium (Mg), sulfur (S), potassium (K), chlorine (Cl), phosphorus (P), calcium (Ca), and a few others. Scores of elements needed only in minute quantities are the trace elements (or microminerals) such as manganese (Mn), iodine (I), iron (Fe), copper (Cu), zinc (Zn), fluorine (F), cobalt (Co), selenium (Se), and many others.

E. Water.--Water is critical to life as we know it . . . in part because virtually all biochemical reactions occurring in living tissues can occur only when reactants are dissolved in water. The importance of water results also from special chemical properties of the molecule:

(1) Polarity: Figures 3-11 and 3-12 in your text show the structure of the water molecule and how the one oxygen and two hydrogen molecules are positioned relative to each other. Notice that water is an electrically polar molecule . . . with the oxygen region being more negative and the hydrogen ends more positive. So, the hydrogens of a water molecule attract the oxygen of other water molecules. The result is a very regular arrangement of water molecules in space. This arrangement is most apparent in frozen water (ice). The polarity of the water molecules causes water to expand, rather than to contract! Hence, ice cubes float in a glass of tea . . . and beneath the ice of a frozen pond there is usually a zone of denser, cold liquid water in which aquatic organisms can live until the overlying ice thaws.

(2) Cohesion and Adhesion: Water molecules attract each other as well as other polar substances and surfaces due to hydrogen bonding. Hence, water "beads" into droplets on the waxed surface of a car's hood and water "climbs" (by capillary action) hundreds of feet through vascular bundles to tree tops.

(3) Thermal Properties: Water has the capacity to store large amounts of heat. Its specific heat is much greater than that of many materials. Hence, water is a great insulator. The temperature of ocean water changes very little compared to changes in temperature of the air above the ocean . . . so, aquatic organisms live in surroundings largely protected from great temperature changes.

(4) Solvent Properties: Because of its polarity, water is a very good solvent. Many polar salts and ionic compounds dissolve in water to form aqueous solutions. Polar nonionic compounds, such as sugars and alcohols, also can dissolve in water because water can form hydrogen bonds with them. These water-soluble compounds are said to be hydrophilic (or water-loving).

F. <u>Ionic Dissociation</u>.--When ionic compounds, such as ordinary table salt (NaCl), are placed into a polar solvent, such as water, the cations (Na^+) and anions (Cl^-) dissociate, that is, they separate from each other. Water molecules, which are attracted electrostatically to both Na^+ and Cl^-, surround each ion. Dissociation is maintained in this way, because cations cannot get close enough to anions (and vice versa) to reform an ionic bond. The action of acids, bases, and buffers involves ionic dissociation.

(1) Acids and Bases: Acids and bases are special cases of dissociation of ionic compounds. <u>Acids</u> are compounds whose dissociation releases one or more hydrogen cations (H^+). The strength of an acid is directly related to the number of hydrogen cations released into solution. <u>Bases</u> are compounds that can accept (bond with) the hydrogen ions donated by an acid. For example, ammonia (NH_3) reacts with H^+ to become an ammonium cation (NH_4^+). The acidity of a solution is determined by the concentration of hydrogen ions in the solution and is expressed by a <u>pH value</u>. A low pH (less than 7) indicates an acidic solution whereas pH values above 7 are considered basic or alkaline. Neutral pH is approximately 7. The pH scale is logarithmic; that is, every unit change in pH value indicates a 10-fold change in concentration of hydrogen ions in solution.

(2) Buffers: For many solutions, the pH changes greatly with the addition of an acid or a base. Such fluctuations in acidity can cause problems for reactions that occur only in a narrow pH range. Other solutions, however, have the ability to resist change in pH despite addition of an acid or a base. Such solutions are said to be "buffered." Buffering can be accomplished by adding to a solution a salt of a weak acid (e.g., sodium bicarbonate, $NaHCO_3$). Living organisms developed biological <u>buffer systems</u> long before chemists understood the principles through lab experiments.

TERMS TO UNDERSTAND: Define and identify the following terms. How do they apply to the field of biology as presented in Chapter 3?

element

atom

molecule

compound

protons

neutrons

electrons

nucleus

atomic mass

atomic number

isotope

half-life

orbital

valence

covalent bond

polar

nonpolar

ionic bond

anion

cation

salt

hydrogen bond

van der Waals forces

hydrophobic interaction

organic compounds

inorganic compounds

macrominerals

microminerals

cohesion

surface tension

hydrophilic

hydrophobic

specific heat

heat of vaporization

heat of fusion

ionic dissociation

acid

base

pH

buffer

CHART EXERCISES

A. Complete the chart and equations below to show mass and electrical-charge characteristics of the three types of subatomic particles that compose atoms.

PARTICLE	MASS	CHARGE
		positive
Neutron		
	zero	

Atomic mass = (the number of _____) + (the number of _____).

Atomic number = (the number of _____) OR (the number of _____).

B. Elements cycle between the living and nonliving segments of Earth. Complete the following chart for the elements carbon, hydrogen, oxygen, and nitrogen to indicate (1) the primary reservoir of each element in the physical world, (2) the chemical form of these elements in their reservoir, (3) the chemical form in which these elements enter living organisms, and (4) the chemical form in which these elements leave living organisms.

ELEMENT	RESERVOIR	CHEMICAL FORM IN RESERVOIR	CHEMICAL FORM FOR ENTERING ORGANISMS	CHEMICAL FORM FOR EXITING ORGANISMS
Carbon				
Hydrogen				
Oxygen				
Nitrogen				

DISCUSSION QUESTIONS

1. Why is an understanding of basic chemistry important to the study of biology?

2. Discuss the evidence for the statement that the vast majority of matter is empty space.

3. a. What is the relationship between temperature and the density of water?

   b. What is the ecological importance of this relationship?

4. Given the atomic number and the atomic weight of an element, how can you determine the number of protons, neutrons, and electrons in an atom of that element?

5. a. What is a hydrogen bond?

   b. How does the strength of hydrogen bonds compare to the strength of other kinds of chemical bonds discussed in Chapter 3?

   c. What are three situations in which hydrogen bonds play an important biological role?

TESTING YOUR UNDERSTANDING:

For each of the test items below, three of the lettered alternatives are true and the other is false. Determine which alternative is false and write its letter in the blank to the left of the question number. On the blank line below alternative D, write the corrected version of the false statement.

_____ 1. A. An atom is the smallest unit that possesses the characteristics of an element.
B. A molecule consists of two or more atoms of the same or different elements bonded together.
C. An ion is the smallest unit that possesses the properties of a compound.
D. Water is an example of a compound whose molecules include two atoms of hydrogen and one atom of oxygen.

Correction: _____

_____ 2. A. The nucleus of an atom has no net electrical charge because equal numbers of positively charged and negatively charged particles are located in the nucleus.
B. Neutrons are found in the nucleus of an atom.
C. Electrons are negatively charged particles located in orbitals.
D. Protons are positively charged particles located in the nucleus.

Correction: _____

_____ 3. A. The mass of an electron is so small compared to the mass of a proton that an electron's mass is designated as zero.
B. The mass of a neutron is equal to the mass of an electron.
C. The atomic mass of an element is the sum of the number of protons and neutrons.
D. For any electrically balanced atom, the number of protons equals the number of electrons.

Correction: _____

_____ 4. A. An atom having an atomic number of 12 has 12 protons.
B. An atom having an atomic number of 19 has 19 electrons.
C. An atom having an atomic mass of 16 has the sum of the number of protons and neutrons equal to 16.
D. An atom having an atomic mass of 32 has the sum of the number of protons and electrons equal to 32.

Correction: _____

_____ 5. A. The isotopes of a particular element differ from each other only in the number of neutrons.
B. Isotopes of a particular element have different atomic numbers.
C. Isotopes of a particular element have different atomic masses.
D. All isotopes of a particular element have the same number of electrons.

Correction: _____

_____ 6. A. Deuterium is an isotope of hydrogen.
   B. Radioactive isotopes decay to stable isotopes by emitting neutrons from the nucleus.
   C. Because all isotopes of the same element behave the same chemically, radioisotopes can be used as tracers in biochemical reactions.
   D. The half-life of carbon-14 is longer than the half-life of uranium-238.

   Correction: _____

_____ 7. A. All electrons of an atom orbit the nucleus at the same distance from the nucleus.
   B. Because the exact location of an electron is very difficult to determine, electrons are said to occur in electron charge clouds.
   C. The innermost orbital contains a maximum of two electrons.
   D. The second orbital can contain up to eight electrons.

   Correction: _____

_____ 8. A. The chemical behavior of an atom is determined by the number of electrons in its outermost orbital.
   B. The number of electrons that an atom must gain or lose to complete its outer orbital is called its valence.
   C. An atom with its outermost orbital full of electrons is highly reactive.
   D. Electrons located in inner orbitals possess less energy than electrons in outer orbitals.

   Correction: _____

_____ 9. A. A covalent bond is approximately twenty times stronger than an ionic bond.
   B. Covalent bonding involves sharing of electrons by two or more atoms.
   C. Ionic bonding involves the gain or loss of electrons by atoms.
   D. Bonds in which electrons are shared unevenly between the atoms involved produce nonpolar molecules.

   Correction: _____

_____ 10. A. Because water is a polar molecule, water is a very good solvent for polar salts and ionic compounds.
   B. Anions are the negatively charged ions of a dissolved salt.
   C. Cations are attracted to the positively charged region of a polar molecule.
   D. Hydrogen bonds are weaker than covalent or ionic bonds.

   Correction: _____

_____ 11. A. Hydrophobic substances dissolve readily in water because they form hydrogen bonds with water molecules.
   B. In hydrogen bonding, the positively charged hydrogen nucleus is attracted to the negatively charged region of another atom.
   C. Cohesion of water molecules is due largely to the effects of hydrogen bonds between the water molecules.
   D. Hydrogen bonds are important in determining many of the properties of proteins, carbohydrates, and nucleic acids.

   Correction: _____

_____ 12. A. Van der Waals forces are very weak interactions between molecules due to interactions of electrons with nuclei.
B. Hydrophobic interactions do not involve attractions of molecules, but instead involve repulsion of nonpolar molecules from molecules of a polar solvent.
C. Sugars dissolve easily in water because sugar molecules form covalent bonds with water molecules.
D. Ionic compounds, such as NaCl, dissolve in water by the process of ionic dissociations.

Correction: _____

_____ 13. A. Acids are compounds which can liberate a hydrogen ion.
B. A pH value indicates the log of the concentration of hydrogen ions in solution.
C. An amphoteric substance can act as either an acid or a base.
D. Stronger acids dissociate more completely than do weaker acids.

Correction: _____

_____ 14. A. Acidic solutions have pH values less than 7.
B. A solution with pH of 4 is twice as acidic as a solution with pH of 2.
C. Buffer systems are used widely in both the lab and in biological organisms to maintain stable pH.
D. Most biochemical reactions in living cells occur at pH levels near 7.

Correction: _____

_____ 15. A. The field of science known as organic chemistry focusses on compounds containing carbon.
B. The four elements that are most abundant in living organisms are also the four most abundant elements in Earth's crust.
C. Plants obtain carbon primarily from gaseous carbon dioxide.
D. Carbon dioxide and water can be combined into sugars by plants using the process of photosynthesis.

Correction: _____

_____ 16. A. The carbon in carbon-containing compounds ultimately is oxidized into carbon dioxide.
B. For oxygen to be usable by either terrestrial or aquatic organisms, oxygen must be dissolved in water.
C. Nitrogen is an element that is found in all protein molecules.
D. Most plants and animals can use nitrogen obtained directly from the atmosphere.

Correction: _____

_____ 17. A. Sodium and chlorine are examples of trace elements or microminerals.
B. Nitrogen-fixing bacteria play a critical role in converting gaseous nitrogen into nitrogen compounds that can be used by plants and animals.
C. Urea, uric acid, and ammonia are waste products of nitrogen metabolism.
D. Water is the most abundant compound in living tissues.

Correction: _____

_____ 18. A. Ice floats in water because the density of water is greater below 0° C than above 0° C.
B. Water striders can walk on water without sinking because the force exerted on the water surface by their body mass is less than the strength of the hydrogen bonds between water molecules at the surface.
C. A substance that has a high heat capacity is a good temperature buffer.
D. The specific heat of water is greater than the specific heat of ethyl alcohol.

Correction: _____

_____ 19. A. The amount of heat required to evaporate a standard amount of liquid at its normal boiling point is its heat of fusion.
B. A calorie is the amount of heat required to raise the temperature of one gram of water by one degree Celsius.
C. Heat capacity is the amount of heat required to raise the temperature of a given quantity of a substance by one degree Celsius without changing its phase.
D. Density is mass per unit volume.

Correction: _____

_____ 20. A. Water has a specific heat of one calorie.
B. The heat of vaporization of water at 100° C is 540 calories per gram.
C. One kilogram equals 2.2 pounds.
D. One liter equals 1.06 gallons.

Correction: _____

CHAPTER 4--ORGANIC MOLECULES IN LIVING SYSTEMS

CHAPTER OVERVIEW

In the preceding chapter, we were introduced to four biologically important elements: carbon, hydrogen, oxygen, and nitrogen. These elements are the primary components of four categories of macromolecules (carbohydrates, lipids, proteins, and nucleic acids) found in the vast diversity of living organisms. This chapter presents the basic chemical characteristics of these various biomolecules and outlines special properties of various categories of biomolecules that allow them to perform their biological functions.

In Chapter 4, we also examine the question of how life arose from the non-living. Earlier views held that life continually developed from non-living matter whereas, in the modern interpretation, all life comes from pre-existing life. Several experiments that examined how the first biomolecules came into being and how the earliest cells formed are summarized.

TOPIC SUMMARIES

A. <u>Unity Within Diversity</u>.--The diversity of life--from bacteria to giant sequoia trees to kangaroos--is united by its underlying chemical composition. All organisms consist of several categories of biomolecules that are organized into different forms and sizes but arranged according to the same rules of chemistry. The major categories of organic compounds are <u>carbohydrates</u>, <u>lipids</u>, <u>proteins</u>, and <u>nucleic acids</u>. Molecules in these various categories serve several general functions:

(1) They may contribute to body structure.
(2) Some contain great amounts of chemical energy and can, therefore, serve as fuels.
(3) Some large molecules often contain genetic information.
(4) Some catalyze biochemical reactions.

B. <u>Construction of Organic Compounds</u>.--All organic compounds contain carbon, and virtually all also include hydrogen and oxygen. Proteins contain these three elements plus nitrogen (and, often, other elements as well). Atoms of these elements readily share electrons, and consequently they form covalent bonds with each other.

Carbon has a valence of four. Carbon reacts to fill its outer shell with eight electrons by sharing four from atoms of other elements. For example, a carbon atom can bond with four hydrogen atoms (each with a valence of one) . . . or with two oxygen atoms (each with a valence of two) . . . or with some other combination of elements to achieve a complete outer shell.

Carbon, of course, is not the only element with a valence of four. Silicon, for example, is a very abundant element in Earth's crust and also has a valence of four. However, silicon does not bond to form long chains and rings as readily as carbon. Also, the oxide of silicon (SiO_2) corresponding to that of carbon (CO_2) is a solid (quartz) rather than a gas (carbon dioxide). The importance of carbon chains and rings as well as the gaseous nature of carbon dioxide will become apparent in later chapters.

C. <u>Chemical Formulas</u>.--Formulas of different types provide different kinds of information about a compound. An <u>empirical formula</u> indicates only the number of atoms of each element in a molecule, whereas a <u>structural formula</u> expresses the spatial position of each atom in the molecule. For example, the empirical formula for glucose is $C_6H_{12}O_6$; structural formulas showing various chain and ring arrangements for glucose are shown in Fig. 4-6 of the text.

Each different structural organization of a molecule is a different isomer of that molecule. Structural isomers, such as the sugars glucose and fructose (Fig. 4.1), are compounds in which atoms are linked differently; they are distinguishable from each other because they have different chemical behavior. Stereoisomers of a compound have the same bonding linkages and have similar chemical behavior. They differ only in spatial positioning of atoms or atomic groups; see text Fig. 4-2 for examples of stereoisomerism in molecules of alanine, an amino acid. These molecules are mirror images of each other. A primary biological significance of stereoisomerism is their stereospecificity--some enzymes "recognize" one epimer but not the other. We'll see more about this in Chapter 8.

D. Carbohydrates.--This class of macromolecules includes mainly the sugars and starches. They are made of carbon, hydrogen, and oxygen. Almost all carbohydrate molecules have twice as many hydrogen atoms as oxygen atoms. The number of carbon atoms is often, but not always, the same as the number of oxygen atoms. Carbohydrate molecules consist either of straight chains (Fig. 3.1A) or of rings of carbon atoms (Fig. 3.1B). To most carbon atoms are attached -OH (hydroxyl) groups. To one of the carbon atoms, either an aldehyde or a keto group is attached.

Single monomer molecules of a carbohydrate are called monosaccharides, such as glucose, a six-carbon sugar. Monosaccharide molecules can polymerize by condensation reactions in which one monosaccharide bonds to another; in the process a water molecule is formed from a hydrogen from one monomer and a hydroxyl group from the other. For example, two glucose molecules condense to form one maltose molecule (Fig. 4.1).

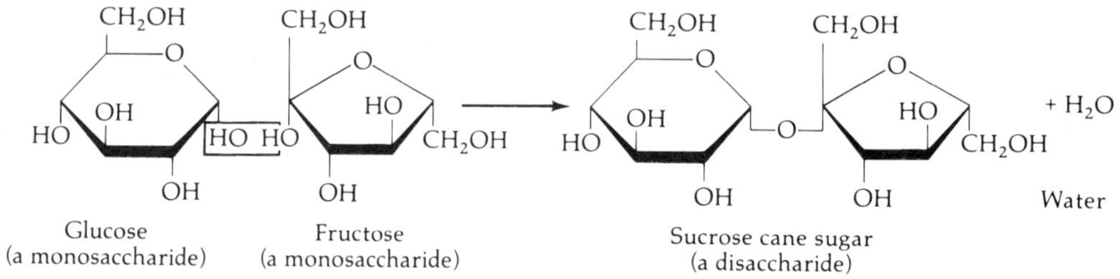

Fig. 4.1: Condensation reaction between monosaccharides to form a dissacharide

A sugar consisting of two monosaccharides is a disaccharide. A carbohydrate chain can grow to great size by continued addition of monomers. Examples of such large polysaccharides include cellulose, glycogen, and starch. Cellulose serves as a strong structural molecule, primarily in the cell walls of plants. Glycogen (in animals) and starch (in plants) are energy storage molecules that can be used as a fuel to drive the organism's metabolism.

E. Lipids.--Like carbohydrates, lipids also are made of carbon, hydrogen, and oxygen, but the ratio of hydrogen to oxygen is not 2:1 in lipids as it is in carbohydrates. Lipids dissolve in organic solvents, but only slightly or not at all in water. Lipids have a variety of functions:

(1) Lipids often serve as a chemical means to store energy; most of us have greater amounts of body fat (adipose tissue) than our energy demands require.
(2) Such body fat can insulate and pad the body of an organism and can contribute to body contour (in humans, figure or physique).
(3) Some vitamins and hormones are classified into a category of lipids called steroids.

(4) **Membranes** surrounding cells and many cellular organelles are made largely of **phospholipid** molecules. Molecules of this special group of amphipathic lipids are relatively long, with one end being polar (water soluble) and the other end non-polar (not water soluble). We'll examine the importance of phospholipids in Chapter 6.

The simple lipids are called **triglycerides**. These molecules are formed by **condensation reactions** of the hydroxyl groups of three **fatty acid** molecules with the hydrogens of the three hydroxyl groups of a **glycerol** molecule (Fig. 4.2).

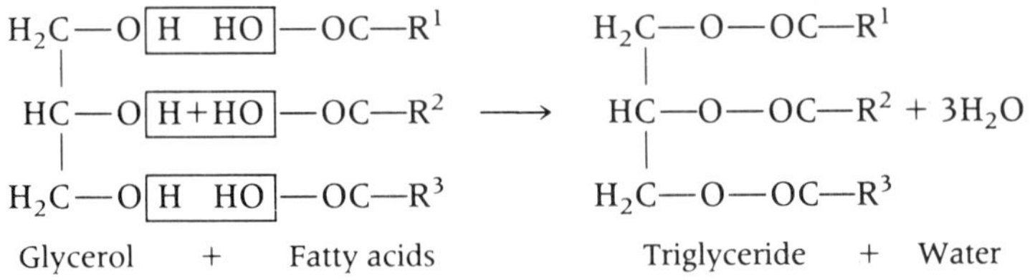

$$H_2C-O\boxed{H \;\; HO}-OC-R^1 \qquad\qquad H_2C-O-OC-R^1$$
$$HC-O\boxed{H+HO}-OC-R^2 \longrightarrow HC-O-OC-R^2 + 3H_2O$$
$$H_2C-O\boxed{H \;\; HO}-OC-R^3 \qquad\qquad H_2C-O-OC-R^3$$

Glycerol + Fatty acids → Triglyceride + Water

Fig. 4.2: Condensation reaction between glycerol and fatty acids to form a triglyceride

F. **Proteins**.--Proteins differ from carbohydrates and lipids in the presence of nitrogen in addition to carbon, hydrogen, and oxygen. The nitrogen is found as the central atom in the **amino group** ($-NH_2$). Although proteins often are immensely large molecules, they are built of only about 20 kinds of small molecules called **amino acids**. Amino acids are recognized by a generalized formula: The next-to-last carbon of the carbon chain has attached to it (1) an amino group, (2) a **carboxyl group** (-COOH), (3) a hydrogen, and (4) the rest of the chain (or residue) symbolized by the letter R.

Amino-acid monomers bond with each other by condensation reactions between the OH of the carboxyl group of one amino-acid molecule and an H from the amino group of another amino acid molecule (Fig. 4.3). The bond formed by such a condensation reaction is a **peptide bond**. Hence, molecules formed by such linkage of amino acids are called **peptides**. A moderately long string of amino acids is a **polypeptide**. **Protein** molecules can be made of one or more polypeptide.

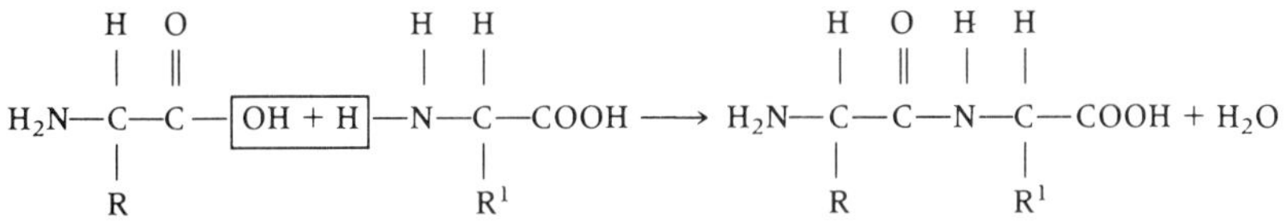

$$H_2N-\underset{R}{\underset{|}{C}}-\overset{O}{\overset{\|}{C}}-\boxed{OH + H}-N-\underset{R^1}{\underset{|}{C}}-COOH \longrightarrow H_2N-\underset{R}{\underset{|}{C}}-\overset{O}{\overset{\|}{C}}-N-\underset{R^1}{\underset{|}{C}}-COOH + H_2O$$

Fig. 4.3: Condensation reaction between amino acids to form a peptide

Proteins have four structural levels. The linear sequence of amino acids is the protein's **primary structure**. **Secondary structure** describes portions of the molecule as coiled (helical) or not coiled (non-helical). **Tertiary structure** refers to the three-dimensional pattern of folds and turns of the coiled portion of the molecule. **Quaternary structure** applies to molecules consisting of two or more polypeptides and refers to how the polypeptide chains are arranged in space. All levels of protein structure are determined by a protein's primary structure.

Proteins have diverse roles in living organisms. **Structural proteins**, such as keratin, collagen, and elastin, usually are strong molecules having an elongate (fibrous) conformation. In contrast, proteins serving as enzymes, as hormones, and in immunological defense, have a more rounded (globular) shape. We'll see, too, that certain proteins play a critical role in expressing the genetic code.

By associating other chemical groups (<u>prosthetic groups</u>) with the <u>simple proteins</u> (those consisting only of amino acids) discussed above, we expand the diversity of form and function of proteins. These various <u>conjugated proteins</u> will be examined in the chapters pertaining to their specific functions.

G. <u>Nucleic Acids</u>.--The detailed structure and functions of nucleic acids are treated in Chapters 11 and 16. At this point we should note that nucleic acids are the largest of the biomolecules. <u>Nucleotide</u> monomers bond into long nucleic-acid chains. The nucleotides include carbon, hydrogen, oxygen, nitrogen, and phosphorus arranged in a ring structure. Nucleotide sequence in DNA (<u>deoxyribonucleic acid</u>) and RNA (<u>ribonucleic acid</u>) constitutes the genetic code. Nucleic acids truly are informational molecules!

H. <u>Origin of Life</u>.--One of the matters that has long concerned humans is the question of how life began. From previous chapters, we know that Earth existed a long time before life came about in this planet. In this chapter we have examined several major groups of macromolecules, all of which are produced by living organisms. But if there were no living organisms around to produce the first biomolecules, then where did the first biomolecules come from?

As recently as only a few centuries ago, this paradox was not an intellectual problem. Ancient philosophers believed that life arose spontaneously from such things as decaying flesh and from pond ooze. This concept of <u>spontaneous generation</u> was discredited in the 18th and 19th centuries by the experiments of Spallanzani and of Pasteur. Spallanzani, for example, observed that bacteria grew in nutrient broth that was exposed to air and to flies. He conducted experiments in which he boiled broth and then sealed the container so that air and flies could not contact the broth. In such sealed flasks there was no bacterial growth. He speculated that new life did <u>not</u> spontaneously arise in broth, but instead that microscopic bacteria were transported to the broth by wind currents and larger animals like flies.

This, the beginning of the demise for spontaneous generation, led to the paradox of how the first biomolecules were produced if there existed no biological organisms to produce the biomolecules. Several experimenters have suggested plausible abiotic means by which the first biomolecules came into being. The scenario of <u>chemical evolution</u> can be briefly summarized: The atmosphere of early Earth probably was charged with the energy of lightning which ionized atmospheric gases into simple organic molecules such as methane (CH_4) and ammonia (NH_3). Many of these molecules dissolved in the ocean waters to form a <u>dilute soup</u>. In this solution, simple organic molecules continually reacted to form progressively more complex organic molecules, such as amino acids, the building blocks of proteins. Indeed, in experiments by Miller and Urey, amino acids spontaneously polymerized into <u>proteinoids</u>, large molecules very similar to proteins. Polymerization of simple lipids might have produced the first membranes. Some scientists think that minute clay particles might have served as surfaces on which polymerization and assembly of collecitons of macromolecules occurred. Some of the researchers who have investigated the origin of life include Oparin, Miller, Urey, and Cairns-Smith; be sure you learn which parts of this view of chemical evolution were contributed by each of these scientists. You should realize that many questions remain to be answered before the full story of chemical evolution is known.

TERMS TO UNDERSTAND: Define and the identify the following terms:

empirical formula

structural formula

symmetrical tetrahedron

structural isomers

stereoisomers

stereospecificity

functional group

carboxyl group

organic acid

carbohydrate

hydroxyl group

saccharide

-ose

condensation reaction

hydrolysis reaction

lipids

saturated fatty acid

unsaturated fatty acid

amphipathic

phospholipid

steroid

protein

amino acid

amino group

peptide bond

polypeptide

primary structure

secondary structure

tertiary structure

quaternary structure

denatured

oligomeric protein

conjugated protein

prosthetic group

enzyme

coenzyme

nucleic acid

nucleotide

spontaneous generation

biogenesis

chemical evolution

proteinoids

CHART EXERCISES

A. The chemical characteristics of many organic molecules are determined by associated groups of atoms called functional groups. Complete this chart to give the name and formula of the functional group for each category of organic macromolecule, and give an example of such a molecule from this chapter of the text.

CATEGORY	FUNCTIONAL GROUP		EXAMPLE
	NAME	FORMULA	
Organic acid			
Alcohol			
Amino acid			

B. A variety of structural compounds is classified as carbohydrates. These large molecules are made of monomers. Complete the chart to indicate (1) the identity of the monomer molecules, (2) the chemical group to which the monomers belong, and (3) the natural situations in which these carbohydrates occur.

STRUCTURAL COMPOUND	MONOMER	CLASSIFICATION OF COMPONENTS	OCCURRENCE IN NATURE
Cellulose			
Chitin			

Glycogen

Starch

C. Complete this chart for the four listed categories of macromolecules. Indicate which of the four elements listed are important components of each type of molecule. Give the general formula that indicates the ratio of these elements for each category. What functional group(s) are responsible for the chemical characters of each type of molecule.

MACROMOLECULE	ELEMENTAL COMPONENTS				GENERAL FORMULA	FUNCTIONAL GROUP(S)
	C	H	O	N		
Carbohydrates						
Lipids						
Proteins						
Nucleic Acids						

PEOPLE TO KNOW: What is the significant biological contribution of each of these individuals as indicated in Chapter 4?

Abbe Lazzaro Spallanzani

Louis Pasteur

John Tyndall

A. I. Oparin

Stanley L. Miller

Harold C. Urey

DISCUSSION QUESTIONS

1. Explain what is meant by the phrase "unity within diversity." Give an example.

2. What are the four general functions of organic compounds in living organisms? Provide one example of each function to demonstrate that you understand these functions.

3. Distinguish between the two forms of isomerism. Provide examples of the two types. What is the biological significance of isomerism?

4. What is the relationship between condensation and hydrolysis reactions? In what classes of organic compounds do they occur? Give an example.

5. Discuss the diverse roles of lipids in living organisms.

6. What is meant by the paradox of the "principle of biogenesis" and the apparent abiotic origin of life on Earth?

TESTING YOUR UNDERSTANDING:

For each of the test items below, three of the lettered alternatives are true and the other is false. Determine which alternative is false and write its letter in the blank to the left of the question number. On the blank line below alternative D, write the corrected version of the false statement.

_____ 1. A. Lipids and carbohydrates are two classes of macromolecules that can serve as energy-rich fuels.
B. The valence for a carbon atom is four.
C. All protein molecules contain atoms of carbon, hydrogen, oxygen, and nitrogen.
D. The monomers of polysaccharides are amino acids.

Correction: _____

_____ 2. A. The structural formula of a molecule indicates both the number of atoms of each element in the molecule and the spatial distribution of those atoms.
B. The empirical formula indicates only the spatial positioning of atoms in a molecule.
C. Isomers are compounds that have the same empirical forms, but different structural formulas.
D. Two molecules that are mirror-images of each other are called stereoisomers.

Correction: _____

_____ 3. A. The chemical behavior of a molecule is largely determined by the chemical properties of the functional group of that molecule.
B. Stereoisomers can be distinguished from each other on the basis of the direction in which they rotate a beam of light.
C. The carboxyl group of an organic acid is an effective hydrogen ion acceptor.
D. Alcohol molecules have at least one hydroxyl group.

Correction: _____

_____ 4. A. The dioxide compoud of carbon (CO_2) is a gas at room temperature, whereas that of silicon (SiO_2) is a solid at room temperature.
B. Molecules of carbohydrates possess a -COOH group at one end of their carbon chains and an $-NH_2$ group at the other end.
C. One of the reasons that carbon, rather than silicon, is the central element of organic compounds is that carbon atoms bond to form chains more readily than do silicon atoms.
D. The polymerization reactions that bond amino-acid molecules into polypeptides are called condensation reactions.

Correction: _____

_____ 5. A. Unlike nucleic acids, proteins, and carbohydrates, fatty acid molecules do not polymerize to form complex lipids.
B. In some, but not all, carbohydrate molecules the number of carbon atoms is the same as the number of oxygen atoms.
C. A hexose, such as glucose, is a carbohydrate molecule containing six carbon atoms.
D. A hydrolysis reaction is one in which polymers are split into monomers with the accompanying release of water molecules.

Correction: _____

_____ 6. A. Chitin, a polymer of amino sugars, occurs widely in the cell walls of plants.
B. The animal starch glycogen has a highly branched molecular structure.
C. Plant and animal starches are energy storage molecules rather than structural molecules.
D. Cellulose, a polymer of thousands of glucose molecules, cannot be digested by many of the organisms that consume it.

Correction: _____

_____ 7. A. Simple lipids and simple carbohydrates are constructed only of carbon, hydrogen, and oxygen.
B. Triglycerides are simple lipids that are produced by reactions of glycerol and fatty acids.
C. Any fatty acid whose carbon chain includes any double bonds between carbon atoms is referred to as saturated.
D. The amphipathic properties of certain triglycerides make them suitable for the construction of biological membranes.

Correction: _____

_____ 8. A. Soaps are the sodium or potassium salts of fatty acids.
B. The group of lipids that contains cholesterol and several hormones is the steroids.
C. Most biochemical reactions are catalyzed by molecules classified as proteins.
D. Both functional groups of an amino acid donate hydrogen ions during polymerization reactions that produce proteins.

Correction: _____

_____ 9. A. Acidic amino acids possess two amino groups and one carboxyl group.
B. Peptide bonding occurs between the hydroxyl group of the acidic end of one amino acid and a hydrogen of the basic end of another amino acid.
C. The polypeptide chains of a protein molecule are held together by disulfide bonds.
D. The primary structure of a protein refers to the amino acid sequence of that protein.

Correction: _____

_____ 10. A. The helical structure of segments of a protein molecule is stabilized by hydrogen bonds between carboxyl and amine groups in successive turns of the helix.
B. Quaternary structure is a property of proteins composed of two or more protomers.
C. Proteins which function as enzymes are usually fibrous rather than globular.
D. Most protein molecules are water soluble.

Correction: _____

_____ 11. A. Hemoglobin is an example of a nucleoprotein.
B. Conjugated proteins are proteins to which prosthetic groups are attached.
C. Nucleotides polymerize into nucleic acids, large molecules whose central role is to encode genetic information.
D. Chromatin is a nucleoprotein that is formed by bonding deoxyribonucleic acid with protein molecules.

Correction: _____

_____ 12. A. The principle of biogenesis states that all life comes from pre-existing life.
B. Experimental evidence suggests that the chemical evolution of life occurred primarily in the atmosphere of early Earth.
C. Experiments conducted by Spallanzani and by Pasteur discredited the belief that life could spontaneously arise from inanimate materials.
D. The first life on planet Earth probably arose by a process of spontaneous generation.

Correction: _____

_____ 13. A. The atmosphere of early Earth probably included gases such as methane, ammonia, water, and hydrogen.
B. Amino acids are among the several kinds of organic molecules that were probably found in the oceans of early Earth.
C. The energy required to convert atmospheric gases into organic molecules dissolved in the oceans of early Earth probably came from several sources including lightning and radiation.
D. Organic molecules probably are still being produced today in the same ways they were before life first evolved.

Correction: _____

_____ 14. A. According to one theory of the origin of life, clay particles probably provided the surfaces on which organic molecules aggregated for assembly into primitive cells.
B. Oparin conducted experiments indicating that simple organic acids can be produced by energizing an atmosphere of methane, ammonia, hydrogen sulfide, and water.
C. The heating of amino acids causes them to polymerize into proteinoids, large molecules that resemble proteins.
D. Many investigators favor the following sequence of steps in the evolution of cells: amino acids to proteinoids to proteins to microspheres to cells.

Correction: _____

CHAPTER 5--CELL STRUCTURE

CHAPTER OVERVIEW

In the organizational hierarchy of biology, the cell is the least level that is considered to be alive. For many species, the organism consists of only one cell, whereas for other species, many trillions (or more) of cells make up the whole organism. Cell size varies tremendously, from microscopic to macroscopic.

The cell theory is one of several unifying principles of biology. This theory states that all living things are made of cells and that cells can arise only from pre-existing cells. Emergence of this theory followed a series of technological developments in optics--the development and refinement of the microscope. The newfound ability to see very small objects led to description of many new species to kingdoms (Plantae, Animalia) already known from macroscopic species, to recognition of new kingdoms of organisms (Monera, Protista, Fungi), and to description of "non-living" forms (viruses, plasmids, viroids, prions).

Despite the diversity of types of cells, all are composed of subsets of a comparatively few kinds of intracellular organelles. Cell structure is complex. The remainder of Chapter 5 describes the structure of eukaryotic (Protista, Fungi, Plantae, Animalia) and prokaryotic cells (Monera) and outlines the functions of various organelles of these cells.

TOPIC SUMMARIES

A. <u>Cell Theory</u>.--The invention of magnifying lenses in the 17th century (by Robert Hooke and Anton van Leeuwenhoek) allowed biologists to observe and study a previously unknown and unsuspected realm of biology--a realm occupied by microorganisms. Microscopic study (by Matthias Schleiden and Theodor Schwann among others) of the larger organisms already known to biology revealed that they, too, are made of cells; such organisms are referred to as being <u>multicellular</u>. Rudolf von Virchow's studies indicated that new cells result from division of existing cells. The <u>cell theory</u> emerged from the findings of microscopic study of living organisms. This theory, one of the great unifying principles of biology, states that (1) all living organisms are made of cells and (2) all cells arise from pre-existing cells.

B. <u>Survey of Life in Regard to Cellularity</u>.--The scheme of classifying living organisms that is used in your text recognizes five kingdoms (Chapter 37). All species in kingdoms <u>Monera</u> (bacteria, etc.) and <u>Protista</u> (amebae, paramecia, etc.) and some species in kingdoms <u>Fungi</u> (yeasts, etc.) and <u>Plantae</u> (many algae, etc.) are <u>unicellular</u>. Despite consisting of only single cells, many of these species still exhibit a remarkable degree of specialization, with portions of the cell having functions resembling the roles of eyes, mouths, fins, etc. of <u>multicellular</u> species. All multicellular organisms (including some plants, fungi, and all members of kingdom Animalia) begin life as a single cell which differentiates to become <u>specialized</u> for specific functions.

Although all organisms are made of cells, these cells vary tremendously in size and shape. Indeed, most are microscopic, averaging about 0.01 mm or 10 micrometers. Others are quite long and thin; some human nerve cells, for example, extend from the hand to the spinal cord. A bird egg, such as you ate for breakfast this morning, is but a single cell which has stored within it all the nutrition the developing chick needs until hatching. A primary factor determining size and shape of a cell is the need for adequate exchange of materials (nutrients, wastes, gases, etc.) between the cell and its surroundings. The <u>ratio of cell surface area to cell volume</u> is a means of expressing exchange capacity of a cell or for other objects.

C. <u>Subcellular Entities</u>.--Four types of biological entities that are organized below the cellular level are known. All are smaller than the smallest known cells, and they are not made of organelles (more on organelles below) as are cells.

(1) <u>Viruses</u> are merely small amounts of nucleic acid housed in a protein envelope; all viruses require association with living cells in order to produce more virus particles. Depending on the type of virus, viral infection can produce an array of diseases.

(2) <u>Plasmids</u> are simpler than viruses; plasmids consist of bits of DNA lacking a protein coat and exist only within living cells.

(3) <u>Viroids</u> are known thus far only from plant cells. They are very small bits of RNA without a protein coat. Viroids produce a variety of diseases in their host plants.

(4) <u>Prions</u> also can be pathogenic to their hosts. Prions consist of small amounts of protein without any associated nucleic acid.

D. <u>Cell Structure</u>.--Despite the diversity of species and of cell types, all cells are constructed along variations of the same general plan. Your text examines cell structure for three primary cell types: eukaryotic animal and plant cells and prokaryotic cells. Before proceeding, however, let's present a few of the characteristics that distinguish eukaryotic from prokaryotic cells. Whether they are plant, animal, fungal or protist, all <u>eukaryotic cells</u> share at least the following features:

(1) A <u>nucleus</u>, in which <u>chromosomes</u> of DNA are located, is delimited from other parts of the cell by a membrane, the <u>nuclear envelope</u>.

(2) The rest of the cell's contents, the <u>cytoplasm</u>, is the site of most of the cell's metabolic activities.

(3) The cytoplasm contains a number of kinds of small <u>organelles</u> which are enclosed by membranes.

(4) The outer boundary of the cell is defined by the <u>cell membrane</u>.

In contrast, <u>prokaryotic cells</u>

(1) <u>lack</u> a membrane-delimited nucleus,

(2) possess <u>no</u> membrane-bounded organelles,

(3) do <u>not</u> have multiple chromosomes,

(4) have DNA that is <u>not</u> associated with proteins, and

(5) have <u>cell walls</u> made of polysaccharides <u>other</u> than cellulose.

Other differences will become apparent as we survey cellular structure below. Refer to Table 5-1 in your text for a convenient summary of cell features according to kingdom.

E. <u>Eukaryotic Animal Cells</u>.--The primary structures (and their functions) composing eukaryotic animal cells are outlined below:

(1) The cell membrane encloses the contents of a cell and regulates movement of most materials between the cell's interior and the surrounding environment. This membrane and others that surround certain organelles are made largely of phospholipid and protein molecules (more on this in Chapter 6).

(2) The nucleus is usually the most prominent cellular organelle. Most animal cells possess only one nucleus. The nuclear envelope, which separates nuclear contents from the cytoplasm, is a double-membrane structure with pores and plugs (annuli) which help regulate diffusion of large molecules to and from the cytoplasm. Chromatin, a threadlike network of DNA that is complexed with proteins, carries the cell's genetic code; chromatin condenses into chromosomes in preparation for cell division (more on this in Chapter 12). The other major feature of a nucleus is the nucleolus, a spherical body involved in the production of ribosomes and of ribosomal RNA.

(3) The endoplasmic reticulum (ER) is a system of membrane-bounded channels coursing through much of the cytoplasm. The ER is continuous with the outer membrane of the nuclear envelope and, in places, it merges with the cell membrane. ER serves as sites of synthesis of a variety of macromelocules and as a system for transporting materials within the cell and exporting materials from the cell. The association of ribosomes with ER (a situation referred to as rough ER) indicates the occurrence of protein synthesis in that part of the ER. Synthesis of steroids and other lipids occurs in smooth ER, ER that has no associated ribosomes. Ribosomes are granular structures composed of ribonucleoprotein particles.

(4) A Golgi apparatus consists of a stack of flattened vesicles associated with endoplasmic reticulum. These apparati are sites of collecting, refining, and sorting of materials produced by the ER. Secretory vacuoles containing materials to be secreted (released) from the cell pinch off from the Golgi stack and move to the periphery where they merge with the cell membrane to release their contents. Some vesicles (microbodies) formed from a Golgi apparatus, however, remain within the cell; such microbodies contain an array of enzymes for processes such as digestion (lysosomes) and oxidation (peroxisomes).

(5) Mitochondria are rounded, elongate structures bounded by two membranes. These organelles are the sites of cellular respiration in which energy-yielding reactions occur. The energy released from oxidation of sugar (glucose) is chemically stored as ATP (adenosine triphosphate). We'll examine this set of reactions in some detail in Chapter 9.

(6) A pair of small cylindrical bodies, the centrioles, is located near the nucleus. Their role relates to the movement of chromosomes along the mitotic spindle during cell division.

(7) Cilia (short and numerous) and flagella (long and few) are elongate appendages that extend from the cell membrane into the surrounding environment. Movements of these structures in a coordinated fashion serve to propel the cell through a fluid medium and to establish at the cell surface currents in the surrounding medium. The microscopic structure of eukaryotic cilia and flagella includes microtubules arranged in a "9+2" pattern. Microtubules are heterodimers of the proteins alpha and beta tubulin. Movement of cilia and flagella involves chemical interaction of these tubulins with dynein and ATP.

(8) The cytoplasm is highly structured. The array of organelles is held in place with the microtrabecular lattice, an irregular network of fine interconnected fibers running through the cytoplasm. The cytomatrix includes three other systems of filaments in addition to the microtrabecular lattice: microtubules involved in maintaining cell shape, microfilaments involved in movements associated with cell motility, cell division and embryonic development, and intermediate filaments whose role is not yet understood.

F. <u>Eukaryotic Plant Cells</u>.--Being eukaryotic cells, animal and plant cells are certainly more similar to each other than either is to a prokaryotic cell. Features shared by plant and animal cells include the nucleus, mitochondria, Golgi apparatus (called <u>dictyosomes</u> in plants), ER, ribosomes, cilia, and flagella. These structures have corresponding functions in both plant and animal cells. Notably absent from plant cells are centrioles. <u>Cell walls</u> and <u>plastids</u> are key features in plants, but are absent from animal cells. Although <u>vacuoles</u> occur in some animal cells, they are prominent features of plant cells (Fig. 5.1):

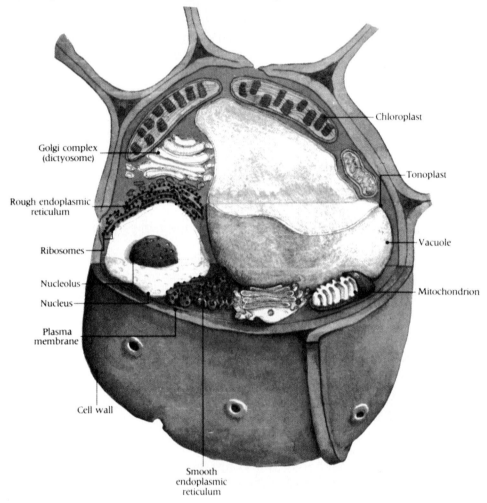

Fig. 5.1: Drawing of generalized plant cell

(1) Plant cells are bounded by a cell membrane (as are animal cells) and by a <u>cell wall</u> which encloses the cell membrane and all cell contents. Cellulose is the primary structural polysaccharide which composes the rigid cell wall. Tiny pores in the cell wall facilitate movement of materials into and out of a plant cell.

(2) <u>Plastids</u> resemble mitochondria in that both are enveloped by membranes and both contain DNA unlike that in the cell nuclei. Functions of these two organelles, however, are quite different. <u>Chromoplasts</u> are a category of plastids which contain pigments. For example, <u>chloroplasts</u> contain green pigment molecules, <u>chlorophyll</u>; chloroplasts serve an indispensible role in plant <u>photosynthesis</u>, the biochemical pathways in which light energy is converted to chemical forms usable by living organisms. <u>Amyloplasts</u>, an example of unpigmented plastids called <u>leucoplasts</u>, store energy in the form of starch.

(3) <u>Vacuoles</u> are fluid-filled cavities that are surrounded by a membrane, a <u>tonoplast</u>. The <u>cell sap</u> that fills a vacuole is a watery solution consisting of various materials such as amino acids, sugars, pigments (e.g., the colorful <u>anthocyanins</u>), and wastes. Vacuoles often occupy up to 90% of the volume of a cell. The degree of wilting or rigidity of a plant often relates to the amount of vacuolar space filled by water. In some plants, vacuoles work to rid cells of excess water.

G. <u>Prokaryotic Cells</u>.--This cell design exists only in kingdom Monera (Fig. 5.2). Prokaryotic cells probably originated earlier than eukaryotic cells. Such cells are simpler in design than are eukaryotic cells. Several prokaryotic characteristics are listed above at the beginning of section D. Additional features of prokaryotic cells include the following:

(1) Chlorophyll or other pigments, if present, are <u>not</u> contained in membrane-limited organelles.

(2) If cilia or flagella are present, their design is different than the "9+2" structure of eukaryotes.

(3) Cell membrane structure differs from eukaryotes in that the component lipids are fewer in number and are chemically simpler.

(4) Materials such as sugars and lipids are stored in <u>storage granules</u> rather than in membrane-bounded organelles.

(5) Although ribosomal function in both prokaryotic and eukaryotic cells is similar, prokaryotic ribosomes are smaller than are those in eukaryotes.

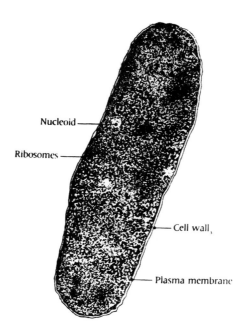

Fig. 5.2: Drawing of generalized prokaryotic cell

TERMS TO UNDERSTAND: Define and the identify the following terms:

cell

tissue

cell theory

Monera

Protista

Fungi

Plantae

Animalia

specialization

differentiation

virus

plasmid

viroid

prion

eukaryotic

prokaryotic

cell membrane

nucleus

cytoplasm

nuclear membrane

annulus

pore complex

chromatin

chromosome

nucleolus

rough endoplasmic reticulum

smooth endoplasmic reticulum

ribosome

polysome

Golgi apparatus

dictyosome

microbody

lysosome

phagocytosis

mitochondrion

centriole

cilium

flagellum

basal body

kinetosome

9+2 pattern

microtubule

tubulin

microtrabecular lattice

cytoskeleton

microfilament

cell wall

plastid

chloroplast

middle lamella

primary wall

secondary wall

lignin

stroma

grana

thylakoid

vacuole

anthocyanins

CHART EXERCISES

List, in the appropriate columns, characteristics of viruses that argue for or against their being living organisms.

LIVING FEATURES	NON-LIVING FEATURES
1.	1.
2.	2.
3.	3.
4.	4.
5.	5.

Considering the information in the chart above, decide whether viruses are living or non-living; write your summary argument in the space below.

PEOPLE TO KNOW: What is the significant biological contribution of each of these individuals as indicated in Chapter 5?

Robert Hooke

Anton van Leeuwenhoek

Matthias Schleiden

Theodor Schwann

Rudolf von Virchow

MATCHING

A. Write the letter of the appropriate organelle function into the blank next to the organelle that performs that function in a eukaryotic cell.

_____ 1. dictyosome

_____ 2. pore complex

_____ 3. nucleolus

_____ 4. cilia

_____ 5. lysosomes

_____ 6. chromatin

_____ 7. cell membrane

_____ 8. endoplasmic reticulum

_____ 9. mitochondrion

_____ 10. rough ER

A. carry materials which digest macromolecules

B. tangled strands of DNA which encode genetic information

C. sites of protein synthesis

D. separates cell contents from extracellular materials

E. network of channels in cytoplasm which transport materials within cell

F. stacks of flattened vesicles involved in secretion of cell products

G. regulates movement of some materials between cytoplasm and nucleoplasm

H. sites where energy-releasing metabolic reactions occur

I. produces ribosomes

J. hair-like structures which create currents at cell surface

B. Write the letter of the description (right column) into the blank next to the appropriate molecule (left column).

_____ 1. lignin

_____ 2. tubulin

_____ 3. chlorophyll

_____ 4. granulose

_____ 5. cellulose

A. a polymer of aromatic alcohols which reinforces cell walls

B. pigment found in a certain kind of plastid

C. primary component of plant cell walls

D. a storage carbohydrate in prokaryotes

E. globular proteins that form microtubules

DISCUSSION QUESTIONS

1. Distinguish between these four levels of subcellular organization: biomolecules, macromolecules, supramolecular complexes, and organelles.

2. Label the accompanying diagram of a generalized cell. For each structure, indicate next to its name whether it is found in only prokaryotic cells (P), only in eukaryotic cells (E), or in both types of cells (B).

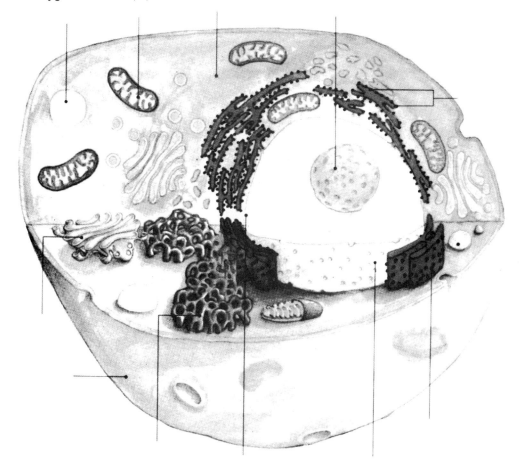

3. Distinguish between the following terms than can be used to categorize organisms: unicellular, multicellular, and acellular. Give an example of an organism for each category.

4. Summarize the diversity of size and shape of cells known to biology. What critical factor limits the maximum size of a cell?

TESTING YOUR UNDERSTANDING:

For each of the test items below, three of the lettered alternatives are true and the other is false. Determine which alternative is false and write its letter in the blank to the left of the question number. On the blank line below alternative D, write the corrected version of the false statement:

_____ 1. A. The cell is the smallest level of biological organization that possesses the characteristics of life.
B. Schleiden and Schwann reported that all organisms are made of cells.
C. There was disagreement among 19th century scientists regarding the means by which new cells are formed.
D. The cell theory states that new cells come from pre-existing cells by a process that resembles crystal formation.

Correction: _____

_____ 2. A. The cell theory contradicts the probable abiotic origin of life.
B. Unicellular organisms occur in only three of the five recognized kingdoms of life.
C. Organelles are assembled from supramolecular complexes which are held together by a variety of noncovalent bonds.
D. Cells of unicellular species lead lives that are independent of other such cells, whereas cells of multicellular species are dependent components of the larger organism.

Correction: _____

_____ 3. A. All cells of living species are microscopic.
B. Some of the smallest known cells are bacteria.
C. As cells increase in size, the ratio of cell surface area to cell volume decreases.
D. Changes in the surface area-to-volume ratio affect the effectiveness of the process of diffusion in delivering materials to the cell interior.

Correction: _____

_____ 4. A. Viruses are classified as non-living due to their inability to reproduce in the absence of host cells.
B. Plasmids, which are entities made of DNA, occur in many different types of bacterial cells.
C. The type of nucleic acid found in viroids is RNA.
D. A prion consists merely of a small bit of DNA surrounded by protein molecules.

Correction: _____

_____ 5. A. All eukaryotic and prokaryotic cells possess a cell membrane made of lipids.
B. Many of the organelles of prokaryotic cells are enclosed by membranes.
C. Bacterial cells lack a nucleus.
D. Polysaccharides are the primary chemical components of cell walls regardless of the kingdom into which the cell is classified.

Correction: _____

_____ 6. A. Prokaryotic cells are usually larger than eukaryotic cells.
B. The plasmalemma of a eukaryotic cell is composed of outer, middle, and inner layers.
C. The cell membrane can be regarded as a leaky container that regulates movement of materials into and out of the cell.
D. The endoplasmic reticulum connects to the nuclear envelope.

Correction: _____

_____ 7. A. Chromatin consists of DNA.
B. Ribosomes are produced inside of the nucleus.
C. The chromosomes of a eukaryotic cell are located within the nucleolus.
D. Synthesis of proteins by the endoplasmic reticulum requires the presence of ribosomes.

Correction: _____

_____ 8. A. Clusters of ribosomes are called polysomes.
B. Smooth endoplasmic reticulum is the site of synthesis of steroids and other lipids.
C. In plants, the Golgi apparatus is called a dictyosome.
D. Mitochondria are vesicles that pinch away from the surface of a Golgi apparatus.

Correction: _____

_____ 9. A. The sequence of events by which proteins are exported from a cell includes action of a Golgi apparatus.
B. A number of "storage diseases" in humans have been traced to presence of defective lysosomes.
C. The process of phagocytosis involves activity by lysosomes.
D. Mitochondria are organelles in which protoplasts are stored.

Correction: _____

_____ 10. A. The inner membrane of a mitochondrion is characterized by inwardly projecting plates called cristae.
B. The mitotic spindle develops from the outer membrane of a mitochondrion.
C. The cilia of a eukaryotic cell are composed of several proteins including tubulins and dynein.
D. Flagella and cilia function to propel a cell or to establish currents in the medium surrounding the cell.

Correction: _____

_____ 11. A. The basal body of a prokaryotic cell has fibrils which are arranged in the "9+2" pattern.
B. The spatial positioning of organelles with the cytoplasm is largely determined by the microtrabecular lattice.
C. Microtubules are long, thread-like molecules that function to maintain the shape of a cell.
D. Microfilaments are involved in the movements of cell division.

Correction: _____

_____ 12. A. A primary difference between animal and plant cells is the presence of a cellulose cell membrane in plant cells but not in animal cells.
B. The secondary cell wall of many plant cells contains a reinforcing polymer called lignin.
C. Some plastids function as storage structures.
D. Plastids that contain pigments are called chromoplasts.

Correction: _____

_____ 13. A. Mitochondria and chloroplasts are similar in that both are bounded by a double membrane.
B. Most of the space within plant cells is occupied by vacuoles.
C. The stacks of flat structures within a Golgi apparatus are called thylakoids.
D. Contractile vacuoles are important in maintaining water balance within a cell.

Correction: _____

_____ 14. A. The chlorophyll of prokaryotes is not contained within chloroplasts.
B. Prokaryotic cells lack endoplasmic reticulum.
C. Prokaryotes possess one chromosome which is located in the nuclear zone.
D. Ribosomes of prokaryotes and eukaryotes have similar functions and are of similar size.

Correction: _____

_____ 15. A. In eukaryotic cells, DNA is located in the nuclei, the chloroplasts and the mitochondria.
B. In eukaryotic cells, RNA is located only in the cytoplasm.
C. A cell wall characterizes cells in three kingdoms of organisms.
D. Photosynthesis is not known for any species in kingdoms Fungi and Animalia.

Correction: _____

CHAPTER 6--MEMBRANES AND CELL WALLS

CHAPTER OVERVIEW

Each cell of all living organisms is separated from its surroundings by a membrane. Furthermore, such cells contain an array of structures and organelles, many of which are themselves separated from their cytoplasmic surroundings by membranes. These membranes serve several purposes, the most important being to regulate movement (in both directions) of materials between the cell's interior and exterior.

Chemically, membranes are composed mainly of phospholipid and protein molecules. The phospholipids are arranged in a bilayer, with the hydrophilic parts of the molecules located at the membrane surfaces and the hydrophobic ends in the interior. Protein molecules are attached to and/or embedded within the phospholipid bilayer. A primary role of membrane proteins is to assist in movement of other molecules and other materials across the membrane.

Membranes are selectively permeable, that is, they allow some materials to cross but prevent others from passing through. Movement of materials occurs in a variety of ways including simple diffusion, facilitative diffusion, active transport, endocytosis, and exocytosis.

TOPIC SUMMARIES

A. <u>Functions of Membranes</u>.--The plasma membrane (<u>plasmalemma</u>) is the structure that physically separates the cell contents from the outside world. It is, therefore, a structure that serves a protective role. Yet, the cell membrane is not a complete barrier to passage of materials; a cell could not conduct many (if any) of its metabolic activities were raw materials unable to enter or products unable to leave the cell. Additionally, the cell membrane interacts with the cytoskeleton to determine cell size and shape and to maintain the physical integrity of the cell.

Keep in mind that the plasmalemma is not the only membranous structure in a cell. For eukaryotes, at least, many organelles are separated from the rest of the cytoplasm by membranes. Examples of such organelles include mitochondria, chloroplasts, endoplasmic reticulum, nuclei, vesicles and others. The functions of membranes surrounding these organelles are quite similar to those of the cell membrane.

B. <u>Concepts Related to Molecular Movements</u>.--In one way or another, movements of materials across a biological membrane can be explained on a molecular basis. Therefore, we will briefly examine several topics related to molecular movement. A concept fundamental to understanding passage of materials across membranes is that all molecules are in constant motion. Several factors affect the rate of molecular movement: Energy content of molecules is greater at higher temperatures. Correspondingly, molecules of gases move faster than those of liquids and those of liquids move faster than those of solids. Additionally, rate of movement decreases with increasing molecular weight.

1. Diffusion--Whether in a biological or a physical setting, molecules of a substance tend to move from areas of greter concentration to areas of lower concentration (that is, down a <u>concentration gradient</u>). The textbook example involves opening a perfume bottle (an area of great concentration of perfume molecules) in a room containing no or few of these molecules. Given enough time, the perfume molecules will disperse (<u>diffuse</u>) evenly throughout the room to achieve an <u>equilibrium</u> condition. To achieve such a uniform distribution of molecules in the room, it might seem that the molecules have minds of their own and that they intentionally seek out and move in directions where fewer molecules are located. Rest assured, however, that

molecular motion is completely random, not directional. Even at equilibrium, molecules are still in motion, yet concentration does not change because equal numbers of molecules are moving in opposite directions.

    2. Osmosis and Permeability--Diffusion occurs in gaseous, liquid and solid mediums; in each case the process continues until the molecules are evenly dispersed within the available space. <u>Osmosis</u> is a special case of diffusion in which a partition (called a membrane in biological systems) divides the available space. As in diffusion, molecules move randomly to achieve equal concentrations on each side of the membrane. However, characteristics of the membrane determine whether molecules will be able to pass through the membrane to achieve equilibrium. Membranes that allow indiscriminant passage of molecules are said to be <u>fully permeable</u>, those permitting no passage are <u>impermeable</u>, and those which are selective are <u>semipermeable</u> or <u>differentially permeable</u>.

    Biological membranes are differentially permeable. The ability of molecules to pass through such a membrane depends on factors including size, electrical charge and lipid solubility of the dissolved (<u>solute</u>) molecules. Some solute molecules are not allowed passage. So, how can equilibrium concentrations be attained? In biological systems, most reactions occur in aqueous (watery) solutions; the water molecules serving as the <u>solvent</u> are, however, small enough to pass through membranes. An alternate means of reaching equilibrium is for solvent (water) rather than solute molecules to pass. This is the means by which osmosis operates: Water moves from a chamber of low solute (and, therefore, high solvent) concentration across the membrane to a chamber of high solute (low solvent) concentration.

    Several terms are used to indicate relative solute concentrations. A solution with greater solute concentration is <u>hypertonic</u> to one with lesser solute concentration. Conversely, a <u>hypotonic</u> solution has a relatively lower solute concentration. Hence, water moves from a hypotonic to a hypertonic solution. Solutions with equal concentrations of dissolved molecules are <u>isotonic</u> solutions; no net change occurs between chambers containing solutions that are isotonic to each other.

    Solute concentrations on opposite sides of cell membranes (or between the interior of an organism and the outside world) are not often equal. Therefore, solvent and solute molecules are continually moving back and forth.

C. <u>Membrane Structure</u>.--The <u>unit membrane</u> that surrounds cells and many organelles consists of molecules arranged in a <u>bilayer</u>. Lipid and protein molecules are the primary components of biological membranes. Historically, two similar models have been proposed to describe membrane structure; the fluid-mosaic model has supplanted the model proposed by Davson and Danielli in 1925.

    1. Lipids--The primary lipid components of membranes are phospholipids, cholesterol and glycosphingolipids; prokaryotes lack cholesterol in their membranes. As you learned in Chapter 4, phospholipids are relatively long molecules having a polar hydrophilic head and a nonpolar hydrophilic tail. Phospholipids are arranged in two layers of molecules such that the tails of molecules of inner and outer layers contact each other. The heads of the outer layer face outwardly and those of the inner layer are directed inwardly towards the cytoplasm (Fig. 6.1). The internal hydrophobic region (formed by the tails) effectively prevents passage of polar molecules either direction across the membrane. How then do molecules of water, one of many critical polar molecules, pass across? See part D below for the answer.

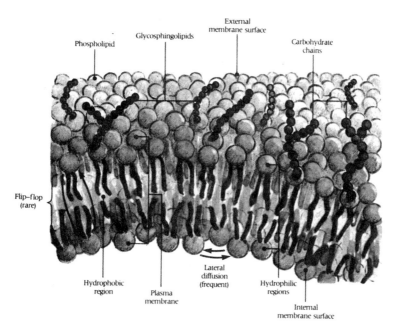

Fig. 6.1: Drawing of fluid-mosaic model of a biological membrane

Lipid molecules are not rigidly fixed into position in a membrane. Rather their positions are fluid (hence the name of the model). Molecules often shift sideways from place to place within either inner or outer sheets of the bilayer Such shifting is important for a cell to carry out its activities in an everchanging environment. Only occasionally, however, do lipid molecules move from one monolayer to the other.

2. Proteins--Although they are less abundant as membrane components than are lipid molecules, protein molecules perform most membrane functions. Membrane proteins are classified either by their chemical structure or by their location relative to the membrane surfaces. Chemically, proteins may be either <u>globular</u> or consist of a single chain of amino acids forming an <u>alpha helix</u>. Attached to ends of proteins protruding through the membrane surfaces are simple sugars; these protein-monosaccharide complexes, called <u>glycoproteins,</u> are an important part of the system by which a cell can "recognize" molecules, cells, or other materials.

Protein molecules can occur on the surface of and/or embedded within the membrane. <u>Intrinsic proteins</u> are largely embedded within the bilayer; their affinity for the interior of the membrane correlates with the presence of a nonpolar residue. (Check back to Chapter 4 to refresh your memory about protein residues.) Polar proteins are hydrophilic; such proteins that attach to the inner or outer surfaces are <u>extrinsic proteins</u>. It is such surface proteins that are involved in cell recognitions. Interconnections of various membrane proteins play an important role in determining and maintaining cell shape.

D. <u>Movement Across Membranes</u>.--Three distinct processes facilitate movement of materials across membranes. Factors that determine the particular process by which a molecule crosses pertain to electrical charge of the molecule, molecular size, and its lipid solubility. The processes differ in regard to involvement of protein carrier molecules and expenditure of energy.

1. Simple diffusion. Many nonpolar, lipid-soluble molecules cross membranes down a concentration gradient freely by <u>simple diffusion</u>. This is a process that does not require the cell to spend energy.

There are some polar, hydrophilic molecules (e.g., water) that also can cross membranes freely, as if they were diffusing. It appears, however, that rather than diffusing, water molecules pass through very narrow pores that are lined by hydrophilic molecules. The narrow diameter of these pores keeps larger molecular-weight molecules from passing.

2. Facilitated diffusion. This is a second process for moving molecules down a concentration gradient; as in diffusion, the cell does not invest energy in this process. Molecules moved by facilitated diffusion are too large to pass through the pores used by water and/or carry charges that are not compatible for passing through the hydrophobic interior of the membrane. Movement is facilitated by association of the molecule awaiting transport with a protein carrier molecule, called a permease. The carrier-passenger complex then diffuses to the opposite side of the membrane where the passenger is released and the carrier is available to combine with another passernger. Because the number of carrier molecules is limited, there is a maximum rate of movement that cannot be exceeded even if more passenger molecules are available.

3. Active transport. Many types of molecules and ions, independent of size and polarity, need to be moved from areas of lower concentration to higher concentration. As with facilitated diffusion, carrier molecules are required. Additionally, the cell must spend energy to move against the concentration gradient. Examples of active transport include accumulation of glucose within a cell and the sodium-potassium pump involved with transmission of nerve impulses.

4. Endocytosis and exocytosis. Some large molecules and other materials (such as viruses and bacteria) are too large to cross membranes by diffusion, facilitated diffusion or active transport. Such materials enter a cell by endocytosis, a process in which the flexible cell membrane surrounds the particle. The membrane that has engulfed the particle then pinches away from the plasmalemma to become a vesicle surrounded by cytoplasm. Be sure to review the endocytic cycle.

E. Cell Walls.--The cell membranes of plants, fungi, and bacteria are enclosed by a rigid carbohydrate cell wall. The cellulose cell wall of plants can be up to three layers thick. Communication between cells occurs through pores called plasmodesmata. Cell walls of fungi are made of chitin, whereas a single large molecule of peptidoglycan surrounds a bacterial cell.

TERMS TO UNDERSTAND: Define and the identify the following terms:

colloid

concentration gradient

diffusion

equilibrium

imbibition

permeable

semipermeable

osmosis

osmotic pressure

isotonic

hypertonic

hypotonic

plasmolysis

turgor pressure

unit membrane

Davson-Danielli model

cholesterol

phospholipids

liposomes

glycoproteins

glycophorin

HLA system

intrinsic (integral) protein

extrinsic (peripheral) protein

membrane skeleton

freeze fracturing technique

simple diffuion

membrane channel

active transport

membrane pump

calcium channel

endocytosis

phagocytosis

pinocytosis

exocytosis

receptor

ligand

endocytic cycle

coated pit

clathrin

coated vesicle

endosome

CURL

LDL

CHART EXERCISES

Different kinds of molecules and materials pass through membranes by a number of different processes. Complete the chart to indicate characteristics of materials that pass by each of these mechanisms.

CHARACTER	DIFFUSION	FACILITATIVE DIFFUSTION	ACTIVE TRANSPORT	ENDOCYTOSIS AND EXOCYTOSIS
Size of molecule or particle				
Electrical charge				
Solubility in lipids				
Example of molecule or particle				
Concentration gradient				

PEOPLE TO KNOW: What is the significant biological contribution of each of these individuals as indicated in Chapter 6?

E. Gorter and F. Grendel

Hugh Davson and James Danielli

S. J. Singer and G. I. Nicolson

Ilya Metchnikoff

Michael S. Brown and Joseph L. Goldstein

DISCUSSION QUESTIONS

1. Why do hydrophobic and hydrophilic molecules pass through membranes by different mechanisms?

2. What are the different roles of lipid and protein molecules in biological membranes?

3. Briefly outline the endocytic cycle.

4. Distinguish between <u>colloid</u> and <u>solution</u>.

5. Consider a vessel that is divided into two chambers by a semipermeable partition (see figure below) The partition allows passage of water, but not of glucose. Side A of the vessel contains a 3-molar solution of glucose; a 5-molar solution of glucose fills side B.

   a. Circle one of the terms next to each chamber to indicate concentrations of the solutions relative to each other.

b. Circle one arrow to indicate the net movement of water molecules as they move towards equilibrium.

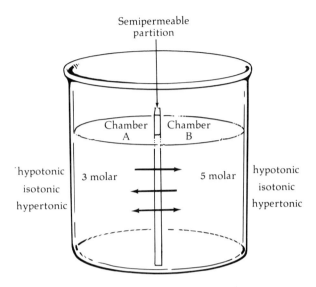

TESTING YOUR UNDERSTANDING:

For each of the test items below, three of the lettered alternatives are true and the other is false. Determine which alternative is false and write its letter in the blank to the left of the question number. On the blank line below alternative D, write the corrected version of the false statement.

_____ 1. A. Molecules in solutions and in colloids are in continuous motion.
B. Many large molecules that are unable to dissolve into solution can remain suspended as a colloid.
C. Movements of individual molecules are random.
D. Protein and lipid molecules usually form solutions whereas sugars and salts usually form colloid systems.

Correction: _____

_____ 2. A. Causes of movements of particles through a fluid medium include random motion of the molecules as well as convection.
B. Net movement of diffusing molecules is from areas of higher solvent concentration to areas to lower solvent concentrations.
C. Even when diffusing molecules have reached equilibrium, molecules remain in motion.
D. Rate of diffusion is greater at higher temperatures.

Correction: _____

63

_____ 3. A. The process in which a structure swells due to absorption of water is called imbibition.
B. Imbibition is an example of active transport.
C. Cell membranes are differentially permeable.
D. Osmosis is a special case of diffusion in which molecules move across membranes.

Correction: _____

_____ 4. A. Hypotonic solutions have greater osmotic pressures than do hypertonic solutions.
B. The process of plasmolysis involves removal of water from the water vacuole of a plant cell.
C. For plasmolysis to occur, a cell must be placed in a hypertonic solution.
D. Much of the structural support of plant tissues results from turgor pressure.

Correction: _____

_____ 5. A. Gorter and Grendel were the first researchers to propose that biological membranes were made of a double layer of lipids.
B. The currently used "fluid-mosaic" model of membranes was proposed by Singer and Nicolson.
C. Both the "fluid-mosaic" and Davson-Danielli models indicated the presence of lipid and protein molecules in membranes.
D. Proteins form a uniform coating on both sides of a membrane's lipid bilayer.

Correction: _____

_____ 6. A. The hydrophobic ends of phospholipid molecules are positioned near the inner and outer surfaces of the cell membrane.
B. Phospholipid molecules can change their locations within the membrane.
C. Phospholipid molecules can shift between inner and outer monolayers of a membrane.
D. Cell membranes of eukaryotes contain about twice as many phospholipid molecules as cholesterol molecules.

Correction: _____

_____ 7. A. The membrane skeleton consists mainly of interconnected membrane-protein molecules.
B. Cell recognition functions of membranes involve extrinsic proteins found at the outer surface of the membrane.
C. Extrinsic proteins frequently aid hydrophobic molecules in crossing the membrane.
D. Carrier molecules are often known as membrane permeases.

Correction: _____

_____ 8. A. Although phospholipid molecules are relatively free to move about within the membrane, protein molecules are held in fixed positions.
B. The protein content of a membrane varies according to cell type.
C. Most extrinsic proteins have polarized amino acid residues.
D. The part of a protein molecule that is embedded in the membrane interior usually is nonpolar.

Correction: _____

_____ 9. A. To conduct the processes of diffusion and facilitated diffusion, a cell must spend energy.
B. Active transport involves use of carrier molecules.
C. Active transport can move materials against a concentration gradient.
D. Materials cannot diffuse across a membrane from an area of lower concentration to an area of higher concentration.

Correction: _____

_____ 10. A. Water molecules move through tiny pores in the membrane.
B. Membrane channels are pores in membranes through which molecules pass.
C. Use of a membrane channel requires a passenger molecule to join with a permease.
D. Membrane channels are involved in export of materials from, and import of materials into, a cell.

Correction: _____

_____ 11. A. Potassium ions leave a cell by the process of facilitated diffusion.
B. For materials moved by facilitated diffusion, the rate of movement across a membrane continues to increase as concentration increases.
C. The rate of movement by active transport increases only to a certain rate even if the concentration of passenger molecules is increased.
D. Facilitated diffusion can move materials only from areas of higher concentration to areas of lower concentration.

Correction: _____

_____ 12. A. Export of waste molecules is an example of a situation involving active transport.
B. Although active transport of materials requires energy, that energy is not always obtained from ATP.
C. The cellular balance of sodium and potassium must be maintained by active transport because the cellular membrane normally is not permeable to sodium.
D. Movement of glucose and sodium in opposite directions is an example of coupled transport.

Correction: _____

CHAPTER 7--CELL FUNCTIONS

CHAPTER OVERVIEW

In this chapter, we continue our studies of the cellular basis of life. Already, we have seen that all organsims consist of one or more cells, and, regardless of cell type or the species of organism, that all cells have many shared components. It is reasonable, then, to expect that cells, even if in different organs or in different organisms, share some structural and functional features.

Chapter 7 addresses three broad areas of universal biological mechanisms exhibited by cells. Cells communicate (or interact) with each other in a variety of ways. Communication can be between cells that are physically associated with each other, between cells that are broadly separated within the same organism, or between cells of different organisms. In multicellular organisms, cells are the building blocks of which tissues, organs, systems and, ultimately, the entire organism are made. Virtually all types of cells exhibit some sort of motion that occurs due to expenditure of metabolic energy by the cell. Such motility may result in movement of the entire cell from one place to another or may involve movements of various subcellular structures. This chapter surveys these three categories of cell functions and develops a foundation that we'll build upon in later chapters.

TOPIC SUMMARIES

A. Cells As Building Blocks.--Multicellular organisms are built of cells that interact in a coordinated manner. Multicellularity is a complex way of existence which generally involves a division of labors among the cells of an organism. Different cellular functions generally correspond with different cell morphologies. Differentiation is the process by which cells acquire their different characteristics to carry out these different roles. Differentiation of the many specialized tissues of an organism from a single original cell is one of the most intriguing of all biological phenomena. A broad outline of general cell types and tissues of plants and animals follows:

(1) Cell types and tissues of plants. A plant body generally consists of the aboveground shoot system of stems and leaves and the belowground root system. Plant growth occurs in specialized regions called meristems. Growth in length is called primary growth. Growth in diameter, secondary growth, occurs in lateral meristematic tissues called vascular cambium. Three broad tissue types result from growth at such meristems:

The outermost layers of cells which cover underlying tissues are protective tissues. These include sheets of closely-fitted flat cells called epidermis and the periderm, a corky layer of dead box-shaped cells.

One of the three fundamental (or ground) tissues is parenchyma, made of thin-walled vacuolated cells which often contain chloroplasts. Cells composing sclerenchyma and collenchyma are thick-walled and serve to enhance the mechanical strength of plant parts.

The vascular tissues conduct materials within a plant. Xylem tissue, consisting of tracheids and vessel elements, transport water and dissolved minerals from the root system to the rest of the plant. These cells, which are stacked end-to-end in columns and that have perforate end walls, lose their nuclei and cytoplasm as they mature and effectively conduct water even after their deaths. Distribution of food materials from production sites in leaves and stems is the role of phloem tissue. Columnar sieve-tube elements and sieve cells are stacked into sieve tubes in which the cytoplasmic contents of these cells remain. Companion cells usually are associated with phloem tissue; they secrete materials into the sieve tubes. Fig. 7.1 illustrates several types of vascular cells.

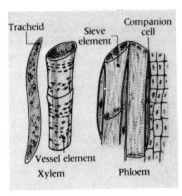

Fig. 7.1: Drawings of several types of vascular cells

(2) Cell types and tissues of animals. The great array of animal tissues can be grouped broadly into four categories. Body surfaces and the inner and outer surfaces of organs, cavities and passageways are lined with a thin covering of epithelial tissues. Cells of epithelium are joined closely to each other; very little intercellular material is present. Subdivisions of epithelial tissues are based upon cell shape and structure. Squamous cells are flattened, columnar cells are elongate, and cuboidal cells are just that--cuboidal.

A primary role of connective tissues is to bind together and to support other tissue types. A considerable amount of intercellular material fills the often large spaces between cells. The hard mineralized portion of bony tissue is an example of an intercellular material. Similarly, the plasma portion of blood is the "filler" between blood cells. Other categories of connective tissue are cartilage and connective tissue proper.

Muscular tissue is characterized by its ability to shorten and, thereby, to cause motion. Muscle is classified by the microscopically-visible presence (striated muscle) or absence (smooth muscle) of streaks or striations. Cardiac muscle of the heart and skeletal muscle possess striations.

Neurons are specialized cells whose role is to transmit electrical nerve impulses. Neurons plus the supporting nerve cells which do not conduct impulses compose nervous tissue. Nerve impulses travel in only one direction in neurons: from a series of short processes (dendrites) to the globular cell body and then away from the cell body along a long process (axon). Transmission of a nerve impulse from one neuron to another occurs across a small gap called a synapse. We'll deal with all of these plant and animal tissue types and their functions in more detail as we progress through the course.

B. Communication between Cells.--Cells that compose multicellular organisms do not operate independently; rather they interact in an integrated fashion that requires intercellular communication. Interacting cells might be in direct contact with each other or they might be separated by considerable distances, the distance depending on the size of the organism.

(1) Junctions between cells. Three functional types of junctions are recognized. Tight junctions exist between cells where cell membranes and/or wall materials of the cells fuse broadly. Such complete fusion causes obliteration of intercellular space. Hence, leakage through tissues whose cells are so joined does not occur. This anti-leak arrangement is

particularly important in situations where indiscriminant passage of materials is not desirable (for example, the lining of the walls of the small intestine).

Another type of intercellular contact, adhering junctions, functions to reinforce tissue structure. This association involves cell-to-cell contact in disc-like patches or plaques (spot desmosomes) or in more elongate band-like arrangements (belt desmosomes). These adhesions involve membrane proteins that anchor to the cytoskeletons.

Although the term gap junction may at first seem nonsensical, it is an appropriate name for an arrangement that involves contact but that also facilitates passage of low molecular weight materials between the joined cells. These junctions are composed of intergral membrane protein molecules whose positioniong defines a narrow tube-like channel piercing the cell membranes.

(2) Signalling systems. Certain types of molecules serve as messengers. The hormones of animal endocrine systems carry information from their sites of production throughout the body to cells located some distance away. Short-distance molecular signalling involves specific surface molecules, most of which are glycoproteins associated with the cells' membranes. Joining of cells involves cell adhesion molecules (CAM's); these molecules not only facilitate cell recognition but also can influence the cytoskeleton of a neighboring cell. CAM's, such as fibronectins and ubiquitin, apparently play a major role in regulating movement of cells during embryonic development and in determining where these cells come to rest.

C. Motion of and Within Cells.--Motility and mobility are not synonyms. Rather motility is motion in living systems that results from expenditure of metabolic energy and which results in the performance of mechanical work. Motility is manifest in many ways; it may involve movement of an entire cell from one place to another or it may refer to movement of parts of cells or to motion of structures within a cell. Your textbook lists at least 13 examples of motility. Such motions are based of one of two general sets of structures and mechanisms:

(1) Movements based on microfilaments. Microfilaments are protein-based structures that are capable of causing contraction. These are the structures that allow certain types of cells to drastically change their shapes and to move from one place to another. Such ameboid movement characterizes amebas (of course), certain other protists, and certain cell types found in multicellular animals (for example, macrophages and blood cells such as leucocytes and platelets). Ameboid movement involves protrusion of a pseudopod, its attachment to the surface by a ruffled membrane, a streaming of cytoplasm into the pseudopod, and a retraction of the remainder of the cell toward the pseudopod.

Contractile microfilaments are made of two primary proteins, actin and myosin, which polymerize into long filaments plus an array of smaller protein molecules variously involved in the contractile process. Actin occurs not only in animal muscle tissue, but in virtually all eukaryotic cells yet investigated. The occurrence of myosin is more limited, being found usually in muscle tissues. The mechanism by which muscles (and probably nonmuscular structures) contract is explained by the sliding filament model in which the amount of tension generated relates to the degree of overlap of thin filaments (actin) and thick filaments (myosin).

Like most other physiological processes, motility is a regulated rather than haphazard event. A key part in motility is the assembly of globular actin monomers in filamentous polymers. The degree of polymerization of actin is governed by an array of regulatory actin-binding proteins. Some of these proteins inhibit actin polymerization (for example, profilin), whereas others seem to limit length of a polymer chain (gelsolin, accumentin). Spectrin promotes cross-linking of actin filaments to each other. The role of vinculin is to attach actin filaments to other structures such as membranes. Tropomyosin facilitates binding of actin filaments to myosin filaments.

(2) Movements based on microtubules. The motion of cilia and flagella probably results from a mechanism in which microtubules, arranged in their characteristic "9+2" pattern, slide past each other. Recent research has demonstrated that microtubules are assembled in a directional manner; polymerization occurs at one end while tubulin subunits are removed from the other end. We'll examine microtubular motility in greater detail in later chapters, especially in the context of cell division.

TERMS TO UNDERSTAND: Define and the identify the following terms:

differentiation

root system

shoot system

meristem

cambium

epidermis

periderm

parenchyma

sclerenchyma

collenchyma

vascular tissues

xylem

tracheids

vessel elements

phloem

phloem cells

sieve plates

sieve tubes

companion cells

plasmodesmata

epithelium

squamous

columnar

cuboidal

connective tissues

muscular tissues

smooth muscle

striated muscle

cardiac muscle

nervous tissue

neuron

axon

dendrite

synapse

tight junctions

microvilli

adhering junctions

desmosomes

gap junctions

hormones

glycoproteins

cell adhesion molecules (CAM's)

mobility

motility

microtubules

microfilaments

actin

tubulin

ameboid movement

pseudopod

mantle

myosin

actin-binding proteins

tropomyosin

directional assembly

CHART EXERCISE

Animal tissues are divided into four major categories: epithelium, musclar tissue, nervous tissue, and connective tissue. Complete the chart below to indicate the functions and special characteristics of the five tissue examples provided.

TISSUE TYPE	MAJOR CATEGORY	FUNCTION	SPECIAL CHARACTERISTICS
Cardiac muscle			
Nerve			
Bone			
Skin			
Smooth muscle			

MATCHING

Match the molecule in the left column with the most appropriate role or description for that molecule in the right column.

	Molecule		Role or Description
____	1. accumentin	A.	the primary component of which thin microfilaments are made
____	2. fibronectin	B.	responsible for "homing" ability of certain cell types
____	3. actin	C.	a signalling ribonucleotide that passes between cells through gap junctions
____	4. profilin	D.	thick protein filaments of muscle tissue
____	5. cyclic AMP	E.	a glycoprotein involved in cell adhesion
____	6. ubiquitin	F.	a protein that binds to the pointed end of an actin polymer
____	7. tubulin	G.	the primary component of which microtubules are made
____	8. protein 110K	H.	inhibits actin polymerization
____	9. myosin	I.	attaches actin to cell membranes

DISCUSSION QUESTIONS

1. Why is it important that cells of an organism be able to communicate with each other? Briefly outline several ways by which such intercellular intraorganismal communication occurs.

2. Distinguish between <u>mobility</u> and <u>motility</u>. Provide a biological example of each.

3. Describe the sequence of events by which ameboid movement occurs.

4. Why is it important that motility processes be regulated? Give examples of specific molecules that are involved in regulating motility.

5. Compare and contrast the structure of plant and animal tissues involved in corresponding activities such as protecting the organism . . . or conducting fluids within the organism . . . or in communication within the organism.

LABELLING EXERCISE

The diagram below represents a longitudinal section through a plant stem. On the basis of information presented in Chapter 7 (and elsewhere in the textbook as needed) identify the specific plant tissues indicated in the diagram below. In the blanks next to the diagram, write (a) the name and (b) the function of the indicated tissue. Select labels from these choices: cambium, epidermis, fundamental tissues, phloem, and xylem.

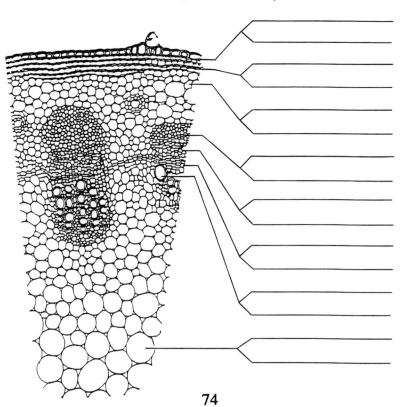

TESTING YOUR UNDERSTANDING:

For each of the test items below, three of the lettered alternatives are true and the other is false. Determine which alternative is false and write its letter in the blank to the left of the question number. On the blank line below alternative D, write the corrected version of the false statement.

_____ 1. A. Tight junctions are frequently found in tissues where indiscriminant movement of molecules between cells is not desirable.
B. The ability of cells to communicate with each other by diffusible substances exists in eukaryotes but is unknown in prokaryotes.
C. The coordinated interaction of the cells in a multicellular body requires that cells be able to communicate with each other.
D. One of the means by which cells can communicate involves exchange of cytoplasm between cells.

Correction: _____

_____ 2. A. Impermeable joints are important in tissues involved in controlled passage of materials.
B. Desmosomes are composed of membrane protein molecules that are anchored to the cytoskeleton.
C. Desmosomes are tube-like structures that allow passage of low-molecular weight compounds.
D. Adhering junctions are involved with the processes of cell support and with the union of adjacent cells.

Correction: _____

_____ 3. A. Most of the specific surface molecules that serve as recognition signals between cells are glycolipids.
B. Messenger molecules that are secreted from cells and that affect other cells at distant locations are called hormones.
C. A role of cell adhesion molecules is to regulate the movements of cells during development.
D. Fibronectins, a group of glycoprotien molecules, are located on the surface of cell membranes.

Correction: _____

_____ 4. A. Differentiaiton is the process by which cells become morphologically and functionally specialized.
B. Composite tissues are composed of two or more types of cells.
C. Meristematic tissues of plants are involved in transport of minerals and water throughout plant bodies.
D. The chloroplasts of plants often are contained in parenchymal cells.

Correction: _____

_____ 5. A. Epidermal cells are flat cells that provide protection for underlying tissues.
        B. Peridermal cells divide to produce cambium.
        C. Trachieds and vessel elements function in their roles of water transport and of lending mechanical support to a plant even after cell death.
        D. Phloem cells are separated by seive plates.

        Correction: _____

_____ 6. A. Direct communication between adjacent plant cells is achieved by structures called collenchyma.
        B. Plant sap is conducted within xylem tissue.
        C. Bone is an example of connective tissue.
        D. The amount of extracellular material in connective tissue is greater than in epithelial tissue.

        Correction: _____

_____ 7. A. Squamous and columnar are types of cells characterizing epithelial tissue.
        B. Nervous tissue is specialized for transmitting electrical impulses.
        C. Muscle tissue can be classified on the basis of its microscopic appearance as well as on the degree of voluntary control that can be exerted on it.
        D. The primary muscle tissue found in the heart lacks striations.

        Correction: _____

_____ 8. A. Motion in a biological system that does not require expenditure of metabolic energy is mobility.
        B. Motility can involve motion of subcellular structures or of entire cells.
        C. Ameboid movement occurs in certain types of cells from kingdoms Protista and Animalia.
        D. Motility results from activity of lipid-based structures called microtubules and microfilaments.

        Correction: _____

_____ 9. A. Microfilaments are made of actin and myosin whereas microtubules are composed of tubulin.
        B. Although actin and myosin have been known for decades to be present in muscle tissue of animals, these molecules are not yet known from other kingdoms.
        C. Pseudopods attach to the surface by regions of plasmalemma called ruffled membrane.
        D. Polymerization of actin occurs more rapidly at the barbed end of these filaments.

        Correction: _____

CHAPTER 8--METABOLISM AND THE TRANSFER OF ENERGY

CHAPTER OVERVIEW

Cells (and, therefore, organisms) are vessels in which a vast array of chemical processes occurs continuously. One segment of these metabolic processes degrades materials brought into the cell to provide the energy and the raw materials from which the other segment of metabolism constructs biomolecules.

Associated with every biochemical reaction is a transfer of energy between reactants, products, and the surroundings. In this chapter, we examine the nature and behavior of energy as described by the Laws of Thermodynamics. We study the connection between energy and metabolic reactions. We see how metabolic reactions are driven by biological catalysts called enzymes. Regulation of enzyme-mediated reactions occurs at two levels: several strategies affect the level of enzyme activity, while other approaches target the rates of enzyme synthesis.

TOPIC SUMMARIES

A. Energy.--Energy occurs in many forms, including light, electricity, chemical, heat, and others. Whatever the form, energy represents a capacity to do work. Stored energy is referred to as potential energy; when that stored energy is put into motion, then the energy has been converted into kinetic energy. For example, a compressed spring contains potential energy which transforms into kinetic energy when the spring is released.

The energy that flows through living organisms originates from various physical sources, primarily the sun. Plants convert into biomass less than 1% of the solar energy that strikes the earth! Plants are the (primary) group of organisms responsible for transferring energy from the physical world into forms that are usable by virtually all other kinds of organisms. Seemingly, all biological activity involves energy transformations. Bioenergetics is the study of energy transformations in living organisms. Therefore, an understanding of biology requires a grasp of the nature and behavior of energy.

B. Energy and the Laws of Thermodynamics.--The nature and behavior of energy in any system is described by the two laws of thermodynamics.

(1) First Law of Thermodynamics. The First Law states that the total energy content of the entire universe remains constant. This implies that energy can neither be created nor destroyed, but it can be changed from one form to another. For example, the chemical energy bound into the hydrocarbons of gasoline in a car is converted into energy of motion, heat, and sound. The energy content of the gasoline oxidized by the car engine equals the sum of the amounts of the various kinds of energy produced as the engine runs.

Chemical reactions either release (exergonic reactions) or absorb (endergonic reactions) energy. Heat energy is measured in units called calories. The heat content of a substance is called enthalpy. The reactants and the products of a reaction usually have different heat contents. If this difference is negative, then the products possess a lesser heat content than did the reactants; therefore, heat energy was released by that reaction (an exothermic reaction). A positive value for change in enthalpy indicates that the products contain more heat energy than did the reactants; heat energy was absorbed by that reaction (an endothermic reaction). In living organisms, these two types of reactions usually are coupled so that heat energy liberated by one reaction is used to drive reactions that require energy.

(2) Second Law of Thermodynamics. <u>Entropy</u> refers to randomness or the disorder of a system. The <u>Second Law</u> states that the entropy of the universe is increasing. A system with an entropy value of zero is perfectly ordered and probably purely hypothetical; in such a system there could be no atomic or molecular motion. At temperatures only slightly above absolute zero, atomic and molecular motion occurs and produces disorder and, thereby, has an increasing entropy value. All chemical and physical processes cause increases in the entropy of a system. Increase in entropy is spontaneous and is not reversible. A corallary of the Second Law, then, is that the total energy content of the universe eventually will degrade into unusable form and all processes in the universe will grind to a halt!

So far as is known, no chemical or other processes are completely efficient. Therefore, some of the transferred energy degrades into unusable form in every reaction; only a part of the transferred energy is still usable. The difference between total energy change and the lost energy is referred to as <u>free energy</u>--energy that is available for future use. Equations 8.3 and 8.4 in your textbook expresses this relationship. Energy-releasing reactions always have a negative ΔG, that is, with values less than zero. A reaction can occur spontaneously only if ΔG is negative. Endergonic reactions can occur only because the needed free energy is supplied from some other source . . . such as a coupled exergonic reaction.

Understanding the concept of <u>free energy</u> may be made easier by considering an analogy familiar to us all--a comparison of gross wages versus net wages. Our salaries are taxed so that our take-home pay always is somewhat less than our total wages or salaries. Think of the gross pay you receive from your employer as the total change in energy (ΔH in textbook equation 8.4), of your take-home pay as the available or free energy (ΔG; economists call this "disposable income"), and of the deducted taxes as the entropy (ΔS; although our tax dollars are usable by others, an individual might still consider them lost) of the system.

C. <u>Metabolism</u>--Metabolism can be thought of as the sum total of all chemical reactions that occur within a cell or organism. These many thousands of reactions can be grouped into three sets of processes:

(1) All cells need <u>energy</u> and simple <u>raw materials</u> in order to build the more complex biomolecules and the myriad of structures made from such molecules. <u>Catabolism</u> is that set of metabolic reactions which breaks down the materials which an organism takes in to liberate energy and the required chemical building blocks.

(2) Raw materials taken in by an organism are the source of all of the chemical components of that organism. Yet, many (or perhaps, most) of the chemicals forming an organism are different than the materials that entered the organism. It is apparent, then, that organisms are capable not only of breaking down materials, but also of constructing new and different molecules from those breakdown products. This segment of metabolism which involves <u>biosynthesis</u> is referred to as <u>anabolism</u>.

(3) The processes of catabolism and anabolism are integrated and coordinated by an array of <u>regulatory mechanisms</u>. We'll examine some of these at the end of this chapter.

D. <u>Metabolism and Chemical Equilibrium</u>.--Thermodynamics tells us whether a reaction can occur. <u>Kinetics</u> is the branch of physical chemistry that deals with questions concerning how fast a reaction occurs, the <u>reaction rate</u>. Many reactions, such as the <u>reversible reaction</u> expressed by the general equation 8.5 in your textbook, can occur in both forward and backward directions. That is, reactants A and B on the left side of the equation react to form products C and D. The reverse reaction, in which "products" C and D react to form "reactants" A and B, occurs simultaneously. For any reversible reaction, the forward and reverse reactions <u>initially</u> do <u>not</u> occur at the same rate; <u>rate constants</u> indicate the speeds at which these reactions proceed. Reactions can, however, progress to a point where forward and reverse reactions occur at equal rates; the reaction is then said to have achieved equilibrium. The equilibrium constant, K_{eq}, for that reaction indicates the ratio of products to reactants at which equilibrium occurs.

An implication arising from the phenomenon of equilibrium is that all such reactions are, to some degree, wasteful in that not all reactant molecules have been converted into product. Organisms cannot operate if reactions achieve equilibrium. A way of assuring that equilibrium does not occur is to prevent accumulation of products, by such means as consuming the product at an appropriate rate.

E. Metabolism and Catalysts.--Metabolism is concerned mainly with energy in chemical form. Chemical reactions involve motion of atoms and molecules; reactions not only produce products that are different than the reactants, but also involve changes in the energy content of the chemicals involved. We'll compare energy-releasing and energy-storing reactions below, but for now let's examine the concept of activation energy: Many reactions require a bit of "encouragement" in order to get underway. Refer to Fig. 8.1 which depicts the energy content of reactant molecules through the course of the reaction. Notice that reactant molecules must somehow achieve an energy boost to get over the hump before the reaction can begin. How? This extra energy is the energy of activation for that reaction.

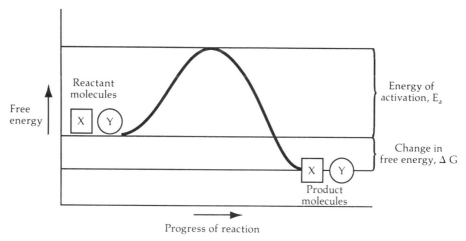

Fig. 8.1: Change in energy content of molecules during course of reaction

There are several ways to provide activation energy. All of these ways involve increasing the frequency with which the reactant molecules collide with each other. Increasing the temperature causes molecules to move more rapidly and, therefore, increases the likelihood of collisions. A general rule of thumb is that reaction rates double or even triple with every 10 degree Celsius increase in temperature. Another logical way to increase molecular collisions is to increase the concentrations of the reactants by adding more reactant molecules to the reaction vessel. In biological systems, however, a different approach is commonly used: catalysts are used to dramatically speed up reactions by lowering activation energies

F. Enzymes--Biological catalysts are known as enzymes. The names of enzymes can usually be recognized by the presence of the suffix -ase which is added either to the name of the substrate or to a word describing the type of reaction being catalyzed. For example, proteases are enzymes which promote reactions that break down protein molecules, and a dehydrogenase drives a reaction which removes a hydrogen atom from a molecule.

Enzymes are characterized as follows: All are proteins. They operate under the mild and relatively constant conditions of temperature, pressure and pH that occur in living organisms. They are remarkably specific--most enzymes will catalyze only one or a few reactions which involve only certain reactant molecules. Because of their tremendous efficiency, one enzyme molecule can catalyze reactions of thousands or even millions of molecules per minute. Like other catalysts, enzymes are not consumed nor permanently altered by the reactions they catalyze. Remember, also, that the reactions catalyzed by enzymes would indeed occur in their absence, but at a tremendously slower rate. . . perhaps too slowly to keep an organism alive.

Over a century ago, Fischer and Ehrlich proposed a mechanism, the template theory (lock-and-key theory), to explain how enzymes operate to lower activation energies: Each enzyme molecule (E in the equation below) has on its surface an active site where specific reactant molecules (substrate; S in the equation) attach. Molecules united in such a way are called the enzyme-substrate complex (ES in equation 8.1). When the substrate molecule(s) attaches, it induces a change in the shape of the active site of the enzyme; refer to the induced fit model (Fig. 8.2).

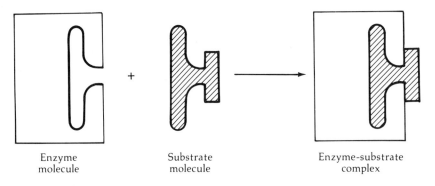

Enzyme molecule   Substrate molecule   Enzyme-substrate complex

Fig. 8.2: Induced fit model of forming the enzyme-substrate complex

These changes result in the proper alignment of the reactive groups of the enzyme with the effect of bringing reactant molecules close together in proper orientation. When associated in this activated enzyme-substrate complex (ES' in equation 8.1), various bonds of the reactant molecules are strained and weakened; this achieves a lower activation energy barrier than can occur in the absence of the enzyme. As the substrate molecules react, the enzyme molecule is released to re-enter the catalytic cycle. The following equation (equation 8.8 in your textbook) symbolically represents the steps involved in a reaction catalyzed by an enzyme:

$$E + S \longleftrightarrow ES \longleftrightarrow ES' \longrightarrow E + products \quad (8.1)$$

Already we have noted that enzymes are rather "picky" in terms of which reactions they can catalyze. The degree of specificity seen depends upon the enzyme being considered: Acetylcholinesterase recognizes only acetylcholine, whereas lipases will attack the ester linkages present in virtually any kind of fatty acid. One of the consequences of enzyme specificity is the phenomenon of competitive inhibition, a situation in which an enzyme is "fooled" into recognizing and binding, at the active site, with a molecule that closely resembles the appropriate substrate molecule. Such competition by counterfeit substrates causes a slower reaction rate, with the amount of slowing related to the relative concentrations of the competitor and substrate molecules. Noncompetitive inhibition is another way of slowing reaction rate. In this situation, the inhibitor molecule or atom attaches to the enzyme molecule at a location other than the active site and causes the enzyme molecule to change in shape. Such shape alteration prevents substrate molecules from binding at the active site, regardless of the amount of substrate present. Many poisons, such as cyanide, operate by noncompetitive inhibition.

One of the means of regulating enzyme activity involves attachment of effector molecules at regulatory sites, sites different than either the active site or that occupied by a noncompetitive inhibitor. Allosteric effectors cause changes in the shape of an enzyme molecule; some molecules (allosteric activators) induce changes that increase the affinity of the enzyme molecule for its substrate and, thereby, increase reaction rate. Conversely, allosteric inhibitors cause changes that decrease enzyme-substrate affinity.

Many enzymes require "assistance" by other molecules in order to drive a reaction. These non-protein assisting molecules are called coenzymes. An enzyme (apoenzyme) combined with its coenzyme forms an active holoenzyme. The functional role of some coenzymes, such as flavin mononucleotide (FMN) and nicotinamide-adenine dinucleotide (NAD), pertains to shuttling hydrogen atoms between various compounds; we'll spend lots more time with FMN and NAD in the upcoming chapters on cellular respiration and photosynthesis. Cofactors are inorganic coenzymes, the best-known of which are metal ions such as magnesium, calcium, zinc, and copper.

G. Metabolism and Oxidation-Reduction Reactions.--The reactions of metabolism involve movements of hydrogen atoms and electrons between compounds. Reactions involving such shuttling are usually coupled, such that a hydrogen liberated in one reaction is used in a related reaction. These coupled reactions are called oxidation-reduction (redox) reactions. We are interested in oxidation-reduction reactions because movement of electrons between atoms or molecules is accompanied by movement of energy.

Textbook Table 8-1 lists various chemical events that embody oxidations and reductions. Oxidation is any reaction in which an atom or molecule loses an electron; two ways by which oxidation can occur are by adding oxygen or by removing hydrogen. Conversely, a reduction is any reaction in which an atom or molecule acquires an electron; two ways by which reduction can occur are by removing oxygen or by adding hydrogen. In any redox reaction, one reactant becomes reduced while another becomes oxidized.

H. Control of Cellular Metabolism.--The preceding paragraph hints at the existence of means of regulating cellular metabolism. An effective regulatory mechanism controls rates of enzyme reactions, prevents waste of materials and energy, and must function efficiently and continuously. Known regulatory mechanisms focus on controlling either the catalytic activity of enzymes or the rates at which enzymes are synthesized.

Most metabolic processes require a series of chemical steps called a metabolic pathway. Most steps in a pathway are mediated by enzymes specific to each particular step. In much the same way as a chain's weakest link determines the strength of the entire chain, the slowest step (rate-limiting step) in a metabolic pathway forms a bottleneck that determines the rate at which the entire pathway can proceed. Rate-limiting steps usually occur early in a pathway.

(1) Control by altering enzymatic activity. Feedback inhibition is a common means of slowing a metabolic pathway. This occurs when the chemical product of the pathway is an inhibitor of an enzyme that catalyzes a reaction of that pathway. Such inhibitory molecules operate by allosterically lowering the enzyme's affinity for its substrate.

Altering the supply of an enzyme's substrate is an effective means of regulating a metabolic pathway. This strategy can work at substrate concentrations below the concentration at which the population of enzyme molecules is saturated. Changes in substrate concentration beyond the saturation level do not affect reaction rates. This relationship is depicted in textbook Figure 8-11. Two other methods of regulating enzyme activity pertain to the actions of hormones and to coupling mechanisms; we'll examine these more closely in later chapters.

(2) Control by affecting enzyme synthesis. Much of this sort of control occurs at the level of the gene, a stretch of DNA which codes for synthesis of enzymes. Substrate molecules (or in some cases, molecules that very closely resemble substrate molecules for a specific enzyme) can induce the gene for that enzyme to produce molecules of that enzyme. An opposite approach may also be used: production of an enzyme can be repressed at elevated concentrations of molecules produced by a reaction or pathway catalyzed by that enzyme.

Certain hormones seem also to work at a level that affects rates of enzyme synthesis. One other means of regulating enzyme activity pertains to activation of molecules that are precursors of an enzyme; placed into the proper chemical surroundings, these "pre-enzymes" are converted into functional enzyme molecules.

TERMS TO UNDERSTAND: Define and the identify the following terms:

energy

work

thermodynamics

bioenergetics

universe

system

equilibrium

calorie

enthalpy

exothermic

endothermic

free energy

coupled reactions

catabolism

anabolism

kinetics

equilibrium constant

activation energy

catalyst

enzyme

active site

induced-fit model

competitive inhibition

noncompetitive inhibition

effectors

allosteric effect

coenzyme

apoenzyme

holoenzyme

cofactors

oxidation

reduction

rate-limiting step

feedback inhibition

cooperativity

induction

repression

derepression

PEOPLE TO KNOW: What is the significant biological contribution of each of these individuals as indicated in Chapter 8?

Emil Fischer

Paul Ehrlich

Willard Gibbs

Christian Anfinsen

Daniel Koshland

DISCUSSION QUESTIONS

1. Discuss the three primary functions of metabolism.

2. The rate at which a chemical reaction proceeds depends on the frequency of collisions of the reactant molecules. Describe three ways in which the probability of such collisions can be increased.

3. The actions of enzymes can be inhibited in a variety of ways. Describe how the mechanisms of competitive inhibition and noncompetitive inhibition differ.

4. What is the role of allosteric effects in regulating enzyme activity? Provide examples.

5. In the context of this chapter, what is the role of certain vitamins to cellular metabolism?

6. What are the implications of the Second Law of Thermodynamics for the future of the universe?

7. What is the relationship of the following terms: endergonic, endothermic, reduction, electrons?

8. Using kinetics, explain how preventing the accumulation of the product of a reaction can be an effective means of regulating a metabolic reaction.

9. Repression and induction are important processes in regulating enzyme-mediated reactions. Compare these two processes.

10. What is meant by the process of coupling of reactions? What is the value of such an arrangement of reactions? Provide examples of materials shuttled between coupled reactions.

TESTING YOUR UNDERSTANDING:

For each of the test items below, three of the lettered alternatives are true and the other is false. Determine which alternative is false and write its letter in the blank to the left of the question number. On the blank line below alternative D, write the corrected version of the false statement.

_____ 1. A. Catabolic reactions represent the degradative portion of metabolism.
B. Catabolic reactions provide the energy and the raw materials required for biosynthesis.
C. Metabolism can be properly defined as the sum of all the chemical reactions occurring within a cell.
D. Unlike kinetic energy, potential energy always involves matter in motion.

Correction: _____

_____ 2. A. Activation energy is the energy that must be added to the reactant molecules in order to initiate that chemical reaction.
B. Decreasing the temperature of a reactant system is one way of supplying the needed activation energy.
C. The addition of an appropriate catalyst to a reactant system is an effective means of supplying activation energy.
D. Energy of activation can be achieved by any factor that increases the frequency of collisions of reactant molecules.

Correction: _____

_____ 3. A. Biological catalysts are known as enzymes.
B. All enzymes are proteins.
C. Although many catalysts are permanently altered during the reactions they promote, they generally are not consumed during that reaction.
D. Catalysts do not cause reactions to occur that would not eventually occur in the absence of the catalyst.

Correction: _____

_____ 4. A. The rate of a chemical reaction increases by a factor of about 2 or 3 for every 10 degree Celsius increase in temperature.
B. Most biological catalysts operate over broad ranges of temperature, pressure and pH.
C. According to the rules of nomenclature, an enzyme promoting the breakdown of cellulose could correctly be called cellulase.
D. Enzymes are specific--that is, each type of enzyme is selective in which reaction(s) it will catalyze.

Correction: _____

_____ 5. A. A reactant molecule involved in a reaction catalyzed by an enzyme attaches to the active site of the enzyme molecule.
B. Competitive inhibitor molecules attach at the active site of an enzyme.
C. The template theory explaining the phenomenon of enzyme specificity was proposed by Fischer and Ehrlich about 100 years ago.
D. The induced fit model explains how competing molecules can inhibit enzyme-driven reactions.

Correction: _____

_____ 6. A. The conformation of an enzyme's active site changes when the enzyme and reactant molecules are associated into an enzyme-substrate complex.
B. Many poisons operate by a mechanism of noncompetitive inhibition of enzymes.
C. Allosteric effector molecules attach to an enzyme at the regulatory site which is different than the active site.
D. The association of an allosteric effector with an enzyme always decreases the affinity of that enzyme for its substrate.

Correction: _____

_____ 7. A. Many enzymes must be associated with coenzymes in order to carry out their roles as catalysts.
B. Although apoenzymes are always proteins, coenzymes are not proteins.
C. Coenzymes frequently operate as donor-acceptors of chemical groups such as hydrogen atoms.
D. The importance of vitamins in human bodies relates to the fact that several inorganic cofactors derive from vitamin molecules.

Correction: _____

_____ 8. A. A reduction can be defined as the acquisition of an electron by an atom or molecule.
B. Loss of an electron has the effect of decreasing the net electrical charge of an atom or molecule.
C. Because oxidation reactions release energy, they are referred to as exergonic.
D. Reduction reactions that absorb heat energy are endothermic.

Correction: _____

_____ 9. A. Solar energy is the primary physical source of energy that enters the living world.
B. Entropy refers to heat content of a system.
C. The Second Law of Thermodynamics states that the entropy of the universe is increasing.
D. A chemical reaction with a negative enthalpy change is an exergonic reaction.

Correction: _____

_____ 10. A. A stick of dynamite possesses greater entropy before it explodes than after it explodes.
B. In biological systems, the energy released by an exergonic reaction usually is used by an endergonic reaction.
C. With each passing moment, the entropy of the universe increases.
D. The maximum state of entropy in an organism occurs at the time of death.

Correction: _____

_____ 11. A. The concept of free energy allows accounting for the inefficiency of reactions: some of the energy released in a reaction is lost as heat.
B. Increases in entropy are accompanied by decreases in free energy.
C. A reaction can occur spontaneously only if free energy has a positive value.
D. The letter G symbolizes free energy, whereas S denotes entropy.

Correction: _____

_____ 12. A. Reversible reactions, such as those occurring in living organisms, can achieve equilibrium in which forward and backward reactions occur at the same rates.
B. The ratio of the equilibrium constants for the forward and backward reactions of a reversible reaction is the rate constant.
C. Reversible reactions that occur in living organisms generally do not reach equilibrium because the products of the forward reaction are not allowed to accumulate.
D. Kinetics is the branch of physical chemistry that examines rates of chemical reactions.

Correction: _____

_____ 13. A. Most known mechanisms for controlling cellular metabolism focus on the enzymes that catalyze those reactions.
B. Rate-limiting steps of a biochemical pathway usually are positioned early in the sequence of reactions.
C. Elevated concentrations of the end-product of a metabolic pathway often cause that pathway to stop operating.
D. The existence of feedback inhibition suggests that regulatory enzymes have inhibitor specificity but lack substrate specificity.

Correction: _____

_____ 14. A. Regulation of enzyme activity can be accomplished by changing the concentration of substrate molecules.
B. Increasing substrate concentration beyond the level at which enzyme molecules are saturated will increase the rate of that enzyme reaction.
C. Hormones can affect reaction rates by competing with the coenzyme of that enzyme.
D. In some cases, a hormone can be the coenzyme in a reaction.

Correction: _____

_____ 15. A. In repression, the addition of substrate molecules to the system inhibits further production of the enzyme.
B. The conversion of pepsinogen into pepsin is an example of a regulatory mechanism that focusses control at the rate of production of an enzyme.
C. The following is an example of the process of induction: rate of production of an enzyme increases when molecules of its substrate are added to the system.
D. Hormones can control enzyme-regulated reactions at the levels of enzyme activity and of enzyme synthesis.

Correction: _____

CHAPTER 9--HOW CELLS EXTRACT ENERGY FROM THE ENVIRONMENT

CHAPTER OVERVIEW

We can classify organisms on the basis of their food and energy sources: Those that can synthesize the organic materials they need from simple inorganic compounds are autotrophs. Heterotrophs are very limited in what sorts of materials they can synthesize, so they require many preformed organic compounds in their food. We'll examine synthesis by autotrophs in the next chapter, but here (in Chapter 9) we'll see that <u>all</u> organisms follow more or less the same pathways for releasing energy and nutrients from organic compounds.

The sugar glucose is the key molecule for entering the catabolic segment of metabolism. All species carry out glycolysis (the first ten steps in the breakdown of glucose) in which a small amount of chemical energy is liberated in usable form. Three alternate routes exist for the product of glycolysis, pyruvate. The pathway followed varies according to species and reflects the availability of oxygen in the immediate surroundings. If little or no oxygen is present, then either anaerobic fermentation or lactate respiration occurs. The citric acid cycle and oxidative phosphorylation occur in the presence of oxygen; such aerobic oxidation of glucose is <u>much</u> more efficient in production of molecules of the energy currency, ATP.

The reactions involved in these metabolic pathways are numerous and often complex. We will not concentrate on specific reactions, but instead try to develop the "big picture." Focus your attention on the "summary equations" which show beginning and ending materials and leave out most of the specifics that occur in between. Concentrate on (1) the fates of the carbon molecules of the original glucose molecules, (2) on coenzyme molecules in their role of shuttling hydrogen ions, and (3) transfer of energy between compounds.

TOPIC SUMMARIES

A. <u>Using Energy and Carbon Sources to Classify Organisms</u>.--As we have seen in earlier chapters, all organisms need energy to carry out life processes. We can divide organisms into groups based on the source of this energy. <u>Autotrophs</u> ("self-feeders") obtain their energy directly from the physical environment, whereas <u>heterotrophs</u> ("other feeders") derive their energy from tissues of other organisms. Be sure to study the classification of organisms summarized in textbook Table 9-1.

(1) Autotrophs. Sunlight is the energy source for most autotrophic species, the <u>photosynthetic autotrophs</u>. For photosynthetic bacteria, sunlight is trapped by pigment molecules called <u>bacteriochlorophyll</u> and it provides the energy to reduce carbon dioxide (an inorganic carbon compound) to a sugar (an organic carbon compound). You should recall from Chapter 8 that reduction reactions require an input of energy . . . hence the value of sunlight. The following summary equation generalizes the events of <u>photosynthesis</u> in many kinds of <u>bacteria</u> (note that photosynthesis in plants and cyanobacteria is different . . . see Chapter 10):

$$2H_2X + CO_2 \xrightarrow{\text{sunlight}} (CH_2O) + 2X + H_2O$$

where H_2X is some <u>reducing agent</u> and $2X$ is the oxidized form of the reducing agent. For some photosynthetic bacteria, water (H_2O) is the reducing agent and molecular oxygen (O_2) is the oxidized form of the reducing agent. Some other bacteria use a different approach by obtaining hydrogens from hydrogen sulfide (H_2S) to produce elemental sulfur (S) or from molecular hydrogen (H_2) to produce methane gas (CH_4).

Some other types of bacteria (<u>chemosynthetic autotrophs</u>) operate without sunlight. Some of these, the <u>chemolithotrophs</u>, rely instead on the energy stored in the chemical bonds of reduced <u>in</u>organic compounds such as methane, ammonia, sulfate, nitrate, ferric iron, manganese and others. Those bacteria which require organic compounds for energy sources are referred to as <u>chemoorganotrophs</u>.

(2) Heterotrophs. Organisms of this group can<u>not</u> synthesize their own organic compounds <u>solely</u> from <u>in</u>organic materials as can the autotrophs. Animals are probably the heterotrophs with which you are most familiar, although fungi and many species of kingdoms Monera and Protista also are heterotrophic. These organisms require some sort(s) of preformed organic compounds as sources of energy and nutrients. These foods usually are the tissues (either living, dead or decomposing) of other organisms. Keep in mind that great variation in food requirements exists from one species to another.

B. <u>Storing Energy in Chemical Bonds</u>.--Energy enters the biological world by specialized metabolic pathways that <u>only</u> autotrophic species can conduct--the various processes of <u>photosynthesis</u> and <u>chemosynthesis</u>. All organisms, autotrophs and heterotrophs alike, carry out processes (<u>glycolysis</u> and <u>respiration</u>) by which such chemically-stored energy is released to drive the many energy-requiring metabolic processes. The summary reactions of these two sets of metabolic pathways are the opposites of each other:

$$\text{(reaction going from left to right)}$$
$$\text{photo/chemosynthesis}$$
$$CO_2 + H_2O + \text{energy} \longleftrightarrow (CH_2O)_n + O_2$$
$$\text{glycolysis \& respiration}$$
$$\text{(reaction going from right to left)}$$

$(CH_2O)_n$ represents a carbohydrate (usually a sugar such as glucose) which contains both energy and carbon.

All chemical bonds contain energy, but certain types of bonds between certain elements contain particularly high amounts of energy. Although such energy-rich bonds occur in a variety of compounds, two compounds are of particular biological importance (Fig. 9.1): adenosine diphosphate (ADP) and adenosine triphosphate (ATP):

Adenosine diphosphate (ADP)   Adenosine triphosphate (ATP)

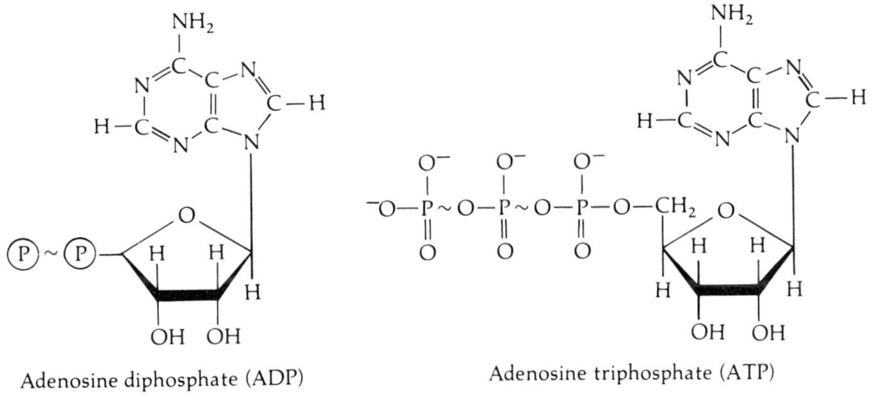

Fig. 9.1: Structural formulas for ADP and ATP

These energy-rich phosphanhydride bonds, denoted by the ~ symbol in the structural formulae above, are those joining the phosphate groups to each other. Photosynthesis and chemosynthesis produce these bonds to store energy (generally by converting ADP to ATP by the addition of inorganic phosphate, P_i) for later release of energy as glycolysis and respiration break these bonds (converting ATP to ADP and P_i).

Ordinary phosphate bonds contain less energy than phosphanhydride bonds. The two major types of chemical reactions that can transform ordinary bonds to high-energy ones are dehydrations (where a water molecule is removed) and oxidations (which you already know about from Chapter 8 . . . if you do not recall what sorts of electron, hydrogen, or oxygen movements occur in oxidation reactions, go back and learn it now!).

Let's expand the energy-releasing direction of the summary equation above to include those molecules that produce the high energy bonds:

$$C_6H_{12}O_6 + 6O_2 + 36ADP + 36P_i \longrightarrow 6CO_2 + 36ATP + 6H_2O$$

The efficiency of energy release in this pathway is relatively high. A mole of glucose contains 686 kcal of energy. Depending on the version of glycolysis and respiration used, 36 to 38 moles of ATP (representing 263 to 277 kcal) are produced from a mole of glucose for 38% to 40% efficiency, respectively.

We need to make one other point before examining specific steps in the energy-releasing process. Conversion of inorganic phosphate to organic phosphate is coupled with the uptake of oxygen. Such reaction coupling is apparent from the name of an important segment of the process: oxidative phosphorylation.

C. The Four Stages of Glucose Metabolism.--The metabolic pathway in which glucose is oxidized into carbon dioxide, water, and energy is divisible into four segments. We'll consider each stage separately below:

(1) Glycolysis. The first stage, which involves 10 reactions catalyzed by 10 enzymes, occurs in the fluid matrix of the cytoplasm. The name for this stage, glycolysis, indicates exactly what happens in this series of reactions: glucose molecules are split (lysed). The summary equation is:

$$C_6H_{12}O_6 + 2ADP + 2P_i + 2NAD^+ \longrightarrow 2 \text{ pyruvate} + 2ATP + 2NADH + 2H^+$$

The individual steps in glycolysis are shown in the textbook in Table 9-2 and Figure 9-5. Significant events in glycolysis are (a) splitting of one 6-carbon molecule of glucose into two 3-carbon molecules of pyruvate, (b) a net gain of 2 ATP molecules per molecule of glucose, and (c) two hydrogens from a molecule of glucose reduce two molecules of the coenzyme NAD^+ to form two molecules of NADH. In stage 4, these and other molecules of NADH will be oxidized to produce more ATP.

Seemingly, all organisms conduct glycolysis. However, the fate of pyruvate produced by glycolysis varies depending on the species considered. In yeasts (a group of fungi), pyruvate ferments to become ethanol (the type of alcohol that is suitable for drinking--grain alcohol). Carbon-dioxide gas produced by fermentation is important to brewers and bakers. The reaction sequence involves conversion of pyruvate into the 2-carbon molecule, acetaldehyde, and the release of carbon dioxide gas. Acetaldehyde then enters the following reaction:

Fate 1: acetaldehyde + NADH + H^+ ----> ethanol + NAD^+

A second of fate of pyruvate produced in glycolysis is used in situations of low oxygen content, in cells of animals (especially in muscle tissue), plants, and microorganisms. Pyruvate is reduced to lactate as follows:

Fate 2: pyruvate + NADH + H^+ ----> lactate + NAD^+

We will see below that the role of oxygen in cellular respiration is merely to carry any of the hydrogens stripped from glucose and, in fates 1 and 3 for pyruvate, from pyruvate as well. Fate 2 of pyruvate does not require oxygen because the the NAD^+ produced by reduction of pyruvate to lactate cycles back to glycolysis where it oxidizes glucose; this is the phenomenon known as the Pasteur effect. Fate 2 is the anaerobic version of glycolysis.

Aerobic oxidation of the various reduced coenzymes represents the third fate available to pyruvate. This approach is much more efficient than anaerobic respiration; it yields 36 or 38 ATP molecules per glucose molecule as compared to only 2 ATP in fates 1 and 2 above. Stages 2, 3 and 4 of glucose metabolism describe this strategy.

(2) Stage 2: Pyruvate oxidation. This stage occurs in the presence of oxygen. One of the three carbons of pyruvate is eliminated as carbon dioxide gas; the other two carbons are converted into the molecule acetyl coenzyme A (acetyl CoA):

2 pyruvate + $2NAD^+$ + 2 CoA ----> 2 acetyl CoA + $2CO_2$ + 2 $NADH_2$

The major events of stage 2 are: (a) release of one carbon as carbon dioxide gas, (b) transfer of two hydrogens from pyruvate to form two $NADH_2$.

(3) Stage 3: Citric-acid cycle. Acetyl CoA is the form in which organic carbon enters this stage in which all of its carbons eventually are converted into carbon dioxide gas. Each acetyl CoA molecule combines with a 4-carbon oxaloacetate molecule to form a new 6-carbon molecule of citric acid (hence, the name of stage 3). The next two steps involve oxidations (by the oxidized forms of NAD and FAD) of citric acid into a 5-carbon and then a 4-carbon molecule which eventually is converted into a new oxaloacetate molecule. This new oxaloacetate molecule binds a different acetyl-CoA molecule and then re-enters the cycle. In the process, two carbons are liberated as carbon dioxide gas and the coenzymes NAD and FAD are reduced. The summary equation is:

2 acetyl CoA + $4H_2O$ + 2ADP + $2P_i$ + 6NAD+ + $6H^+$ + 2FAD + 2GDP ---->

$4CO_2$ + 2ATP + 2CoA + 6NADH + $2FADH_2$ + 2GTP

(4) Stage 4: Oxidative phosphorylation. In stages 1, 2, and 3, all carbons of the original glucose molecule are liberated as carbon dioxide gas. All that remains is to re-oxidize the various reduced coenzyme molecules and, in the process, to produce ATP molecules. The summary equation of these events, which occur inside of mitochondria, is:

10NADH + $2FADH_2$ + $6O_2$ + 32ADP + $32P_i$ ---->

$12H_2O$ + 32ATP + $10NAD^+$ + $10H^+$ + 2 FAD

ATP molecules are produced by transferring the accumulated total of 24 hydrogen atoms (and their electrons) from the reduced coenzymes to the final electron acceptor, oxygen molecules.

Before getting to the oxygen molecules, the electrons are passed by a series of oxidation-reduction reactions through a sequence of compounds called cytochromes.

The Mitchell hypothesis supplies the explanation of how oxidation of the reduced coenzymes (textbook equation 9.16) is coupled with the phosphorylation of ADP into ATP (textbook equation 9.17). This hypothesis also imparts functional meaning to the double-membrane construction of mitochondria. ATP formation requires an active pumping of hydrogen ions from the interior of a mitochondrion outwardly across the inner mitochondrial membrane. For every hydrogen moving down the respiratory chain, three protons are displaced outwardly. Such pumping establishes a proton gradient with higher concentration outside than inside of the inner membrane. Protons move back into the mitochondrion interior by diffusing through channels in the respiratory assemblies of the membrane. This inward surging of protons provides the energy needed to bond a third P_i group to ADP. The key mechanism of the Mitchell Hypothesis is chemiosmosis, a process that we'll encounter again in the next chapter in the context of how chloroplasts are involved in generating ATP molecules.

A few comments regarding our accounting system of ATP production are in order here. Traditionally, we have thought that three molecules of ATP are formed for each pair of hydrogens passing from NADH to oxygen, and that only two molecules of ATP form from passage of each pair of hydrogens from $FADH_2$. However, in light of newer understandings of oxidative phosphorylation, we now realize that these yields of ATP are theoretical maxima that perhaps are attained only rarely. Review the last several paragraphs of textbook Chapter 9 to learn of the various complications that prevent us from determining the exact number of ATP molecules produced per molecule of glucose.

TERMS TO UNDERSTAND: Define and the identify the following terms:

adenosine triphosphate

autotrophs

heterotrophs

photosynthetic autotrophs

chemosynthetic autotrophs

photolithotrophs

photoorganotrophs

bacteriochlorophyll

stromatolites

chemolithotrophs

chemoorganotrophs

photosynthesis

glycolysis

respiration

hydrolysis

oxidative phosphorylation

fermentation

Pasteur effect

Krebs cycle

citric acid cycle

cytochromes

Mitchell hypothesis

proton pump

respiratory assemblies

chemiosmosis

respiration

CHART EXERCISES

The chart below offers you a convenient way to summarize the major events of the various metabolic pathways discussed in Chapter 9. Fill in the chart with the names or formulas of the organic, inorganic and coenzyme reactants, and products of the four listed metabolic pathways. In the "occurrence" column, write the groups of organisms and the tissue types in which that pathway occurs.

METABOLIC PATHWAY	REACTANTS			PRODUCTS			OCCURRENCE
	ORG	INORG	COENZ	ORG	INORG	COENZ	
Glycolysis							
Fermentation							
Anaerobic lactate respiration							
Aerobic respiration							

PEOPLE TO KNOW: What is the significant biological contribution of each of these individuals as indicated in Chapter 9?

Peter Mitchell

Sir Hans Krebs

Louis Pasteur

Fritz Lipmann

Herman Kalckar

DISCUSSION QUESTIONS

1. Respiration refers to processes on both the cellular level (as presented in this chapter) and the gross level (processes of breathing). What, if any, is the connection between these two respiratory processes?

2. The difference between the definitions of autotrophy and heterotrophy are clear-cut. Can you provide examples of organisms that cannot clearly be designated as either autotrophic or heterotrophic?

3. The degree of conversion of glucose molecules into ATP molecules is a good measure of respiratory efficiency. Compare the efficiency of the three pathways that pyruvate molecules produced by glycolysis can follow.

4. The summary equation for the fate of NADH molecules in oxidative phosphorylation contains coupled endergonic and exergonic components. Divide this equation into the two equations that show (a) oxidation of NADH and (b) phosphorylation of ADP. Indicate which reaction is endergonic and which is exergonic.

5. What is the role of osmosis in production of ATP molecules?

6. Discuss reasons why it is not possible to determine precisely the number of ATP molecules produced by the oxidation of a molecule of glucose.

TESTING YOUR UNDERSTANDING:

For each of the test items below, three of the lettered alternatives are true and the other is false. Determine which alternative is false and write its letter in the blank to the left of the question number. On the blank line below alternative D, write the corrected version of the false statement.

_____ 1. A. Autotrophs are able to make all needed organic components from simple inorganic molecules, whereas heterotrophs require preformed carbon-containing compounds as sources of energy and biosynthetic building blocks.
B. Photolithotrophs are autotrophic.
C. Many bacterial organisms utilize bacteriochlorophyll to trap the energy needed to reduce carbon dioxide into organic carbon-containing compounds.
D. For species that use hydrogen sulfide as a reducing agent, waste products of chemosynthesis include methane and water.

Correction: _____

_____ 2. A. Stromatolites are the remains of ancient photosynthetic bacteria.
B. Humans can be properly classified as anaerobic chemoorganotrophs.
C. The earliest organisms probably were anaerobic because there was very little, if any, atmospheric oxygen.
D. Glycolysis and respiration represent the net reverse of the chemical processes of photosynthesis.

Correction: _____

_____ 3. A. The reaction where a terminal phosphate group is split from ATP is a phosphorylation reaction.
B. When it is broken, the terminal phosphanhydride bond of ATP releases about 12 kcal of energy per mole of ATP.
C. An ATP molecule contains two energy-rich phosphanhydride bonds.
D. ADP can be considered the discharged form of the energy currency of living organisms.

Correction: _____

_____ 4. A. One completely oxidized molecule of glucose produces six molecules of carbon dioxide.
B. During glycolysis, one molecule of glucose is split into two molecules of pyruvate.
C. During glycolysis, two molecules of NAD^+ are reduced, but no ATP molecules are produced.
D. No carbon dioxide is produced during glycolysis.

Correction: _____

_____ 5. A. The pathway in which pyruvate is transformed into acetaldehyde is the aerobic pathway.
B. During the fermentation pathway, the three carbon molecules of pyruvate become the carbons of carbon dioxide and ethanol.
C. No oxygen is needed in fermentation because the oxidized electron-carrier molecules (NAD^+) that are produced during fermentation become re-reduced to NADH in glycolysis.
D. Practical applications of the fermentation process include the baking and brewing industries.

Correction: _____

_____ 6. A. Both the lactate and the fermentation pathways are less efficient than is aerobic respiration.
B. The net ATP production in the lactate and the fermentation pathways is two molecules of ATP per molecule of glucose.
C. Fermentation occurs in many kinds of yeasts, whereas the lactate pathway commonly occurs in muscle tissue and in red blood cells.
D. The oxidation reactions of the lactate pathway involve molecular oxygen.

Correction: _____

_____ 7. A. In the presence of oxygen and the appropriate enzymes, pyruvate is converted into acetyl coenzyme A.
B. Acetyl coA is a two-carbon molecule.
C. The process of pyruvate oxidation produces two energy-rich phosphanhydride bonds.
D. One of the carbon atoms of pyruvate is liberated as carbon-dioxide gas.

Correction: _____

_____ 8. A. Of the six carbons of a glucose molecule entering glycolysis, only four carbons remain to enter the citric acid cycle.
B. The first step of the Krebs cycle involves formation of a 6-carbon citric-acid molecule from molecules of acetyl CoA and oxaloactate.
C. During the citric-acid cycle, several different coenzyme molecules become oxidized.
D. Each turn of the Krebs cycle produces four carbon-dioxide molecules for each molecule of glucose that entered glycolysis.

Correction: _____

_____ 9. A. No ATP molecules are produced by the citric-acid cycle.
B. The reactions of the citric-acid cycle occur in the cytosol outside of the mitochondria.
C. NADH molecules that became reduced within a mitochondrion are worth more ATP molecules than are NADH molecules that were reduced outside a mitochondrion.
D. The process of oxidative phosphorylation is about 40% efficient.

Correction: _____

_____ 10. A. The cytochromes are the final acceptors of the hydrogens that enter oxidative phosphorylation.
B. Cytochromes are conjugated proteins whose attached prosthetic groups are iron-containing porphyrins.
C. The production of ATP in living organisms requires the presence of intact mitochondria.
D. A proton gradient maintained by a proton pump is required for the synthesis of ATP.

Correction: _____

_____ 11. A. Because of the distribution of hydrogen ions across the inner mitochondrial membrane, the interior of the mitochondrion is more acidic than is the region just outside of the inner membrane.
B. The energy required for ATP synthesis comes from the diffusion of protons across the inner mitochondrial membrane.
C. The mechanism of chemiosmosis explains the formation of ATP molecules in chloroplasts.
D. ATP production occurs in both chloroplasts and in mitochondria.

Correction: _____

_____ 12. A. Phosphorylation is an endergonic process.
B. Glycolysis is an exergonic process.
C. The hydrogen atoms that enter glycolysis as parts of glucose leave oxidative phosphorylation as parts of water molecules.
D. The oxygen of the water molecules produced by respiration are the same oxygen atoms that were part of the glucose molecules that entered glycolysis.

Correction: _____

CHAPTER 10--CAPTURING THE SUN'S ENERGY: PHOTOSYNTHESIS

CHAPTER OVERVIEW

Virtually all energy contained within the biological world ultimately came from the sun. Only certain organisms (primarily green plants and many bacteria and protists) can directly utilize solar energy. This chapter deals with the structural, physical, chemical, and metabolic aspects and processes of using solar energy to fix carbon, that is, to incorporate carbon (from carbon dioxide) into biomolecules that the organism can use in a myriad of ways.

Photosynthesis is the name of this set of metabolic pathways. Photosynthesis occurs in chloroplasts, organelles which are masterfully adapted to carry out this process. Chloroplasts contain pigment molecules that trap light energy. This light energy drives the light reactions, in the first segment of photosynthesis. The light reactions produce the energy (in the form of ATP) and the reducing power (as NADPH) that is required to fix carbon in the other segment of photosynthesis, known as the Calvin cycle. Some interesting variations in the specific metabolic pathways correlate with the cellular structure of leaves and the environmental stresses affecting the plants.

TOPIC SUMMARIES

A. <u>Light and Energy</u>.--As you already know, light is a form of energy. Light is the primary form in which energy enters the living world. The source of this light energy is the sun. <u>Much less than 1%</u> of the solar energy striking Earth's atmosphere is captured by the photosynthesic structures (e.g., leaves and certain other structures) of photosynthetic organisms (primarily plants).

To understand photosynthesis, we need to review some basic light physics. Light represents the visible (visible to humans) segment of the overall spectrum of electromagnetic energy. This <u>electromagnetic spectrum</u> extends from short-wavelength cosmic rays to long-wavelength radio waves. Light occupies a narrow band of this spectrum, from about 380 to 760 nanometers. The <u>energy content</u> of photons of light varies according to wavelength, such that violet light is more energetic than is red light. We perceive certain small sections of this region of the spectrum as the various <u>colors</u>; these run from red at the longer-wavelength end through orange, yellow, green, blue, indigo to violet at the short end of the visible spectrum. Light which includes all visible wavelengths appears white; an optical prism separates white light into its component bands. Each type of molecule has <u>specific light-absorbing capabilities</u> ... with the color of a molecule or object being those wavelengths that are <u>not</u> absorbed, but instead reflected. Each species of molecule, then, has its own absorption spectrum which scientists can use as a "fingerprint" when attempting to identify chemicals present in a solution.

The process of photosynthesis involves certain types of molecules (<u>pigment molecules</u>) that absorb light energy, with the wavelengths absorbed varying with type of molecule. Light energy so captured drives the endergonic process of photosynthesis. The rest of Chapter 10 is devoted to outlining the essential steps of photosynthesis.

B. <u>Overview of Photosynthesis</u>.--Photosynthesis is the primary means by which carbon and energy pass from the physical world into the biological world. It is the metabolic pathway in which two simple inorganic substances (carbon dioxide and water) are converted into a third inorganic chemical (oxygen) and, most importantly, into organic form as the carbohydrate molecule glucose. A reaction such as this, in which carbon dioxide is converted into an organic compound, is referred to as <u>carbon fixation</u>. A summary equation is:

$$6CO_2 + 6H_2O \longrightarrow 6O_2 + C_6H_{12}O_6$$

This equation should look familiar . . . it is the <u>reverse</u> of the summary reaction for glycolysis and aerobic respiration.

C. <u>Capturing the Energy of Sunlight</u>.--Several kinds of <u>photosynthetic pigment molecules</u> play the key role in transferring light energy into chemical energy in the biological world. In the previous chapter, we were introduced to bacteriochlorophyll found in the photosynthetic bacteria. Now we encounter an array of similar molecules found in plant tissues. These pigments include primarily <u>chlorophylls a and b,</u> large molecules consisting of a <u>porphyrin ring</u> with a central atom of <u>magnesium</u>. The absorbance spectrum for chlorophyll *a* (textbook Fig. 10-7) shows that it best absorbs blue and red light, as suggested by chlorophyll's green color.

Light energy in other regions of the visible light spectrum are not necessarily wasted. A variety of <u>accessory pigments</u> occurs in many plant species; these include several kinds of <u>carotenoids</u> and <u>xanthophylls</u> which are responsible for the yellows, oranges, reds, and browns seen in plant tissues. Transfer of energy into such pigment molecules involves the <u>excitation of an electron</u> in that molecule to a higher energy state. Electrons that are excited are more easily donated to other molecules than are unexcited electrons. Excited pigment molecules, then, become electron donors and, thereby, function as <u>reducing agents</u>.

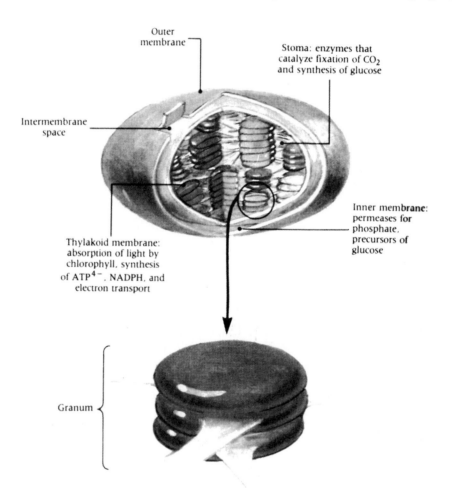

Fig. 10.1: Structure of a chloroplast

Chloroplasts are the sites where photosynthesis occurs (Fig. 10.1). These are organelles surrounded by double membranes. The interior of a chloroplast contains isolated green bodies (grana) positioned in a dense fluid matrix (stroma). The grana consist of stacks of 10 to 20 thylakoids, which are flat photosynthetic membranes. The coloration of the thylakoids correctly implicates them as the sites where the photosynthetic pigment molecules reside. This arrangement of photosynthetic membranes presents an enormous surface area packaged into a relatively small volume.

Pigment molecules are arranged into photosynthetic units in which one chlorophyll molecule serves as the reaction center and the other pigment molecules (antenna pigments) are positioned to efficiently pass energy (in the form of electrons) to the reaction center. Closer examination reveals two kinds of reaction centers, differing in the number of chlorophyll a and b molecules present, the type of protein to which they are bound, and the wavelengths of light best absorbed.

D. The Reactions of Photosynthesis.--Photosynthesis involves two major stages, the first requiring light (the light reactions) and the next not directly requiring light (the reactions of carbon fixation).

(1) Light reactions. The objectives of the light reactions are (a) to produce ATP through steps involving trapping of light energy, and (b) to generate the reducing agent, NADPH. Two versions of the light reactions exist: cyclic and noncyclic photophosphorylation.

Cyclic photophosphorylation occurs in all photosynthetic cells; it is the only method of photophosphorylation in most prokaryotes. This process uses Photosystem I. Within the thylakoid membrane, an excited electron passes from the reaction center chlorophyll molecule through a series of electron carriers, including iron sulfide and plastoquinone. The last carrier, plastocyanin, passes the now energy-poor electron back to the reaction-center chlorophyll molecule; this reduces the chlorophyll molecule and prepares it (actually, one of its electrons) for another pass through the cycle. ATP production involves a version of the chemiosmotic mechanism operating in mitochondria: Using energy released from the electron at the carrier plastocyanin, protons are pumped across the membrane to establish a proton gradient. Energy provided by the return of protons across the membrane drives the phosphorylation reaction, summarized as follows:

$$ADP + P_i + \text{light energy} \longrightarrow ATP$$

Noncyclic photophosphorylation accomplishes the following: (a) produces ATP, as does the cyclic approach above, but additionally it also (b) generates reducing power in the form of NADPH and (c) produces free oxygen. Both Photosystems I and II are involved in this pathway which operates in green plants and certain bacteria. Noncyclic photophosphorylation consists of two distinct steps, both of which require light.

In the first step, light energizes and liberates a pair of electrons from the Photosystem-II chlorophyll molecule. This pair of electrons works its way through the sequence of electron carriers in PS II, releasing energy at each transfer. The transfer from plastoquinone in PS II pumps the protons that drive ATP synthesis. Meanwhile, the original PS-II chlorophyll molecule needs to replace these two electrons: The immediate source of the replacement electrons is protein Z; protein Z replaces its "donated" electron pair by splitting a molecule of water and comandeering its two hydrogens. This releases molecular oxygen.

Getting back to the electron pair displaced from the PS II chlorophyll molecule . . . These wait at the end of the PS-II electron-transport chain until step 2 of noncyclic photophosphorylation in which the chlorophyll in PS I is energized by light. The PS-I chlorophyll then loses a pair of its own electrons; these travel through the PS-I series of electron carriers and eventually join a molecule of $NADP^+$ to reduce it to NADPH. The PS II electron pair then complete their journey by reducing the PS-I chlorophyll molecule. It should be apparent why this is a noncyclic pathway: the original pair of electrons from PS II does not immediately (if ever!) return to the PS II chlorophyll molecule.

The ATP and NADPH produced in the light reactions above are critical reactants in the ensuing light-independent reactions in which carbon dioxide is converted into sugar:

(2) Reactions of carbon fixation. This set of reactions fixes carbon. Carbon dioxide joins with the 5-carbon compound ribulose 1,5-biphosphate (RuBP) in a carboxylation catalyzed by RuBP carboxylase. The resulting 6-carbon compound is very unstable and quickly splits into two 3-carbon molecules of phosphoglyceric acid (PGA). These molecules cycle through a series of steps (known as the Calvin cycle) which use ATP and NADPH from the light reactions to produce a variety of carbohydrates, lipids, and proteins as well as molecules of RuBP which re-enter the cycle.

Photorespiration is an interesting pathway which sometimes accompanies the Calvin cycle. RuBP carboxylase can catalyze two opposing reactions--the one above which fixes carbon dioxide and another in which it oxidizes the same substrate. The latter is a seemingly wasteful process in which RuBP is diverted from the Calvin cycle. Not only is less carbon fixed, but some already-fixed carbon is released and lost! The role of RuBP at any given time is largely determined by the relative concentrations of carbon dioxide and oxygen.

E. Alternative Strategies for Fixing Carbon Dioxide.--The Calvin cycle as presented above reflects one of two approaches for fixing carbon dioxide. Plants in which carbon dioxide joins a 5-carbon compound which quickly breaks into 3-carbon molecules are referred to as C_3 plants. In C_4 plants, carbon dioxide joins instead with a 3-carbon acceptor molecule to form a 4-carbon intermediate. This 4-carbon intermediate loses a carbon dioxide and then enters the Calvin cycle as PGA. A separate cyclic reaction sequence forms a new 3-carbon compound that can accept another carbon dioxide.

Interestingly, leaf structure differs according to whether a species follows the C_3 or the C_4 route. Leaves of both types generally have single cell layers of epidermis on upper and lower leaf surfaces. The undersurface usually is perforated with tiny openings (stomata) through which gases are exchanged with the environment; in particular, carbon dioxide needed for photosynthesis enters the leaf through stomata and water can be lost from the plant through these openings. Sizes of stomatal openings are regulated by pairs of guard cells; these cells swell to open a stoma due to osmotic influx of water. Chloroplasts are found primarily in cells of the leaf interior, especially in the mesophyll cells. The considerable amounts of space occurring between spongy mesophyll cells allow circulation of air within the leaf. Coursing through leaves are bundles of vascular tissues (phloem and xylem).

Beyond this general description, C_3 and C_4 leaf structure differs. C_3 plants have leaves with (1) vascular bundles more widely separated and covered with thin sleeves of mesophyll, (2) spongy mesophyll away from the vascular bundles having chloroplasts, and (3) mesophyll rather than bundle-sheath cells being the primary sites of photosynthesis. As you can see in textbook Figure 10-26, C_4 leaves have (1) vascular bundles close together and surrounded by large bundle sheaths, (2) Kranz arrangement of mesophyll ringing the sheaths, and (3) spongy mesophyll away from these bundle rings with few chloroplasts--consequently, both mesophyll and bundle-sheath cells photosynthesize.

These anatomical differences reflect differences in photosynthetic pathways which, in turn, correlate with the different environmental pressures. C3 plants generally are not subjected to as much water and heat stress as are C4 plants, and if they are, they shift to photorespiration. If C4 plants kept their stomata open during the heat of the day, they would dehydrate. Hence, their stomata close during hot times. This solves the water loss problem, but causes a shortage of carbon dioxide required for photosynthesis. Examine textbook Figure 10-21 to see how C4 plants solve this problem: Carbon dioxide is taken into the leaf during cooler times and is stored by binding it to phosphoenolpyruvate (PEP) in mesophyll cells; malic acid carries the carbon from the carbon dioxide molecule into the adjacent bundle-sheath cell where the carbon dioxide is liberated to enter the RuBP carboxylation and Calvin cycle pathways. This C4 approach is highly effective and has been in use for millions of years. Be sure to read in your text about one other variation on these themes--CAM metabolism.

F. A Final Summary Equation.--The summary equation for photosynthesis presented in part above omits one important reactant--water. Recall that water is the source for electrons needed in noncyclic photophosphorylation. Note, too, that the reactant water molecules in the following equation are not the same water molecules as the product water:

$$6CO_2 + 12H_2O + \text{light energy} \longrightarrow C_6H_{12}O_6 + 6O_2 + 6H_2O$$

TERMS TO UNDERSTAND: Define and the identify the following terms:

photosynthesis

chlorophylls

electromagnetic spectrum

absorbance spectrum

action spectrum

photons

carbon fixation

porphyrin ring

pyrrole ring

carotenoids

xanthophylls

chloroplast

stroma

grana

thylakoids

photosystem I

photosystem II

cyclic photophosphorylation

noncyclic photophosphorylation

plastocyanin

plastoquinone

carboxylation

reductive pentose phosphate cycle (Calvin cycle)

epidermis

cuticle

stomata

guard cells

palisade mesophyll

spongy mesophyll

bundle-sheath cells

Kranz anatomy

crassulacean acid metabolism

PEOPLE TO KNOW: What is the significant biological contribution of each of these individuals as indicated in Chapter 10?

Jan Baptista van Helmont

Joseph Priestly

Daniel I. Arnon

Melvin Calvin

M. D. Hatch and C. R. Slack

DISCUSSION QUESTIONS

1. The summary equation (textbook equation 10.2) for photosynthesis includes water molecules as both reactants and products. Explain.

2. Compare photosynthesis with glycolysis and aerobic respiration.

3. Label the structures in the accompanying illustration of a section through a leaf that are important in photosynthesis. Indicate whether the section is through a C3 leaf or a C4 leaf. Compare the anatomical design of the leaves of C3 and C4 plants. How are the photosynthetic pathways different for these kinds of plants?

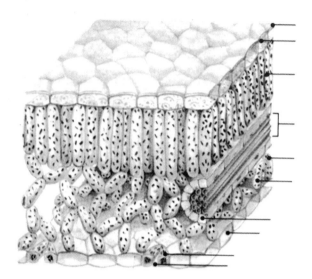

4. Why is the "dark reactions" not an appropriate label for the segment of photosynthesis that fixes carbon?

5. In the light reactions, electrons are passed through a series of electron-carrier molecules. Explain how this is a more suitable arrangement than one in which the electrons are passed directly from a high-energy carrier to a low-energy carrier.

CHART EXERCISES

The chart below offers you a convenient way to summarize the major events of various metabolic pathways discussed in Chapter 10. Fill in the chart with the names or formulas of the organic, inorganic, and coenzyme reactants, and products of the four listed metabolic pathways. In the "occurrence" column, write the groups of organisms, the tissue types, and the environmental situation (where appropriate) in which that pathway occurs.

METABOLIC PATHWAY	REACTANTS ORG	INORG	COENZ	PRODUCTS ORG	INORG	COENZ	OCCURRENCE
Cyclic photo-phosphorylation							
Noncyclic photo-phosphorylation							
C3 photo-synthesis							
C4 photo-synthesis							
Photo-respiration							
CAM metabolism							

TESTING YOUR UNDERSTANDING:

For each of the test items below, three of the lettered alternatives are true and the other is false. Determine which alternative is false and write its letter in the blank to the left of the question number. On the blank line below alternative D, write the corrected version of the false statement.

_____ 1. A. The segment of the electromagnetic spectrum that is visible to humans ranges from about 380 to 760 nanometers.
B. Ultraviolet radiation has a greater energy content than does infrared radiation.
C. The color of an object is the color of the wavelengths of light that the object absorbs.
D. White light consists of a mixture of all visible wavelengths of light.

Correction: _____

_____ 2. A. The absorption spectra for xanthophylls have peaks in the red, orange, and yellow regions.
B. Anoxygenic photosynthesis occurs only in prokaryotes.
C. The role of photosynthetic pigment molecules is to capture light energy.
D. In green plants, carbon dioxide is the source of inorganic carbon that ultimately be fixed into organic carbon.

Correction: _____

_____ 3. A. The light reactions of photosynthesis must occur during the sunlit daytime, whereas the dark reactions can occur only during the darkness of nighttime.
B. The light reactions generate ATP.
C. The light-independent reactions generate organic carbon compounds.
D. The dark reactions consume NADPH.

Correction: _____

_____ 4. A. A characteristic shared by molecules of chlorophyll and carotenoids is the presence of alternately positioned double and single bonds.
B. The absorbance spectra of chlorophyll and xanthophyll are similar.
C. The metal atom that is found in chlorophyll is magnesium.
D. A chlorophyll molecule consists, in part, of a porphyrin ring.

Correction: _____

_____ 5. A. The presence of several different types of photosynthetic pigments in one leaf allows use of a greater portion of the light striking the leaf.
B. Light striking a leaf may be absorbed into the leaf, reflected from the leaf, or transmitted through the leaf.
C. Some of the light that a leaf absorbs is re-radiated from the leaf as fluorescent light.
D. That portion of incident light that is transmitted through the leaf is the light that is used to drive photosynthesis.

Correction: _____

_____ 6. A. Pigment molecules such as chlorophyll *a* are attached to the thylakoid membranes by long hydrophobic side-chains.
B. The various types of chlorophyll molecules differ from each other only in which metallic element is present in the center of the porphyrin ring.
C. Bacteriochlorophyll differs from plant chlorophylls in the presence of four additional hydrogens in one of the pyrrole groups.
D. Carotenoid pigment molecules consist of two 6-carbon rings linked by of straight multiple-carbon chain.

Correction: _____

_____ 7. A. Chlorophyll molecules are located within the external membrane of a chloroplast.
B. The enzymes of the light reactions are located within the thylakoid membranes.
C. The enzymes of the light-independent reactions are located in the stroma.
D. Within a photosynthetic unit, the antenna pigment molecules absorb light energy and pass this energy to the reaction center molecule.

Correction: _____

_____ 8. A. Molecules of both chlorphylls *a* and *b* are present in reaction centers of Photosystem II.
B. Cyclic photophosphorylation involves only Photosystem I.
C. Cyclic photophosphorylation occurs in the photosynthetic cells of all photosynthetic species.
D. Like oxidative phosphorylation, photophosphorylation requires the presence of oxygen molecules.

Correction: _____

_____ 9. A. The processes of both cyclic and noncyclic photophosphorylation produce NADPH and oxygen.
B. The cyclic nature of cyclic photophosphorylation refers to the fact that light-energized electrons passing from chlorophyll molecules return to the reaction center after passing through the electron transport chain.
C. The energy released from electrons as they pass the carrier plastoquinone drives the proton pump.
D. The last carrier in the electron transport chain of cyclic photophosphorylation is plastocyanin.

Correction: _____

_____ 10. A. In noncyclic photophosphorylation, the ultimate source of electrons needed to reduce the electron-deficient chlorophyll molecule is water.
B. In noncyclic photophosphorylation, electrons lost from the PS-II chlorophyll molecule pass through the carrier chain and finally reduce $NADP^+$.
C. In noncyclic photophosphorylation, the immediate source of electrons needed to reduce the electron-deficient chlorophyll molecule is protein Z.
D. ATP produced in noncyclic photophosphorylation is the source of the reducing power required to convert carbon dioxide to glucose.

Correction: _____

_____ 11. A. The reaction in which ribulose biphosphate accepts a carbon-dioxide molecule is a carboxylation reaction.
B. Ribulose biphosphate carboxylase is an enzyme that catalyzes a reaction in the reductive pentose phosphate cycle as well as a reaction in photorespiration.
C. Each turn of the reductive pentose-phosphate cycle regenerates two new molecules of ribulose biphosphate that serve as acceptors of carbon dioxide.
D. The metabolic pathway that converts phosphoglyceric acid (PGA) into glucose is an exact reversal of the glycolytic pathway in which glucose is converted into PGA.

Correction: _____

_____ 12. A. Photorespiration is a biochemical strategy used by some plants to conserve carbon dioxide.
B. C4 plants have evolved a mechanism that overcome the negative effects of photorespiration.
C. C4 plants can fix carbon dioxide even at very low concentrations of gaseous carbon dioxide.
D. Oxaloacetate is a key molecule in C4 photosynthesis.

Correction: _____

_____ 13. A. Stomata of C3 plants usually are open during daylight hours.
B. Kranz anatomy characterizes leaves of C3 plants.
C. Stomata open when guard cells become turgid.
D. In leaves of C4 plants, vascular bundles are located very close together, and the bundle sheath cells contain many large chloroplasts.

Correction: _____

_____ 14. A. In C3 plants, mesophyll cells are the primary sites of photosynthesis.
B. Pyruvate is an important molecule in C4 photosynthesis.
C. Phosphoenolpyruvate is the carbon dioxide acceptor in C4 photosynthesis.
D. In C3 photosynthesis, carbon dioxide is actively pumped into cells against a concentration gradient.

Correction: _____

_____ 15. A. CAM metabolism is utilized by certain desert plants that are adapted to dry conditions and extreme temperatures.
B. Fossil data indicates that C4 photosynthesis has evolved within the last 100,000 years.
C. Like C4 plants, CAM plants avoid water loss by closing their stomata during the day.
D. In CAM plants, carbon dioxide acquired at night is stored as organic acids which later degrade to release carbon dioxide during the day.

Correction: _____

CHAPTER 11--BIOSYNTHESIS OF CELL CONSTITUENTS

CHAPTER OVERVIEW

In the previous several chapters dealing with metabolism, we have focussed on processes in which glucose is the energy-containing product or is the reactant that is degraded to release energy. In this chapter, we broaden our scope (1) to show that useful energy can be liberated by degrading compounds other than glucose, and (2) to introduce pathways by which various important biomolecules can be synthesized by the cell.

One of the alternate pathways of glucose catabolism (the pentose phosphate shunt) produces the 5-carbon sugar that is critical in the manufacture of ribonucleotides, rather than producing just ATP and reducing power. We see, also, that glucose can be converted into complex molecules used in energy storage or in structural roles. We examine how the energy contained in proteins and lipids can be funnelled into the glycolytic and respiratory pathways to release energy. We touch on processes for building complex lipids and an array of amino acids from simple compounds such as acetyl CoA, carbon dioxide, and other molecules in the now-familiar citric-acid cycle.

Several themes run through this chapter; keep these in mind as you study. The anabolic and catabolic pathways leading to and from some particular molecules are usually not the identical reactions in reverse. Certain key compounds repeatedly occur in the synthetic and degradative pathways of most types of molecules.

TOPIC SUMMARIES

A. <u>Pentose Phosphate Shunt</u>.--<u>Ribose</u> is a pentose, a 5-carbon sugar. Ribose is the sugar molecule of which <u>ribonucleotides</u> (such as ATP, NAD, and NADP) and <u>ribonucleic acids</u> are constructed. The pentose phosphate shunt is the metabolic pathway which produces ribose. Additionally, this pathway <u>generates reducing power</u> in the forms of NADH and NADPH. Because the shunt occurs in the cytoplasm, it generates no ATP molecules.

The pentose phosphate shunt represents <u>a fourth way to oxidize glucose</u>. In the two previous chapters we examined three other pathways related to glucose oxidation: glycolysis, respiration, and oxidative phosphorylation. Sugar enters the shunt as glucose 6-phosphate. An early step removes a carbon dioxide from the hexose to leave a pentose, ribose 5-phosphate. The ribose 5-phosphate is diverted into reactions that synthesize ribonucleotides and their derivatives, deoxyribonucleotides. Other aspects of the shunt involve rearrangements of the carbons of a various 3, 4, 5, 6, and 7-carbon sugar molecules. An important value of this part of the shunt is that many of these sugar molecules are converted into molecules that are also part of glycolytic pathway. Hence, these two pathways are <u>integrated</u>, and many resources of the cell can be directed into the pathways having greater demand.

B. <u>Biosynthesis of Polysaccharides</u>.--Polysaccharides are chemicals whose molecules consist of more than one sugar unit (a monosaccharide). For many polysaccharides, glucose molecules (or molecules derived from glucose) are the building-blocks. <u>Cellulose</u>, the chemical of which wood or cotton is made, is among the most widely occurring molecules in the biological world. <u>Starch</u> is a polysaccharide found widely in plants as an energy storage compound.

In animals, <u>glycogen</u> is a primary form of storing energy. Glycogen, a branched polysaccharide made from glucose monosaccharides, can be broken down quickly into glucose 1-phosphate and then into glucose 6-phosphate which can directly enter either the glycolytic pathway or the pentose phosphate shunt. Glycogen breakdown occurs by phosphorolysis, a process in which phosphoric acid splits the polysaccharide, in the presence of the enzyme phosphorylase. Glycogen synthesis involves the enzyme phosphoglucomutase which converts glucose into glucose 1-phosphate which then polymerizes into glycogen. [Note that although glycogen synthesis and degradation are the net reverse reactions, they are not the identical steps in reverse direction. This should be apparent from the involvement of different enzymes for opposite directions of the apparently same reactions.]

C. Lipids.--Lipids are important in structural and in energy-storing roles. The amount of extractable energy stored per carbon molecule is greater for lipids than for glucose! Naturally, it is advantageous for an organism to be able to extract energy from lipids, such as fatty acids. Oxidation of fatty acids occurs in the mitochondria and involves enzymatic splitting into acetyl groups (containing two carbons) which bind with coenzyme A. These acetyl-CoA molecules feed directly into the citric-acid cycle.

Synthesis of fatty acids occurs in the cytoplasm. This pathway requires large amounts of NADPH (generated by the pentose phosphate shunt) which provide the energy used to build the carbon chains of fatty acids. The synthesis of fatty acids begins with two familiar molecules: acetyl CoA and carbon dioxide. These two compounds join to form the 3-carbon molecule malonyl CoA. The carbon chain then lengthens by addition of acetyl CoA and subsequent loss of carbon dioxide to form the 4-carbon compound butyryl CoA. Subsequent growth in carbon chains is by a net gain of two carbons per step (from adding malonyl CoA and losing carbon dioxide).

In Chapter 6, we described the membranes of cells and of many cellular organelles as being phospholipid bilayers. Growth in membranes occurs by expansion of existing membranes. Probably all phospholipid molecules are synthesized on the membranes of endoplasmic reticulum. Enzymes and many of the reactant molecules are attached to the ER surface, yet the reactions themselves occur in the cytoplasm immediately adjacent to the ER. Product molecules reach the appropriate membranes by budding from vesicles and by other means.

D. Nitrogen-containing Compounds.--Nitrogen is a component of many critical biomolecules, including proteins and nucleic acids. Nitrogen is the most abundant element of the atmosphere. Yet, in its molecular form, nitrogen is not available to most species. Most species must rely on nitrogen-fixing bacteria to convert nitrogen into usable forms. . . much more on this in a later chapter. We'll see below that synthesis of amino acids is one of the few ways of getting inorganic nitrogen (in the form of ammonia) into organic form.

Nitrogen in proteins is contained in the amine group of amino acids. Most amino acids can be degraded into acetyl CoA or into pyruvate, both of which are molecules that can join the post-glycolytic steps of glucose metabolism to yield energy. Such a process in which non-carbohydrate precursors are converted into glucose is called gluconeogenesis. Usually, very little of the energy requirement of an organism comes from degradation of amino acids.

Amino-acid synthesis is a much more important process than is amino-acid degradation. Amino-acid synthesis is linked to the citric-acid cycle as a source of the carbon chains of the amino acids. The reaction that is central to amino-acid synthesis is the reductive amination (adding ammonia in the presence of NADH) of alpha-ketoglutaric acid (a 5-carbon compound) into glutamic acid (an amino acid). Many of the other amino acids can then be manufactured by transamination of glutamic acid. Note that many species cannot synthesize all of the types of amino acids required for metabolism . . . some amino acids must be consumed in that form.

E. Assembly of Cell Structures.--In the paragraphs above, we have briefly mentioned how certain types of biomolecules are synthesized. Now, we should address the issue of how cells assemble these molecular components into cellular structures and organelles. Although scientists are making progress, we do not yet fully understand the processes involved in assembly. We'll touch on various topics pertaining to assembly as we proceed through the textbook, but be sure to read in the textbook about an apparent <u>principle of self-assembly</u> in the last paragraphs of this chapter.

TERMS TO UNDERSTAND: Define and identify the following terms:

biosynthesis

pentose phosphate shunt

ribonucleotides

Embden-Meyerhof pathway

glycogen

phosphoglucomutase

phosphorolysis

phosphorylase

glyoxysome

glyoxylate cycle

gluconeogenesis

transamination

essential amino acids

leguminous plants

self-assembly

DISCUSSION QUESTIONS

1. Very few metabolic pathways are identical in both forward and reverse directions. How is this an advantageous arrangement? Give examples to support your answer.

2. What are at least four metabolic pathways in which acetyl coenzyme A is a key participant? What might be the adaptive value of the same chemical occurring in so many different pathways?

3. How does ribose bridge the gap between such dissimilar molecules as glucose and ribonucleotides?

4. Compare the structures and the functional roles of the polysaccharides glycogen and cellulose.

5. Briefly discuss the major changes that would occur in catabolic pathways during early and then late stages of starvation in an animal that normally respired via the glycolytic and aerobic respiratory pathways.

6. On the accompanying outline of a cell, indicate the location(s) in which each of the following pathways occurs: glycolysis, pentose-phosphate shunt, oxidative phosphorylation, photophosphorylation, citric-acid cycle, fatty-acid synthesis, fatty-acid breakdown, phospholipid synthesis, and amino-acid synthesis.

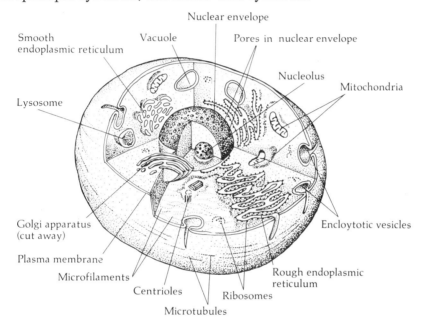

MATCHING: Match the process or pathway in the left column with the appropriate description in the right column.

_____ 1. gluconeogenesis   A. splitting of glycogen by phosphoric acid

_____ 2. Embden-Meyerhoff pathway   B. exchange of alpha-amino group for an alpha-keto group

_____ 3. pentose phosphate shunt   C. production of glucose from a non-carbohydrate precursor

_____ 4. transamination   D. involves glycolysis and aerobic respiration

_____ 5. phosphorolysis   E. leads to production of ribose

TESTING YOUR UNDERSTANDING:

For each of the test items below, three of the lettered alternatives are true and the other is false. Determine which alternative is false and write its letter in the blank to the left of the question number. On the blank line below alternative D, write the corrected version of the false statement.

_____ 1. A. Glucose is a 6-carbon molecule.
B. Carbon dioxide is a 1-carbon molecule.
C. Pentose is a 5-carbon molecule.
D. Malonyl coenzyme A is a 4-carbon molecule.

Correction: _____

_____ 2. A. Butyryl CoA is an intermediate molecule in the synthesis of fatty acids.
B. Ammonia is a reactant in the degradation of glycogen.
C. Glucose 6-phosphate is an intermediate in the synthesis of glycogen.
D. Glucose 6-phosphate is an intermediate in glycolysis.

Correction: _____

_____ 3. A. Phosphoglucomutase catalyzes the conversion of glycogen to glucose.
B. Transaminase is an enzyme involved in the synthesis of other amino acids from glutamic acid.
C. The pentose phosphate shunt produces molecules important in the synthesis of NADH and ATP.
D. Oxidation of fatty acids produces more molecules of ATP per carbon atom than does oxidation of glucose.

Correction: _____

_____ 4. A. Phospholipids are synthesized on membranes of endoplasmic reticulum.
B. The pentose phosphate shunt occurs in the cytoplasm.
C. Fatty-acid breakdown occurs in the mitochondria.
D. Fatty-acid synthesis occurs in the mitochondria.

Correction: _____

_____ 5. A. Although glycogen forms in most animal cells, it occurs in greatest quantities in muscle and liver cells.
B. Starch is found in most plant and animal cells.
C. Phospholipids are important molecules in membranes of all cell types.
D. Most animal species cannot synthesize the amino acids methionine, lysine, and threonine.

Correction: _____

_____ 6. A. Gluconeogenesis does not commonly occur in cells unless there are no available reserves of carbohydrates or lipids that can be converted to glucose.
B. Fatty-acid synthesis occurs when the supply of nutrients exceeds the energetic demand of the cell.
C. During starvation, amino acids are converted into proteins which then enter energy-releasing catabolic reactions.
D. When the glucose supply of an animal cell is exhausted, the energy reserve that the cell first draws upon is glycogen.

Correction: _____

_____ 7. A. The reaction of malonyl CoA with acetyl CoA is a condensation reaction.
B. The type of reaction by which glycogen occurs is a phosphorolysis reaction.
C. The production of deoxyribonucleic acid relies on products of the pentose phosphate shunt.
D. The production of adenosine triphosphate relies on products of transamination reactions.

Correction: _____

_____ 8. A. In the assembly of some organelles, many of the component protein molecules are synthesized outside of the organelle and then imported into the organelle.
B. Some organelles, such as mitochondria and chloroplasts, replicate themselves during the course of cell division.
C. One of the few ways by which inorganic nitrogen can be incorporated into biomolecules of a living organism is by the process of gluconeogenesis.
D. Reductive amination of alpha-ketoglutaric acid yields glutamic acid.

Correction: _____

CHAPTER 12--HOW CELLS REPRODUCE

CHAPTER OVERVIEW

You already know from earlier chapters that all living organisms are composed of one or more cells. Even multicellular organisms begin their lives as single cells. Growth into a multicellular organism involves production of many cells from the original one cell. Reproduction of cells is also important in several other ways, including the repair of damaged tissues and the production of specialized sex cells whose union results in a new organism of that species.

Because prokaryotic cells have somewhat different construction than eukaryotic cells, there are some differences in their processes of cell division. Likewise, there are differences in the cell-division processes according to the purpose of that division process. For example, growth and repair in eukaryotes occurs by the process of mitosis, whereas production of sex cells involves meiosis. In this chapter, we will examine the cellular and subcellular events that occur in these various versions of cell reproduction.

TOPIC SUMMARIES

A. <u>Cell Division</u>.--All cells, whether prokaryotic or eukaryotic, demonstrate at least some similarities in cell division: A cell that is preparing to divide <u>grows larger</u>. <u>Genetic information is duplicated</u> (or is modified in other ways, depending on the specific cell-division process occurring) in preparation for partitioning into new cells. The parts of the cell not related to genetic information (generally, the cytoplasm and cell membrane) split into two regions that will soon be separated from each other; this process is <u>cytokinesis</u>. Many of the differences between prokaryotic and eukaryotic cell-division processes relate to differences in cell structure and in the nature of the genetic material.

(1) Prokaryotic cell division. When compared to eukaryotic cells, <u>prokaryotic cell division</u> seemingly is a simple matter. As in eukaryotes, the genetic information is encoded by DNA; but prokaryotes have all of their DNA arranged into only one structure--a circular <u>genophore</u>--rather than into several to many non-circular chromosomes. The genophore is located in the nucleoid, a clear region that is not surrounded by a membrane, and it is attached to the cell membrane.

The first of the two main processes in prokaryotic cell division is replication of the DNA and then duplication of the genophore (illustrated in textbook Figures 12-1 and 12-6). Next the two genophores must be separated from each other and then distributed to the new <u>daughter cells</u>. The paired genophore, which attaches to the cell membrane and wall, separates as the cell wall grows between the genophores and thereby pulls them apart. Prokaryotic cell division is rapid, requiring as little as 20 minutes.

(2) Eukaryotic cell division. Two cell division processes, mitosis and meiosis, occur in eukaryotes. For eukaryotes, the process of producing identical daughter cells (<u>mitosis</u>), is much slower (requiring about 24 hours) than the corresponding process in prokaryotes. This greater time intuitively makes sense, considering the greater complexity of eukaryotic cells. Eukaryotes usually possess several to many chromosomes which are located in a membrane-bound nucleus.

The activities of most eukaryotic cells can be viewed as parts of a cycle--the <u>cell cycle</u> (Fig. 12.1). <u>Mitosis</u> is the stage in which cell division actually occurs. Next is a <u>gap</u> or resting phase (designated by G_1). This is followed by a <u>synthesis</u> period in which the DNA and associated proteins of chromosomes are synthesized (stage S). Then, a second gap (G$_2$) precedes mitosis. This cycle repeats continuously over many generations for many cell types, whereas some other types of cells undergo only a few turns of the cycle. Still others are intermittently active and inactive (dormant). Read in your textbook to find examples of cell types which follow these different cycling strategies.

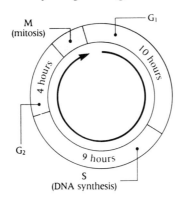

Fig. 12.1: The eukaryotic cell cycle

B. <u>Mitosis in Eukaryotic Cells</u>.--The description above of the cell cycle paints a deceptively simple view of cellular activities. Let's focus now on mitosis, one of the four cycle stages, and examine the <u>major events of division of the nucleus</u>. Although cytologists break mitosis into seemingly discrete stages, the events of cell division are continuous. Keep in mind that the purpose of mitosis is to divide one cell into two genetically-identical daughter cells. The following description summarizes events in animal cells; the process in some protists and most fungal and plant cells differs in various ways.

(1) Interphase. This interval includes three of the four stages of the entire cell cycle (synthesis and the gaps before and after synthesis), and as such probably should not even be recognized as a stage of mitosis. <u>Interphase</u> is not a time of rest, but a time of great metabolic activity including preparation for the next mitotic division. Stage G$_1$ involves dismantling of the structural remnants from the previous mitosis, and it is a time of growth and preparation for the upcoming period of DNA synthesis. During stage S, DNA is synthesized and nuclear chromatin is replicated. Activities during G$_2$ are the continued preparation for the ensuing mitotic division--synthesis of RNA and proteins as well as formation of spindle fibers and other structures that will be used in mitosis.

Mitosis involves two separate series of events--<u>karyokinesis</u>, in which the nucleus and its contents are split, and <u>cytokinesis</u>, in which the remainder of the cell splits. We'll first address karyokinesis:

(2) Prophase. Recall that previously, in stage S of interphase, the genetic information was duplicated in preparation for sending a complete copy to each daughter cell. One of the primary events of prophase is the <u>condensation of chromatin</u> (the fine thread-like material that carries the genetic code) into sausage-shaped <u>chromosomes</u>. Each chromosome has two identical <u>chromatids</u>, and these <u>sister chromatids</u> of each chromosome are joined together by a <u>centromere</u>. These chromosomes are randomly dispersed throughout the nucleoplasm. By the end of mitosis, the sister chromatids of each chromosome pair will be separated, with one chromatid of each pair being sent to each of the two new nuclei.

As chromatin condenses, the nucleolus disappears and the nuclear membrane begins to dissolve. Also, <u>centrioles</u> and the associated <u>radiating fibers</u> move towards opposite poles of the cell, and the mitotic spindle begins to form. These events are related to the mechanism by which the two complete sets of chromosomes will move to opposite ends of the cell.

(3) Metaphase. An early event in metaphase is completion of the spindle, the radiating trackway of microtubules along which chromatids of chromosome pairs will soon separate to move to opposite ends of the cell. Before the chromatids separate, however, the paired chromatids line up neatly along the equatorial plane. Attachment of chromosomes to the spindle is at a point in the centromere known as the kinetochore. Metaphase ends with the chromosomes positioned at the equatorial plane.

(4) Anaphase. Anaphase begins with the division of the centromeres. The now-separated chromatids are quickly pulled towards opposite spindles by ever-shortening microtubules. Simultaneously, another set of microtubules is involved in increasing the distance between the two spindles. An interesting mechanism (microtubular treadmilling) has been proposed to explain how microtubules shorten in one region and lengthen in another with the net effect of moving chromatids poleward; see textbook Figure 12-15 for more details. At the end of anaphase, full sets of single-chromatid chromosomes have collected at opposite poles, spindle fibers lengthen further to continue pushing the poles apart, and cytokinesis begins.

(5) Telophase. Karyokinesis concludes with telophase, which might be thought of as the reverse of the events of prophase. The chromosomes, now present in only single copies, return to chromatin form. The nuclear envelope forms and the nucleolus reappears. Cytokinesis, which we will consider next, rapidly nears completion. Telophase is followed by interphase which prepares for these new cells for the next mitotic division.

C. Cytokinesis in Eukaryotic Cells.--Splitting of the cytoplasm may or may not follow karyokinesis. If each new cell is to have a nucleus with chromosomes, then a division (or cleavage) furrow must form between the two new nuclei. At the furrow, the cell membrane grows inwardly from all sides to meet at the center. In plants, a cell plate formed from vesicles divides the new cells. Some organelles, such as chloroplasts and mitochondria, divide on their own by processes much like fission of prokaryotes. Also, new centrioles develop from old centrioles.

D. Chromosome Structure.--Curious as it might seem, scientists have not yet determined the exact structure of prokaryotic or eukaryotic chromosomes. Of the several models that have been suggested for eukaryotic chromosomes, the following one is best supported by the currently available evidence: Chromatin consists of a string of somewhat spherical nucleosomes measuring 10 nm in diameter. Each nucleosome is made of a strand of DNA of about 140 base pairs in length that is coiled in nearly two turns around a protective core of eight molecules of proteins known as histones. Adjacent nucleosomes are connected by regions of spacer DNA about 60 base pairs long (more about base pairs in Chapter 16). Chromatin fibers, which have a diameter of about 30 nm, seem to represent an orderly folding of polynucleosome strings into helical DNA-protein complexes called solenoids. No models offered to explain the behavior of chromosomes during cell division have yet been substantiated.

E. Ploidy.--Chromosomes occur in sets that include one copy of information for each characteristic exhibited by that organism. The term ploidy indicates the number of sets of chromosomes present in a particular cell. Most cells of the body of most eukaryotic organisms possess two sets of chromosomes; such somatic cells are said to be diploid (symbolized as 2n). One special set of cells, the sex cells, will contain only one set of chromosomes; these gametes are haploid (n). [Some species are exceptional in having more than two sets of chromosomes. . . we'll treat these in appropriate places elsewhere in the textbook.] Haploid cells arise from diploid cells by a special cell-division process called meiosis.

F. <u>Meiosis</u>.--The <u>purpose of meiosis</u> is to reduce chromosomal content by half, either directly or indirectly yielding gametes. This process is known also as <u>reduction division</u>. <u>Gametogenesis</u> refers to the entire process of gamete formation, including the nuclear events of meiosis and other non-nuclear events; the version which produces male gametes (<u>sperm</u>) is <u>spermatogenesis</u> and that yielding female gametes (<u>ova</u>) is <u>oogenesis</u>. A single cell entering spermatogenesis divides into four functional sperm, whereas only one functional ovum and one to three non-functional <u>polar bodies</u> result from oogenesis. <u>Fertilization</u>, the union of two gametes of appropriate sex, restores the full genetic complement.

Two sequences of nuclear and cytoplasmic division occur in meiosis as is indicated by use of numerals I and II in the following summary:

(1) Meiosis I. The two cells produced by this first sequence of steps each contain one set of genetic information. This distribution of genetic information results from splitting of each chromosome pair into corresponding chromosomal homologs, with one homolog for each chromosome pair going to each of the two new nuclei. The major events of each step or meiosis I are outlined below:

Prophase I: This very busy phase has five recognized steps; you should refer to the textbook for details of each. During the <u>condensation stage</u>, the chromatin condenses into visible chromosomes. Next, in the <u>pairing stage</u>, the chromosomes of homologous chromosome pairs line up next to each other in a very precise alignment that matches the homologs gene-by-gene. This alignment is known as <u>synapsis</u>. A <u>synaptonemal complex</u> extending the length of the synapse is visible during this and the following stage. The <u>recombination</u> stage involves the exchange of (usually) corresponding segments of one homolog with the other homolog. Such crossing-over generates new combinations of genetic information. The <u>synthesis stage</u> begins as the synaptonemal complex dissolves and the homologs begin to separate. Yet, the homologs remain attached in a few regions called chiasmata, which are the points of crossing-over. During <u>recondensation</u>, which marks the end of prophase I, the chromosomes condense, the chiasmata move to the ends of the chromosomes, the nucleoli disappear, and the cell envelope breaks up.

Metaphase I: During this phase, homologous chromosome pairs (as <u>tetrads</u>) line up at the spindle midpoint; independent orientation of each tetrad along the spindle results in mixing of maternal and paternal genes. The nuclear envelope disintegrates. Also, centrioles (if present) complete their replication.

Anaphase I: Homologs (as <u>bivalents</u> which are half of each tetrad) move to opposing poles. This phase is another time of mixing of paternal with maternal genes; except by chance alone, there is no mechanism for sending maternal homologs to one pole and paternal homologs to the other.

Telophase I and Interkinesis: The expression of these phases is quite variable according to species. In general, the events that occur are those needed to reorganize the metaphase-I spindle into the two spindles required for the upcoming second meiotic division.

(2) Meiosis II. This sequence of steps partitions the nuclear contents of the products of meiosis I into four nuclei which may, depending on the sex and on other factors, be contained in four independent cells.

Prophase II: Whether this is a prominent phase depends upon the completeness of telophase I and interkinesis . . . that is, whether chromosomes decondensed into chromatin form and whether nuclear envelopes reformed at the end of meiosis I. If these events occurred, then chromatin condenses and the nuclear envelope degenerates in prophase II.

Metaphase II: <u>Chromatid pairs</u> (rather than chromosome pairs as in meiosis I) move to the spindle during metaphase II.

Anaphase II: Chromatid pairs separate and move to opposite poles of the spindle during anaphase II. This event accomplishes the final division of genetic information in gametogenesis.

Telophase II: This stage concludes meiosis. Nuclear envelopes redevelop and chromatids revert to chromatin.

Three points need to be made here. <u>First</u>, although meiosis (a term pertaining to nuclear events) ends with the completion of telophase II, gametogenesis is not necessarily complete; some non-nuclear events need still to occur to position the new nuclei into separate cells or into other situations appropriate for that species or sex. <u>Second</u>, the description of meiosis presented above applies to most animals, many protists, and a few plant species. The many other eukaryotes follow other versions of the process; these variants will be treated in the context of life cycles in Chapter 21. <u>Finally</u>, deviations (such as nondisjunctions) sometimes occur in this intricate process; we'll look at some of these in Chapter 15.

TERMS TO UNDERSTAND: Define and identify the following terms:

fission

budding

mitosis

karyokinesis

cytokinesis

sperm

ovum

zygote

histones

karyotype

homologs

diploid

ploidy

interphase

chromatin

prophase

chromatids

centromere

asters

spindle fibers

metaphase

satellite bodies

kinetochore

anaphase

daughter chromosomes

telophase

division furrow

nucleosome

spacer DNA

cloning

somatic cells

germ cells (gametes)

haploid

meiosis I and II

gonads

gametogenesis

spermatogonia

spermatocyte

oogenesis

polar body

interkinesis

chromomeres

synapsis

synaptonemal complex

crossing-over

chiasmata

tetrads

pollen

nondisjunctions

CHART EXERCISES

Although mitosis and meiosis are alike in some ways, they differ markedly in many other respects. Completing this chart is a useful way to organize your thoughts and to be sure that you have a good understanding of the two processes. Complete the chart below to summarize the essential events in each step of these processes. [Do not rely on textbook Table 12-2 for this information . . . you will learn better if you assemble this information on your own and if you use your own words.]

STAGE	MITOSIS	MEIOSIS I	MEIOSIS II
Interphase			
Prophase			
Metaphase			
Anaphase			
Telophase			

PEOPLE TO KNOW: What is the significant biological contribution of each of these individuals as indicated in Chapter 12?

Rudolf von Virchow

W. S. Sutton

Joachim Hammerling

Daniel Mazia

DISCUSSION QUESTIONS

1. Compare mitosis and meiosis in terms of the purpose(s) of each process.

2. Discuss the differences between plant and animal cells regarding structures involved in spindle formation during processes of cell division.

3. What are the steps of meiosis in which reorganization of genetic information occurs? Discuss the specific events that occur to produce different genetic combinations. Perhaps most importantly, what is the adaptive value of such mechanisms for producing genetic variability?

4. How does meiosis differ from gametogenesis? How does gametogenesis differ according to sex?

5. The illustration below is a sketch of a cell nucleus from an organism whose diploid number is four. Identify and label as many nuclear features as possible. Indicate, as specifically as possible, the stage of cell division demonstrated by this sketch. Present your rationale for determining the stage.

TESTING YOUR UNDERSTANDING:

For each of the test items below, three of the lettered alternatives are true and the other is false. Determine which alternative is false and write its letter in the blank to the left of the question number. On the blank line below alternative D, write the corrected version of the false statement.

_____ 1. A. Karyokinesis occurs in cells of prokaryotic species.
B. Eukaryotic cells generally reproduce by mitosis.
C. Cytokinesis occurs in eukaryotic cells.
D. Fission and budding are means of reproduction used by prokaryotic cells.

Correction: _____

_____ 2. A. Gametogenesis, a process which produces sex cells, involves meiosis.
B. A zygote is a cell that results from fusion of gametes.
C. The cell division process by which a zygote develops into a multicellular organism is mitosis.
D. The zygote of a prokaryotic species carries two copies of genetic information on a structure called a genophore.

Correction: _____

_____ 3. A. Nuclear DNA of eukaryotes is not circular.
B. The DNA of prokaryotes is contained in the nucleoid.
C. The DNA of prokaryotes is complexed with histones to form the genophore.
D. Exonucleases are enzymes that can attack the free end of a DNA molecule.

Correction: _____

_____ 4. A. Completion of cell division in prokaryotes involves the process of cell-wall formation.
B. The separation of duplicate genophores is intimately related to cell-wall formation.
C. In a diploid organism, mitosis produces diploid cells whereas meiosis produces haploid cells.
D. If the haploid number for a species is 72, then that species has a diploid number of 36.

Correction: _____

_____ 5. A. The work of W. S. Sutton demonstrated that chromosomes are the organelles that carry genetic information.
B. The discovery of the processes of mitosis and meiosis supported the claims by von Virchow that all cells come from pre-existing cells.
C. Joachim Hammerling's scientific work produced data that supported the principle of heredity, that is, that like begets like.
D. Gurdon and Brown produced data that suggested that each cell of an organism possesses a full complement of genetic information.

Correction: _____

_____ 6. A. Replication of chromatin occurs during the synthesis stage of interphase.
B. The gap stages occurring before and after the mitosis phase of the cell cycle are periods of greatly reduced metabolic activity.
C. During anaphase and metaphase, DNA is located in structures called chromosomes.
D. Cytokinesis nears completion during telophase.

Correction: _____

_____ 7. A. The nucleolus is visible during metaphase.
B. The nuclear envelope is intact during interphase.
C. Chromatids of the same chromosome pair do not line up next to each other at the metaphase plate of mitosis.
D. Cytokinesis begins during anaphase.

Correction: _____

_____ 8. A. The kinetochore is the point of attachment of spindle microtubules.
   B. The kinetochore is the center for polymerization of spindle microtubules.
   C. The kinetochore is the structure that joins chromatids of a chromosome pair.
   D. The kinetochore is located near the centromere.

   Correction: _____

_____ 9. A. The spindle is a mass of microtubules to which chromatids attach.
   B. The shortening in length of a spindle fiber occurs by disassembly of the fiber at the chromatid end of the fiber.
   C. The location of the division furrow that develops during cytokinesis is influenced by the position of the spindle.
   D. Lengthening of spindle fibers causes the opposite poles to be pushed apart.

   Correction: _____

_____ 10. A. The solenoids of a chromatin fiber are the lengths of DNA that connect nucleosomes.
   B. The spermatocyte is the cell in which the meiotic divisions of spermatogenesis occur.
   C. Although spermatogenesis produces four functional gametes for each diploid cell entering the process, only one functional gamete results from oogenesis.
   D. Chiasmata are the locations at which chromatids of a tetrad are attached.

   Correction: _____

_____ 11. A. Synapsis occurs during prophase I of meiosis.
   B. The synaptonemal complex is a structure that is positioned in the space between homologs during synapsis.
   C. The positioning of chromomeres indicates the sites where crossing-over has occurred.
   D. Crossing-over is a process during which corresponding segments of homologous chromosomes are exchanged with each other with the effect of producing different combinations of genetic information.

   Correction: _____

_____ 12. A. During anaphase I, the four chromatids of each tetrad are separated from each other.
   B. During metaphase I, the bivalents of each homologous chromosome pair line up next to each other at the equatorial plane.
   C. Pairing of homologous chromosomes into tetrads occurs in meiosis, but not in mitosis.
   D. Each chromatid of a tetrad has its own kinetochore.

   Correction: _____

_____ 13. A. During metaphase I, the cell contains four times as much genetic information than will be present in each gamete.
   B. During metaphase of mitosis, the cell contains two times as much genetic information than will be present in each daughter cell.
   C. A primary goal of mitosis is to reduce chromosomal content by one-half.
   D. A primary goal of meiosis is to produce new arrangements of genetic information.

   Correction: _____

_____ 14. A. In meiosis I, paternal chromosomes gather at one pole and maternal chromosomes group at the other pole.
   B. In mitosis, each daughter cell contains complete sets of both maternal and paternal chromosomes.
   C. The processes of both mitosis and meiosis occur in gonads.
   D. Crossing-over does not occur during meiosis II.

   Correction: _____

_____ 15. A. In a version of meiosis characteristic of some fungi, algae and protists, meiotic divisions occur immediately after fertilization.
   B. The versions of meiosis occurring in most animals is called terminal meiosis.
   C. In nondisjunction, cytokinesis does not go to completion; the result is a zygote with three complete sets of chromosomes.
   D. Miscarriages are spontaneous ways by which genetically defective embryos are aborted.

   Correction: _____

CHAPTER 13--GENETICS

CHAPTER OVERVIEW

Humans have applied the principles of heredity for a long time, perhaps from the days when various plants and animals first were domesticated for human use. Only recently, however, have we begun to understand how these principles work. Our scientific understanding of heredity began slightly over a century ago, with the experiments of Mendel and others.

In Chapter 13, we will review some of Mendel's experiments to illustrate the principles of dominance, of segregation, and of independent assortment. We will examine the role of chromosomes and of gametes in heredity, particularly with respect to how new combinations of alleles are introduced into the gene pool. Additionally, we will see the value of using statistics to analyze biological data.

TOPIC SUMMARIES

A. <u>Mendel's Experimental Materials and Approach</u>.--Whether by design or by good fortune, Mendel selected a plant species and a variety of traits that were admirably suited for study of inheritance. The pea plant, *Pisum sativum*, is an easily-grown species that is <u>self-pollinating</u>, that is, pollen (male gametes) from one plant can fertilize the ova (female gametes) of the same plant. A researcher, however, also can manipulate the plants to cause <u>cross-pollination</u> (pollen of one plant fertilizing ova of another plant) to occur. Perhaps of greatest value is that indiscriminant fertilization does not occur by wind-blown pollination. Hence, pea plants are excellent model species because pollination is quite controllable.

Mendel was most fortunate in selecting plant characteristics to study. All traits he studied showed <u>complete dominance</u>, that is, a character was expressed in either one or the other of two alternative ways. He used only <u>true-breeding lines</u>, that is, strains which always expressed a trait in a particular way. Further, the traits he studied were located on different chromosomes . . . the importance of this will become apparent as we proceed. Had the species or traits not been as described above, then Mendel probably would have produced results that would have been considerably more difficult to interpret. Some biological historians have even suggested that Mendel must have "fudged" his data a bit in order come up with the nice figures that he reported!

In his efforts to understand the complexities of inheritance, Mendel devised simple experiments; he studied only one or a few traits at a time. Another important feature of Mendel's experimental approach was that he <u>quantified</u> his findings, that is, he counted and recorded the results. Such as quantitative approach is a critical element to modern scientific studies in any field.

B. <u>Principle of Segregation . . . or . . . Mendel's Hypothesis of Paired Factors</u>.--This principle, which states that the alleles for each trait segregate from each other during meiosis, resulted from study of the pattern of inheritance of individual traits. Such a cross in which only one character is followed is called a <u>monohybrid cross</u>. Mendel crossed true-breeding strains of red-flowered peas with white-flowered peas, and he followed the <u>progeny</u> (offspring) for two generations. These red and white parent plants represented the <u>parental generation</u> (P), and their progeny were the <u>first filial generation</u> (F_1). Mendel then cross-pollinated the F_1 plants to produce the next generation, called the <u>second filial generation</u> (F_2).

In this experiment, both red and white flowers were present in the P and F$_2$ generations, but only red flowers appeared in the F$_1$. Mendel proposed that information determining how a trait is expressed is carried twice in an individual, as "paired factors" that we now refer to as alleles of a particular gene. He also suspected that one allele can surpress the expression of the other allele; the allele that is expressed is the dominant allele and the masked allele is recessive. The recessive allele is expressed only when present twice in an individual (a homozygous condition). Dominant alleles are expressed if present in single copy (heterozygous, one dominant allele with one recessive allele) or double copy (homozygous).

Mendel's practices of counting progeny according to their appearance (phenotypes) and of analyzing data as ratios provided additional evidence suggesting the paired nature of the alleles. The pattern of color inheritance corresponded to the ratios predicted by probability theory for information that is carried in two copies and in which the two copies (alleles) separate from each other into gametes and then reassociate later into new combinations of paired factors (via fertilization to form the zygote). Even though Mendel had a mathematical explanation for his results, he did not have a biological explanation. At the time Mendel was conducting experiments, chromosomes and meiosis had not yet been discovered. The "modern synthesis" of inheritance patterns with the role of chromosomes as carriers of the alleles and with the processes of cell division waited until the turn of the 20th century.

The following chart presents genotypes, phenotypes, and gametes (with proportions of each) for a three-generation monohybrid cross. You should compare this chart with the Punnett square method for determining ratios of gametes and progeny for a monohybrid cross as shown in textbook Figure 13-5.

Parental generation   red parent
phenotype: red
genotype: *RR* (homozygous)
gametes: 100% *R*

white parent
phenotype: white
genotype: *rr* (homozygous)
gametes: 100% *r*

Cross: *RR* x *rr*

F$_1$ generation   phenotype: all progeny (and, hence, all parents in next cross) red
genotype for all: 100% *Rr* (heterozygous)
gametes from one parent: 50% *R* and 50% *r*
gametes from other parent: 50% *R* and 50% *r*

Cross: *Rr* x *Rr*

F$_2$ generation   phenotype: 75% red and 25% white (3:1 phenotypic ratio)
genotypes: 50% *R* x 50% R = 25% *RR*   (red)
50% *R* x 50% *r* = 25% *Rr*   (red)   (1:2:1 genotypic ratio)
50% *r* x 50% R = 25% *rR*   (red)
50% *r* x 50% *r* = 25% *rr*   (white)

C. Testing the Principle of Segregation.--Geneticists use test crosses as an experimental approach to determine the genotype of an individual whose genotype is not known. Mendel used a test cross to determine the genotype of the red-flowered F$_1$ peas in the monohybrid cross above; this cross is diagramed in Punnett-square form in textbook Figure 13-7. He hypothesized that (a) if alleles do indeed segregate independently, and (b) if red is dominant to white, then the red F$_1$ plants do carry the recessive allele for white. So, crossing heterozygous red from the F$_1$ (genotype *R*_, where _ represents the unknown allele) with true-breeding white (genotype *rr*) should produce red flowers and white flowers in approximately equal numbers. His results were as expected!

D. <u>Alternative Modes of Inheritance</u>.--The traits followed in the crosses in parts B and C above operated by <u>complete dominance</u>, wherein all phenotypes of progeny are like that of one parent or of the other. In some other situations, a third phenotype appears in the F_1 and three phenotypes appear in the F_2 of a cross between true-breeding dominant and recessive homozygotes: For example, a cross of red-flowered and white-flowered snapdragons yields an F_1 that is all pink. The F_2 has red and white progeny (as expected) but also an intermediately-colored flower (pink) which represents the heterozygote. The ratio of progeny phenotypes is the same as the genotypic ratio--1 red : 2 pink : 1 white. This mode of inheritance is called <u>incomplete dominance</u>.

There is still another mode of inheritance in which the heterozygote has a phenotype different than either parent. Human blood-type determination is an example of <u>codominance</u>. This is a three-allele system in which alleles A and B dominate allele i. When an individual has both A and B, the effects of both A and B are expressed. Refer to your textbook for more details.

E. <u>Crosses with More than One Trait</u>.--In real life, the processes of gametogenesis and fertilization pass <u>many</u> traits simultaneously; so, our consideration of monohybrid crosses above is unrealistically simplistic. Crosses involving more than one trait, such as <u>dihybrid</u> and <u>trihybrid crosses</u>, illustrate <u>Mendel's second law</u>, the <u>law of independent assortment</u>. This law states that alleles for one trait will segregate independently of alleles for other traits.

For example, in one of Mendel's experiments, alleles for seed color segregated into gametes independently of alleles for seed shape. The independence of this segregation was evident in the F_2 of homozygous dominant individuals (round, yellow seeds) crossed with homozygous recessive individuals (wrinkled, green seeds). The two parental phenotypes as well as two <u>new</u> phenotypes (round, green and wrinkled, yellow) were present in the F_2. The new phenotypes could have resulted in the observed ratios <u>only</u> if alleles for shape had segregated independently from those alleles for color. The individuals possessing the new phenotypes (as well as the gametes that produced them) are referred to as <u>recombinants</u>. This and other mechanisms for recombining genetic information (such as crossing-over during prophase I of meiosis) are primary sources of variation that are critical components of the evolutionary process.

In light of our knowledge of the role of chromosomes in carrying alleles and of how chromosomes behave in meiosis, we realize that Mendel was lucky to have selected traits that were carried on <u>different</u> chromosomes. We know now that characters carried on the same chromosome would not show the same pattern of segregation, because alleles <u>linked</u> on the same chromosome generally go to the same gamete. We'll study inheritance patterns of linked characters in Chapter 14.

To understand how alleles for different characters segregate as described just above, you need to work through the process of determining the allelic make-up of gametes that a set of parents can produce, and then to figure out what genotypes and phenotypes (and in what ratios) result from union of those gametes. Textbook Figure 13-9 presents an excellent example of a dihybrid cross. You should be able to determine genotypic and phenotypic ratios of a cross by both the Punnett-square and the probability-theory methods presented in the textbook. Working genetics problems is probably the best way to be certain that you understand the basics of the rules of inheritance; several sample problems are included in this chapter of the <u>Study Guide</u> to guide your study and to check your understanding.

TERMS TO UNDERSTAND: Define and identify the following terms:

genetics

parental generation (P)

first filial generation (F_1)

second filial generation (F_2)

principle of segregation (Mendel's first law)

alleles

locus

mutation

dominant

recessive

homozygous

heterozygous

phenotype

genotype

Punnett square

monohybrid cross

test cross

intermediate inheritance

incomplete dominance

codominance

dihybrid

recombinants

trihybrid cross

rule of independent assortment (Mendel's second law)

pangenesis

GENETICS PROBLEMS

1. Assume that albinism in a cave-dwelling species of bats is a recessive trait. These bats have normal brown pelage when a dominant allele is present. A cross between an albino bat and a brown bat produced bats which all had brown fur. Crosses between those offspring yielded 13 bats; 12 had brown fur and only one was albino. Using this information, answer the following questions:

   a. What were the genotypes of the two bats of the parental generation?

   b. What were the genotypes of the F_1 bats? What gametes can they produce?

   c. What are the genotypes of the albino F_2 bats and of the brown F_2 bats?

   d. Do the F_2 progeny fit the expected ratio for a monohybrid cross of a homozygous dominant bat with a homozygous recessive bat? What is the expected ratio? Use a chi-square test to test for goodness of fit.

e. If you are uncertain of the genotypes of the brown F2 bats, then design a breeding experiment to determine the genotypes. What would be the results of a test cross if the brown F2 bats were heterozygous . . . or if homozygous?

2. Let's assume that scientists have at long last been successful in capturing two living specimens of the Loch Ness monster, and that zoo keepers have developed a successful captive-breeding program for the monster. Records over several generations have shown that the presence of spikes on the tail is a dominant trait and that a red shine in the eyes is recessive to green eyeshine. The cross of a green-eyed, spike-tailed male with a red-eyed, spikeless female produced a clutch of 15 green-eyed, spike-tailed progeny. Cross-breeding among these 15 monsters produced a generation containing 87 green-eyed, spike-tailed, 32 green-eyed, spikeless, 29 red-eyed, spike-tailed, and 11 red-eyed, spikeless monsters. Using this breeding information, answer the following questions:

a. What is (are) the genotype(s) of the F_1?

b. Use a Punnett square to show (i) the gametes that the F_1 generation can produce, and (ii) to determine the genotypes and phenotypes, and their ratios, in the F_2 generation.

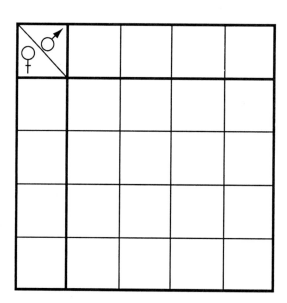

c. What are the expected genotypic and phenotypic ratios for the F$_2$? Do these ratios meet the expected values?

d. Design a cross to determine the genotype of a green-eyed, spike-tailed Loch Ness monster from the F$_1$ generation. How would you change this test cross to determine the genotype of an individual of the same phenotype from the F$_2$?

PEOPLE TO KNOW: What is the significant biological contribution of each of these individuals as indicated in Chapter 13?

August Weismann

Johann Gregor Mendel

Wilhelm Johannsen

R. C. Punnett

Jean Baptiste Lamarck

DISCUSSION QUESTIONS

1. Briefly outline the historical evolution of ideas about the mechanisms of heredity that prevailed before the findings of Mendel and others in the last half of the ninteenth century.

2. Explain how a monohybrid cross followed through a second filial generation can demonstrate the principle of segregation. How can an experiment that follows two or more traits through an F2 generation demonstrate the principle of independent assortment?

3. What practical applications and benefits can you think of that have derived from an understanding of Mendelian genetics?

TESTING YOUR UNDERSTANDING:

For each of the test items below, three of the lettered alternatives are true and the other is false. Determine which alternative is false and write its letter in the blank to the left of the question number. On the blank line below alternative D, write the corrected version of the false statement.

_____ 1. A. An allele is one of the alternate ways in which a gene can exist.
B. In a somatic cell of a diploid organism, each gene is represented by four alleles.
C. The position of a gene on a chromosome is its locus.
D. Many genes can exist in more than two forms.

Correction: _____

_____ 2. A. In a situation of incomplete dominance, a recessive allele will not be expressed in the presence of a dominant allele.
B. New alleles of a gene can arise by mutation.
C. In a situation of complete dominance, a heterozygote expresses the dominant phenotype.
D. In a situation of complete dominance, a monohybrid cross of homozygous dominant and homozygous recessive individuals will produce an F1 generation in which all individuals show the dominant trait.

Correction: _____

_____ 3. A. In a dihybrid cross of homozygous dominant with homozygous recessive individuals, approximately one-sixteenth of the F_2 will express both recessive traits.
   B. In a situation of complete dominance, a dihybrid cross of homozygous dominant with homozygous recessive individuals will yield approximately nine-sixteenths of the F_1 expressing both dominant traits.
   C. In a situation of incomplete dominance, a dihybrid cross of homozygous dominant with homozygous recessive individuals will produce an F_2 in which only about one-sixteenth of the individuals express the homozygous dominant phenotype.
   D. In a situation of incomplete dominance, a dihybrid cross of homozygous dominant with homozygous recessive individuals will produce an F_2 in which about one-fourth of the individuals express the intermediate phenotype for both traits.

   Correction: _____

_____ 4. A. In a test cross, an individual of unknown genotype for a particular trait is crossed with an individual that is homozygous dominant for that trait.
   B. Inheritance in which a heterozygous trait is visibly affected by both alleles is called intermediate inheritance.
   C. Codominance is a situation in which both alleles produce an effect in the heterozygote.
   D. In situations of intermediate inheritance, both the genotypic and phenotypic ratios of the F_2 for a monohybrid cross of homozygous dominant and homozygous recessive individuals will be 1:2:1.

   Correction: _____

_____ 5. A. The inheritance pattern exhibited for blood types in humans is an example of codominance.
   B. The principle of independent assortment explains how recombinant gametes are produced.
   C. In the 9:3:3:1 ratio for the F_2 of a dihybrid cross, the 9 includes those individuals which possess at least one dominant allele for each trait.
   D. The F_2 of a dihybrid cross includes eight genotypes.

   Correction: _____

_____ 6. A. R. C. Punnett coined the term "gene."
   B. Karl Correns was one of several scientists whose experiments confirmed Mendel's findings.
   C. August Weismann was a proponent of the continuity of germinal plasm.
   D. Darwin proposed a hypothesis of pangenesis to account for the inheritance of acquired characteristics.

   Correction: _____

CHAPTER 14--THE PHYSICAL BASIS OF HEREDITY

CHAPTER OVERVIEW

In the previous chapter, we examined general patterns of inheritance without considering the physical basis for passing traits from one generation to the next. In this chapter, we mesh such studies of inheritance with knowledge of cell structure to develop the chromosome theory. We also examine in this chapter the implications of having genes carried on chromosomes. These implications include gene linkage and recombination of alleles by crossing-over. Recombination and various types of mutations are important sources of variation in the genetic make-up of populations.

Another major topic addressed in this chapter is gene interactions. Inheritance of many traits is influenced by more than one gene. Examples of inheritance by gene interactions include epistasis, modifier genes, complementary gene action, and polygenic inheritance.

Additionally, this chapter introduces us to fruit flies, a workhorse of genetic studies. The characteristics of such model species that make them suitable models for genetic studies are presented.

TOPIC SUMMARIES

A. <u>The Chromosome Theory</u>.--Studies of heredity and cell structure by different scientists progressed separately for decades until around 1900. These two fields did not come together until Sutton and Boveri independently perceived that genes were carried on chromosomes. This realization came to be known as the <u>chromosome theory</u>. Movements of chromosomes explain Mendel's <u>laws of independent segregation and of independent assortment</u>.

(1) Independent segregation. It was evident to Mendel that, during gamete formation, the alleles of each gene separate from each other and move to regions that become separate cells. This partitioning of alleles corresponds precisely to the pattern of segregation of homologous chromosome pairs during meiosis. The result is that each of the two alleles for any given gene is present in the gametes in equal numbers: If the parent is homozygous, then all gametes possess the same allele for that gene; if the parent is heterozygous, then the dominant and recessive alleles occur in a 1:1 ratio.

(2) Independent assortment. Consider breeding experiments in which two or more traits are being followed. Mendel observed that all of the traits he studied in dihybrid and trihybrid crosses segregated independently of each other. Independent assortment of multiple traits is apparent when offspring possess characteristics in a <u>different combination</u> than the way the traits were expressed in either parent.

Hence, a cross between a doubly-recessive short-tailed, white mouse and a doubly-heterozygous long-tailed, brown mouse would produce the <u>recombinant</u> short-tailed, brown mice and long-tailed, white mice as well as those offspring resembling both <u>parental</u> phenotypes. The occurrence of all four phenotypes in equal frequency is in agreement with the probability of independent events occurring together. Figure 14.1 below diagrams how alleles for these two traits, each carried on a different chromosome pair, assort independently:

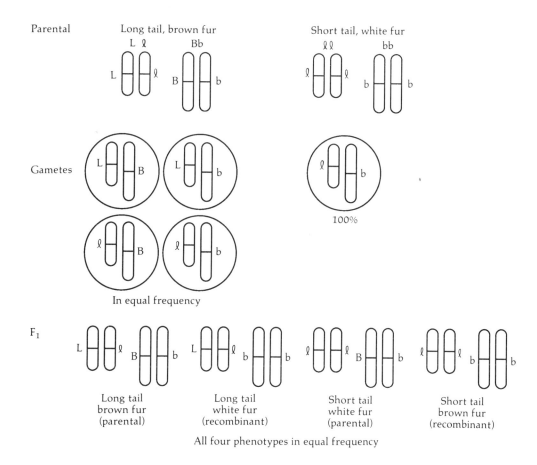

Fig. 14.1: Independent assortment of two alleles carried on different chromosome pairs

B. <u>Why Study Fruit Flies</u>?--A primary model species in the world of genetics is the fruit fly, genus *Drosophila*. What is so magical about these insects? For genetics, the ideal model organism must possess a number of features. These include a short generation time (important since we often follow crosses through several generations). The organisms must be easy to handle and culture. They must possess traits whose alternative phenotypes are easy to read and score. They must be readily available. Fruit flies fit the bill marvelously!

C. <u>Gene Linkage</u>.--"Lady Luck" smiled on Mendel in selecting traits that produced results that indicated independent assortment. However, had he had the misfortune of selecting traits that were carried on the <u>same</u> chromosome, then he would have gotten very <u>different results</u>. Cytological and genetics studies have demonstrated that not just one, but many genes are carried on each chromosome; such genes are <u>linked</u>. It is reasonable to expect that physically-linked stretches of DNA (i.e., a chromosome) normally will remain intact throughout the process of meiosis. Hence, such linked genes would be inherited together, and we would expect crosses to produce progeny whose combinations of traits <u>all</u> resemble their parents; we would expect <u>no</u> recombinant phenotypes to appear as they do in a cross where the genes are located on different chromosomes.

Let's alter the example in part A(2) above. Let's assume here that the genes for tail length and for hair color in mice are linked on the same chromosome. We might <u>expect</u> two phenotypes (short-tailed white and long-tailed brown) in equal frequency. Running the experiment, however, yields the same four F_1 phenotypes as did the cross in which the genes were carried on two different chromosomes. Interestingly, the ratio of these phenotypes is not 1:1:1:1; rather we observe that the parental phenotypes make up, say, about 45% each of the F_1 with the recombinant phenotypes present at only about 5% each. How do we explain these findings? Go back to Chapter 12 and review the events of prophase I in meiosis. Recall that the crossing-over that occurs during synapsis involves exchange of corresponding parts of chromosomes with the result of producing recombinant genotypes and phenotypes. Also, read in textbook Chapter 14 to see how the proportion of recombinant geneotypes in the F_1 can be used to map chromosomes.

D. Gene Interaction.--The traits we have studied so far are controlled by alleles positioned at only one gene locus. Inheritance in many characters is regulated by more than one gene. Such gene interaction operates in several different ways:

    1. Complementary gene action. In some situations, such as the cock's-comb example in the textbook, two genes act together to produce a phenotype that neither gene acting alone can produce.

    2. Epistasis. This interaction involves one gene which masks the effects of another. Consider, for example, inheritance of agouti coat color in a strain of lab mice. Two genes are involved: One gene determines whether coat color will be black (*aa*) or agouti (*AA*). Another gene affects color expression; homozygous recessive (*cc*) produces albino fur because no pigment is produced. A cross between a black (*aaCC*) and an albino (*AAcc*) produces an F_1 which is entirely agouti (*AaCc*). Interbreeding of these F1 agouti produces an F2 with the expected nine genotypes, but only three phenotypes: The presence of both dominant alleles (*A_C_*) yields agouti. Progeny that are homozygous recessive for black and one or more dominant alleles for color (*C_*) are black. Any combination of alleles for the color gene (*AA, Aa, aa*) produces albino because homozygous recessive (*cc*) prevents any color expression; genotypes *A_cc* and *aacc* are albino. Hence, the albino gene can mask any color that is dictated by the color gene.

    Epistasis is suggested by F_2 progeny in which fewer than the expected number of phenotypes occur. In the example above, the ratio was 9 agouti : 3 black : 4 albino.

    3. Modifier genes. In this situation, the way in which one gene is expressed is altered by one or more other genes. Your textbook's example pertains to human eye color. One gene determines whether brown pigment will be deposited in the iris; modifier genes dictate where and how the pigment will be deposited. The result of this interaction is the common eye colors brown, blue, green, hazel, and others.

E. Polygenic Inheritance.--This mode of inheritance involves two to many genes that contribute to expression of a trait. Generally, the types of traits that are inherited in this manner have a continuous distribution rather than the "either-or" expression seen where dominance exists. Examples of polygenic inheritance include the determination of size of tomatoes, degree of spotting in the coats of cattle, and leucocyte count in lab mice, as well as those examples in your textbook. Neither allele of the genes involved exerts dominance, and each gene collaborates with the others in an additive manner. Therefore, the more "tall" alleles that an individual possesses, then the taller that individual will be. Crosses between homozygotes from opposite ends of the continuum will produce progeny that are intermediate for the trait.

F. Influence of Environmental Factors on Gene Expression.--The genotype for a trait determines the potential for expression of that trait. However, an individual's phenotype might differ from that potential . . . much as your performance on an exam might fall short of the level that you feel you are capable of achieving. Environmental factors (such as temperature, nutritional quality, amount of rainfall, etc.) can and do affect how genotypes of many traits are expressed. For example, the acorn that can grow into a mighty oak tree will not live up to its potential if it is not provided with enough water. To continue our exam analogy, external influences such as no sleep on the night before the exam can cause you not to score as well as the amount of studying you did suggested that you should. So, phenotypic expression is a product of genotypic potential and environmental influences. Geneticists measure such influences in terms of penetrance (the proportion of the population with a particular genotype which expresses that genotype) and expressivity (the degree to which an individual expresses the full potential of that genotype).

G. Mutations: Another Source of Genetic Variability.--Even in crosses of known pure-breeding individuals, individuals with a different phenotype (and, correspondingly, usually a different genotype) occasionally appear. These variants often occur by a spontaneous process of gene mutation. Mutation is a natural phenomenon that normally occurs at a low rate, and it is a necessary (and often healthy) part of the evolutionary process. However, excessive mutation rates, such as those caused by over-exposure to radiation and various mutagenic chemicals, can be detrimental and often lethal. Mutations can occur at various levels and in many forms which we will examine later in the context of evolution. Two of the primary modes of mutation involve (a) mutation of one allelic form to another, and (b) alteration of chromosome number and structure.

H. Genetics of Microorganisms.--All of the examples we have used in this chapter thus far have been of eukaryotes. Let us not forget that microorganisms (bacteria, and viruses such as the bateriophages) also possess genetic materials. Indeed, much of what we know of the principles of genetics derived from study of such microbes . . . and the future of genetic engineering seemingly rests with microorganisms.

TERMS TO UNDERSTAND: Define and identify the following terms:

chromosome theory

heteromorphic pair

Mendelian ratios

Drosophila melanogaster

wild type

gene linkage

linkage groups

parental combinations

recombinations

crossing-over

marker gene

map unit

polygenic inheritance

gene interaction

complementary gene action

epistasis

modifier genes

phenocopy

reaction range

penetrance

expressivity

gene mutation

reverse mutation

chromosome mutation

bacteriophage

GENETICS PROBLEMS

1. Answer the questions below on the basis of this scenario: Suppose that you are a horticulturist who is trying to develop a strain of tomatoes that bears larger fruit. You have already determined that inheritance of fruit weight is controlled by three genes, labeled *A*, *B*, and *C*. The *AABBCC* genotype produces 60-gram tomatoes and *aabbcc* produces 30-gram tomatoes. Furthermore, each dominant gene adds 5 grams to fruit weight. Crossing a 60-gram plant with a 30-gram plant yields offspring which all bear 45-gram fruits.

   a. What is the mode of inheritance for fruit size?

   b. What is the genotype of the F_1 produced by crossing an *AABBCC* plant with an *aabbcc* plant?

   c. What are the genotypes and phenotypes of the F_2 produced by crossing F_1 individuals? Use the Punnett square to determine genotypes, and list the phenotypes below the square.

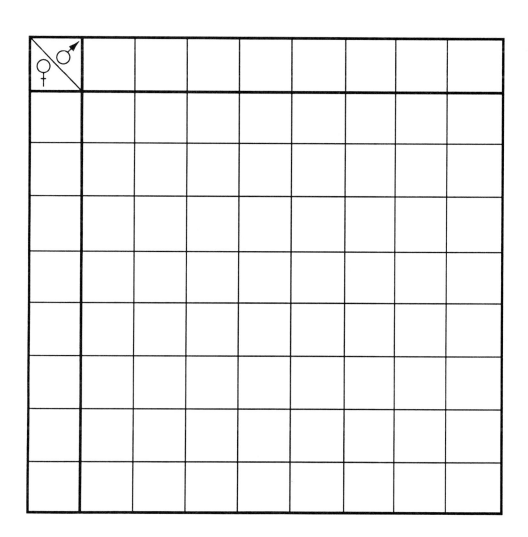

2. Answer the following questions on the basis of the information provided in Chapter 14 of your textbook regarding inheritance of rooster combs. Suppose that a walnut-combed rooster mates with two hens. The mating with a walnut-combed hen produced 6 walnut-combed and 2 rose-combed offspring. Mating with a pea-combed hen yielded offspring in the ratio of 3 walnut : 3 pea : 1 single : 1 rose.

   a. By what mode of inheritance is comb type inherited?

   b. What are the genotypes of the rooster and the two hens?

   c. Use the Punnett square below to determine the genotypes and phenotypes of a cross between a pea-combed rooster and a single-combed hen.

   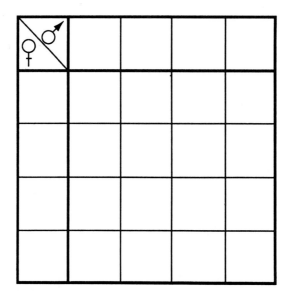

3. Answer the questions below on the basis of the following information. Suppose that tail color and ear size in field mice (genus *Peromyscus*) are inherited by simple dominance. Gray tail color (T) is dominant over white (t), and large ear size (L) is dominant over small (l).

   a. A dihybrid cross of homozygous dominant with homozygous recessive produced 100% gray-tailed, long-eared mice. What is the F_1 genotype?

   What genotypes of gametes can the F_1 produce?

b. Test crosses between mice that are heterozygous for both traits with mice that are doubly-recessive for these traits yielded the following progeny:

gray-tailed, long-eared	42 individuals
gray-tailed, short-eared	9 individuals
white-tailed, long-eared	10 individuals
white-tailed, short-tailed	48 individuals

Are these the phenotypes that would be expected for such a test cross?

If not, then what phenotypes were expected?

How can you account for the results that actually were obtained?

Do you need to return to part A above to revise the types of gametes that were produced by the F_1?

4. What is the role of statistics in biological research? How did Mendel contribute to our modern practice of using statistics in analyzing biological data?

PEOPLE TO KNOW: What is the significant biological contribution of each of these individuals as indicated in Chapter 14?

W. S. Sutton

Theodor Boveri

Elinor Carothers

Thomas Hunt Morgan

William Bateson

E. M. East

Gertrude and Charles B. Davenport

Hugo DeVries

William E. Castle

DISCUSSION QUESTIONS

1. What characteristics are required of a species that is to be used in genetics research? Give several examples of species used for such a purpose.

2. How does gene mutation (as well as other sorts of mutations) contribute to the process of evolution?

3. Summarize the evidence that supports the chromosome theory.

4. How can phenotypic ratios of experimental crosses be useful in determining the mode of inheritance of a trait? In your answer, consider the following: linkage, recombination, gene interaction, and polygenic inheritance.

TESTING YOUR UNDERSTANDING:

For each of the test items below, three of the lettered alternatives are true and the other is false. Determine which alternative is false and write its letter in the blank to the left of the question number. On the blank line below alternative D, write the corrected version of the false statement.

_____ 1. A. W. S. Sutton was an early geneticist who grasped the significance of chromosomes and their role in inheritance.
B. The chromosome theory states that genes are carried on chromosomes.
C. The chromosome theory explains why different alleles of the same trait are inherited as linkage groups.
D. The chromosome theory explains how alleles for genes that are carried on different chromosomes assort independently of each other.

Correction: _____

_____ 2. A. Statistical analysis of the proportion of progeny in each phenotypic class can be a useful means of determining how a trait is inherited.
B. In a situation of complete dominance and in which the traits are on different chromosomes, the expected F_2 phenotypic ratio from a dihybrid cross of a doubly homozygous dominant with a doubly homozygous recessive would be 9:3:3:1.
C. In a situation of complete dominance and in which the traits are on the same chromosome, the expected F_2 phenotypic ratio from a dihybrid cross of a doubly homozygous dominant with a doubly homozygous recessive would be 9:3:3:1.
D. In a situation of epistasis and in which the traits are on different chromosomes, the expected F_2 phenotypic ratio from a dihybrid cross of a doubly homozygous dominant with a doubly homozygous recessive would be 9:3:3:1.

Correction: _____

___ 3. A. Our understanding of gene linkage supports the following statement: If genes for hair color and eye color are carried on the same chromosome, then a cross between a blonde, blue-eyed male and a brunette, green-eyed female could not produce a blonde, green-eyed offspring.
B. In the cross in part A above, a blonde, green-eyed offspring would be called a recombinant.
C. Crossing-over during prophase I is the means by which recombinant genotypes are produced.
D. The technique of mapping chromosomes relies on the production of offspring that possess recombinant genotypes.

Correction: _____

___ 4. A. Epistasis is a mode of inheritance in which two or more genes are involved in determining how a trait will be expressed.
B. Epistatic inheritance can be recognized by a reduction in the number of phenotypic classes present in a progeny generation.
C. In epistasis one gene modifies, but does not mask, the effects of another gene.
D. Complementary gene action is a situation in which two or more genes act together to produce a phenotypic trait that neither gene could produce alone.

Correction: _____

___ 5. A. In cases of polygenic inheritance, the number of phenotypic classes in the F_2 is usually fewer than in inheritance by complete dominance.
B. Many traits that are expressed in a continuous, rather than in a discrete, manner are examples of polygenic inheritance.
C. In polygenic inheritance, most offspring express phenotypes that are intermediate to the phenotypes of their parents.
D. Height and coloration are example of the kinds of traits that are inherited by polygenic inheritance.

Correction: _____

___ 6. A. Expressivity is defined as the percentage of the individuals carrying a gene that actually express the related phenotype.
B. Measures such as penetrance are used to indicate the degree to which environmental conditions can affect the expression of a genotype.
C. A phenocopy is an individual whose phenotype has been altered by the environment in such a way that the phenotype expressed actually resembles that usually associated with some other genotype.
D. The genotype determines the potential range of phenotypes which an individual might express.

Correction: _____

_____ 7. A. Mutations in somatic cells usually are transmitted to subsequent generations of that species.
   B. Gene mutations are one of the sources of genetic variation that have accounted for the present diversity of life on Earth.
   C. Mutation of genes at low frequencies is an event that occurs normally during the processes of cell division.
   D. Although some mutations are detrimental, others are beneficial.

   Correction: _____

_____ 8. A. Not all gene mutations are expressed in the phenotype.
   B. Once a gene has mutated from one allele to another, that same gene cannot mutate back to its original form.
   C. Not only can mutation involve changes at gene loci, but mutation can involve change in chromosome number or chromosome structure.
   D. X-rays and ultraviolet radiation are known mutagens.

   Correction: _____

_____ 9. A. Microorganisms, such as *Escherichia coli*, are excellent experimental subjects for genetics research because their generation time is short.
   B. A reason that bacteria are useful species for genetics studies is that they are haploid; the value of haploidy is that expression of a mutant gene can occur almost instantly.
   C. Bacteriophages are viruses that require bacterial hosts in order to be able to reproduce.
   D. Most alleles of bacterial chromosomes can be referred to as either dominant or recessive.

   Correction: _____

_____ 10. A. Theodor Boveri was one of the cytologists who theorized that genes are carried on chromosomes.
   B. William Castle was an early supporter of the use of *Drosophila* as an experimental species in genetics studies.
   C. The term "epistasis" was first used by A. H. Sturtevant.
   D. The concept of polygenic inheritance was first applied to human genetics by Gertrude and Charles Davenport.

   Correction: _____

CHAPTER 15--SEX AND HUMAN HEREDITY

CHAPTER OVERVIEW

From your own observations, you know that most familiar species are represented by two mating types or sexes. In previous chapters, we have followed breeding experiments that required genetic contributions from two sexes. Yet, we have not examined the genetic basis of sex determination . . . nor have we studied crosses that involve genes carried on the sex chromosomes.

In this chapter, we learn about several systems of sex determination, and we see how phenotypic and genotypic ratios deviate from the "expected" results when genes are carried on the sex chromosomes. This chapter also treats the topic of inherited sex abnormalities and discusses the chromosomal basis for such abnormalities. We'll look also at the basis of several human genetic diseases.

TOPIC SUMMARIES

A. Sex Determination.--Evidence that genes are involved in determining the sex of an individual is provided by an essential difference in the chromosomal make-up of individuals of different sexes: For most animal species, study of karyotypes reveals that most chromosomes occur in pairs having similar size and shape . . . these are the homologous pairs that you studied in Chapter 12. Closer study shows that all chromosomes match into pairs in one sex (the homogametic sex), but that all chromosomes except one or two (depending on the species) match into pairs in the other sex (the heterogametic sex). The unmatched chromosome(s) in the heterogametic sex is(are) the sex chromosomes (gonosomes); the remaining chromosomes that do match into pairs are called autosomes and have no direct role in sex determination. In the homogametic sex, the two sex chromosomes are alike in size and shape and, hence, form a matching pair.

Several different systems of sex determination are known in Kingdom Animalia. You probably are familiar with the XX-XY system of mammals such as in humans in which males are heterogametic (XY) and females are homogametic (XX). This system is reversed in birds, butterflies and some other species, with males being homogametic (ZZ) and females heterogametic (ZW). Grasshoppers show an interesting variation on this theme: Females possess two identical sex chromosomes (XX), whereas males have only one sex chromosome and are designated XO. The insect order Hymenoptera lacks sex chromosomes altogether. In this arrangement, sex is determined by ploidy; females are diploid and males develop from unfertilized eggs and, hence, are haploid.

From the previous paragraph, it might appear that the mere presence or absence of a Y or a W chromosome dictates the sex of an individual. Read in your text for the evidence in fruit flies and in humans that the ratio of the number of X chromosomes to the number of sets of autosomes is a better indicator of sex. Some individuals may possess more or less than the normal number of sex chromosomes, generally as a result of nondisjunction of the sex chromosomes during meiosis.

B. Sex-Linkage.--Many of the differences between sexes are due to the genes that these sex chromosomes carry. Genes for various sex-related traits might be carried on the X but not on the Y chromosome, or vice versa.

Genes that are carried on the X chromosomes are said to be sex-linked. Sex-linkage was discovered by T. H. Morgan when crosses for eye color in fruit flies produced progeny in which all individuals expressing the recessive allele were of the same sex, the heterogametic sex. Reciprocal crosses provided the evidence that genes for sex-linked traits are carried on sex chromosomes. In the crosses we have studied in previous chapters, we were not concerned whether it was the male or the female parent that carried a particular allele . . . this was because the genes for those traits were carried on autosomes. Reciprocal crosses (where alleles carried by each parent are reversed) for autosomal traits yield the same F_1 and F_2 results. Reciprocal crosses for sex-linked traits yield strikingly different results; see the textbook example for crosses of white-eyed and red-eyed *Drosophila* for specific details. Nondisjunction of X chromosomes can lead to exceptional offspring in which a trait expressed in a way normally associated with one sex unexpectedly appears (albeit in very low numbers) in the other sex.

Genes that are carried on the Y chromosome are not represented in the other sex. Such traits are passed on by Y-linked inheritance. These traits can be expressed only in the sex that carries the Y chromosome which, in the case of humans, is males.

Hemophilia and color-blindness are examples of sex-linked traits in humans. One of the few Y-linked traits known for humans is the condition of hairy ears.

C. X-Chromosome Dosage.--Dosage refers to the number of copies of a particular gene or of a particular chromosome in a cell. Nondisjunction during meiosis can result in offspring with other than the normal number of X chromosomes; textbook Table 15-1 shows phenotypic effects of increased X dosage. Increased dosage of genes in autosomes tends to multiply the amounts of certain proteins produced by a cell. However, the amounts of proteins coded by genes on the X chromosome seem not to change regardless of the number of X chromosomes present. Suppression of the expression of extra X genes is accomplished by dosage compensation genes (a type of modifier genes) in *Drosophila* and, in mammals and various other animals, by inactivation via clumping of extra X's into chromatin masses called Barr bodies.

Research has indicated that all X chromosomes (except for one) in any cell become inactivated. Then, in a normal human female who has two X's, which X becomes inactivated? And when does inactivation occur? The Lyon hypothesis (single-active-X hypothesis) proposes that (1) inactivation occurs early in embryonic development, (2) the X that is inactivated for any particular cell is randomly selected and is independent of adjacent cells, and (3) inactivation is permanent. All cells arising (via mitosis) from a particular cell will have the same X inactivated. Because an individual with two or more X's can carry different alleles for a gene on the X, X inactivation can produce individuals with "patchy" (or mosaic) phenotypes. The fur coloration pattern of a calico cat is an excellent example of mosaicism.

D. Sex Abnormalities.--Errors occasionally occur in sex determination and sexual differentiaiton. Gynandromorphy occurs when a sex chromosome is lost from a somatic cell early in development. The way in which male and female characters are expressed in a gynandromorph depends on how early in development that the X chromosome is lost: If lost at the two-cell stage, then left and right halves of the body show traits of the opposite sexes. If an X is lost at a later stage instead, then the offspring presents a mosaic of male and female characteristics.

An array of sex abnormalities has been described for humans. Klinefelter's syndrome, occurring in XXY males, is the most common. Turner's syndrome occurs in XO females. These and other human X and Y abnormalities generally result in individuals having mental deficiencies, physical abnormalities, and infertility.

E. <u>Human Heredity</u>.--Study of human heredity presents special problems. The human life cycle is too long and the number of offspring produced per mating is too low for statistically meaningful data to be gathered. Besides, society would not permit the required manipulation of matings. Hence, other approaches to study of human heredity had to be developed. These acceptable methods include analysis of genealogies (<u>pedigrees</u>), study of identical twins, and study of isolated populations in which considerable amounts of inbreeding occurs.

There are many practical reasons to investigate human heredity. Proper medical treatment (e.g., blood transfusions, Rh disease) relies on understanding of inherited characteristics of blood. Human genetics also assists police in solving crimes, aids in identification of parents of offspring, and permits research of anthropological problems.

F. <u>Chromosomal Analysis</u>.--Much of our knowledge of human genetics has derived from study at the cellular level of the morphology and number of chromosomes. Genes are arranged linearly along chromosomes. Changes in structure and/or number of chromosomes necessarily affect the genetic information content of a cell, and often affect the characteristics of that cell. Chromosomes occasionally break in one or more places; severed sections of chromosomes can be lost (<u>deleted</u>), reattached in reverse order (<u>inverted</u>), or joined to a different chromosome (<u>translocated</u>). Chromosomes generally break at fragile sites; many cases of mental retardation are associated with the "fragile-X syndrome."

Chromosome number also can vary. The presence of complete copies of entire sets of chromosomes is <u>abnormal euploidy</u>. <u>Aneuploidy</u> is the condition of changes in number of one or several chromosomes; presence of an extra 21st chromosome, a condition called <u>trisomy 21</u>, produces the disorder of Down's syndrome. Mapping of the human genome also will allow greater understanding of chromosomal conditions.

Special cytogenetics techniques allow examination of the next generation even before the time of birth. <u>Amniocentesis</u> involves drawing cell-containing fluid from the amniotic cavity surrounding the fetus. Fetal cells then are cultured and examined cytologically for chromosomal abnormalities. Discovery of serious abnormalities might lead to the decision to abort the fetus. The manner in which society makes use of such genetic information has important implications, among them <u>eugenics</u> and the wisdom of <u>abortion</u>.

TERMS TO UNDERSTAND: Define and identify the following terms:

sex chromosomes (gonosomes)

autosomes

homogametic

heterogametic

ZZ-ZW system

XX-XO system

superfemale

sex determination

sex linkage

nondisjunction

Y-linked inheritance

hemophilia

dosage

regulation

dosage compensation genes

Barr body

single-active-X (Lyon) hypothesis

mosaicism

sexual differentiation

gyandromorph

intersex

hermaphrodite

aneuploidy

sex-determined characters

pedigree

karyotype

metacentric

submetacentric

acrocentric

telocentric

deletions

inversions

translocations

abnormal euploidy

trisomy

recombinant DNA

somatic-cell hybridization

chromosome transfer

amniocentesis

eugenics

abortion

anencephaly

CHART EXERCISE

Develop a summary for the sex determination systems listed in the leftmost column. In the columns for sex, indicate the chromosomes which determine that sex and whether that sex is heterogametic or homogametic. In the rightmost column, provide examples of groups of organisms which possess that sex determination system.

SYSTEM	MALE	FEMALE	EXAMPLES OF SPECIES
XX-XY			
XX-XO			
ZZ-ZW			
No sex chromosomes			

PEOPLE TO KNOW: What is the significant biological contribution of each of these individuals as indicated in Chapter 15?

C. B. Bridges

Thomas Hunt Morgan

Murray L. Barr

Mary Lyon and Liane Russell

Sir Archibald Garrod

Yuet Wai Kan and Andree M. Dozy

DISCUSSION QUESTIONS

1. How are the effects of nondisjunction different for the cell division processes of meiosis and mitosis? Give examples of human abnormalities resulting from nondisjunctions.

2. Although genes for many traits are carried on the human X chromosome, very few traits are Y-linked. What is the adaptive value (if any) of such an arrangement?

3. Summarize the modes of inheritance for the following hereditary conditions:

    a. hemophilia

    b. colorblindness

    c. ABO blood groups

4. Describe the karyotype illustrated below. Indicate the number of chromosomes of each morphology (e.g., metacentric, acrocentric, etc.) What is the diploid number for the karyotype shown?

5. How is chromosome mapping important to understanding human heredity?

6. Sometimes biological issues become embroiled in social controversy. Summarize the objective arguments and scientific information regarding the abortion issue.

7. Bleeder's disease (hemophilia) is coded by a recessive, sex-linked allele. Given the preceding information, answer the following questions:

   a. Can a female be a hemophiliac? If yes, how? If no, why not?

   b. If a woman who is heterozygous for the hemophilia gene marries a normal male, what proportion of their children will be hemophiliacs? What sex will their hemophiliac children (if any) be?

   c. If a woman who is homozygous dominant for the hemophilia gene marries a hemophiliac male, what proportion of their children will be hemophiliacs? What sex will their hemophiliac children (if any) be?

8. Describe the type(s) of experimental crosses you would use to determine if a trait is inherited by sex-linkage . . . or by Y-linked inheritance.

TESTING YOUR UNDERSTANDING

For each of the test items below, three of the lettered alternatives are true and the other is false. Determine which alternative is false and write its letter in the blank to the left of the question number. On the blank line below alternative D, write the corrected version of the false statement.

_____ 1. A. The gonosomes of a male fruit fly are an X and a Y.
　　　　B. Male mammals are homogametic.
　　　　C. *Drosophila* possesses four pairs of chromosomes.
　　　　D. Female birds are heterogametic.

　　　　Correction: _____

_____ 2. A. Certain insects, such as the Hymenoptera, do not possess sex chromosomes.
　　　　B. In humans, maleness is determined by the presence of a Y chromosome rather than by the absence of an X.
　　　　C. In fruit flies, femaleness is determined by the ratio of the number of X chromosomes to the number of sets of autosomes.
　　　　D. XXY codes for the male sex in both humans and fruit flies.

　　　　Correction: _____

_____ 3. A. Recessive, sex-linked traits can be expressed in females, but not in males.
　　　　B. Genes carried on the Y chromosome cannot be expressed in female humans.
　　　　C. One way by which an individual can obtain more than two gonosomes is by occurrence of nondisjunction during gametogenesis in one or both parents.
　　　　D. The presence of a trait only in males does not mean that the trait must be transmitted by Y-linkage.

　　　　Correction: _____

_____ 4. A. As with autosomes, increased dosage of the X chromosome causes production of elevated levels of those proteins coded for by genes on those chromosomes.
　　　　B. An individual with XXY gonosome composition has a greater than normal dosage of the X chromosome.
　　　　C. The Lyon hypothesis suggests that the X chromosome to be inactivated is chosen randomly in each cell of a female.
　　　　D. The phenomenon of mosaicism is explained by the single-active-X hypothesis.

　　　　Correction: _____

_____ 5. A. Gynandromorphs are individuals who are part male and part female.
　　　　B. Gynandromorphy occurs by a nondisjunctional loss of an X chromosome during an early mitotic division in an embryo.
　　　　C. Down's syndrome in humans is also referred to as trisomy 21; in this condition, an individual possesses only 21 chromosomes.
　　　　D. Turner's syndrome is a condition that affects only females.

　　　　Correction: _____

_____ 6. A. Aneuploidy refers to conditions in which an individual possesses an abnormal total number of chromosomes.
B. Abnormal euploidy is defined as changes in chromosome counts in multiples of the haploid number for that species.
C. Chromosomes are classified according to position of centromeres and the length of chromosome arms.
D. The centromere of a metacentric chromosome is located at the end of the chromosome.

Correction: _____

_____ 7. A. The change of gene sequence from "xyz" to "xyyz" is an example of translocation of chromosomal structure.
B. Breakage of chromosomes that undergo rearrangements occurs at locations known as fragile sites.
C. Many of the cases of inherited mental retardation are associated with the "fragile-X syndrome."
D. Chromosome breakage does not always lead to chromosomal rearrangements; many breaks spontaneously reunite.

Correction: _____

_____ 8. A. Combining the techniques of karyotyping and amniocentesis allows diagnosis of certain genetic diseases before an infant is born.
B. Knowledge gained from basic genetic research is being applied in many ways, such as genetic counseling and forensic pathology.
C. The amnion is a membranous sac that surrounds a developing embryo.
D. The calico coloration pattern in cats and the condition of colorblindness in humans are both examples of mosaicism.

Correction: _____

_____ 9. A. If hairy ears is a Y-linked trait, then crossing a normal female and a hairy-eared male produces an F_1 with all males having hairy ears and all females with normal ears.
B. Assume that curly tail is a sex-linked trait, and that a cross between a curly-tailed female and a straight-tailed male yielded the following F_1: 50% of females with straight tails, 50% of females with curly tails, 50% of males with curly tails, and 50% of males with straight tails. These results demonstrate that the female parent was homozygous.
C. Consider a situation in which nondisjunction of sex chromosomes occurred during meiosis for both parents. The ratio of the F_1 according to gonosome make-up would be 1 XXXY to 1XX to 1 XY to 1 without sex chromosomes.
D. In a cross between a hemophiliac male and a carrier female, half of the F_1 females will be hemophiliacs and half of the F_1 females will be carriers.

Correction: _____

_____ 10. A. C. B. Bridges determined that femaleness occurs in fruit flies when the X chromosome to autosome ratio is 1.0 or greater.
B. The discovery of sex-linkage is credited to Sir Archibald Garrod.
C. Barr described clumps of chromatin that represent inactivated X chromosomes.
D. Mary Lyon and Liane Russell suggested that the X chromosome that becomes inactiviated is randomly chosen in each cell.

Correction: _____

CHAPTER 16--THE NATURE OF GENES

CHAPTER OVERVIEW

In the preceding chapters on genetics, we have spoken of genes and alleles and their patterns of inheritance. We have determined that genes must be carried on chromosomes. But, we have not yet examined the chemical and physical nature of genes.

In this chapter, we will answer questions such as these: How large is a gene? How many genes can fit onto a chromosome? Of what chemicals are genes made? How, during processes of cell division, are genes duplicated? How do changes in genes occur? How does a gene determine phenotypic expression? In examining these questions and others, we will briefly review some of the landmark experiments that have led to our modern understanding of the nature of genes.

TOPIC SUMMARIES

A. What Is a Gene?--The relatively modern field of molecular genetics provided the understanding required for description of the physical and chemical properties of a gene. Although biologists have been experimenting in genetics for over a century, such a description has developed within only the last few decades. Yet, even before modern molecular techniques were available, geneticists already had learned much about the behavior and other characteristics of the gene. Six such properties of genes follow:

(1) Genes exhibit a great degree of stability from one generation to the next, yet ...
(2) they can be altered (i.e., they possess a degree of mutability).
(3) Genes contain information that governs traits of cells.
(4) Information carried by a gene can be expressed in alternate forms (alleles) which produce distinct phenotypes.
(5) Genes produce copies of themselves (i.e., they replicate) during cell division.
(6) The gene for a specific trait has a specific location (locus) on a chromosome.

B. The Physical Nature of Genes.--Genes are small; even the most powerful electron microscopes cannot enable us to see a gene. The approximate length of a gene ranges from 10 to 100 nm (1 nm = 1 x 10^{-9} meters); the four chromosomes of fruit flies contain up to about 15,000 genes!

Genes are made of nucleic acids, either DNA (deoxyribonucleic acid, in most organisms) or RNA (ribonucleic acid, in several groups of viruses). Our awareness of the chemical make-up of genes resulted from a number of landmark experiments that are summarized below:

(1) Transformation. Experiments by Griffith, and later by Avery, MacLeod and McCarty, demonstrated that DNA is the conveyor of genetic information, at least in certain bacteria. Griffith determined that rats injected with smooth pneumococcus (virulent) would die of pneumonia, whereas injections of rough pneumococcus (nonvirulent) did not produce infection. Interestingly, mice injected with a mixture of living rough cells and heat-killed smooth cells died. Hence, some non-living component of smooth cells could transform the previously-nonvirulent rough cells into a lethal strain. Follow-up experiments by the Avery group determined that DNA was the transforming principle.

(2) Viral replication. Study of the replication of viruses (e.g., bacteriophages) that infect bacterial cells further aided in determining the chemical identity of the genetic material. Hershey and Chase radioactively labelled the sulfur of the protein coat and the phosphorus of

the DNA of <u>virulent bacteriophages</u>. Labelled phage DNA was injected into the host bacterial cells. Completion of the cycle of phage DNA replication and host-cell lysis yielded: (1) old phage ghosts with labelled proteins, and (2) new phages with unlabelled protein coats and with labelled DNA. Examine textbook Figure 16-4 to be sure that you understand how this labelling experiment identified DNA as the genetic material.

(3) Transduction. A second major group of bacteriophages includes the <u>temperate phages</u>. These phages generally do not cause their host cells to lyse. Instead, the phage DNA incorporates with, recombines with, and replicates in concert with the host DNA. Zinder and Lederberg conducted experiments with bacteria (*Salmonella*) and temperate phages. They demonstrated that bits of host-cell DNA often attach to the phage DNA (i.e., prophages) and that this supplemented DNA can be transferred into a new host cell; the new host can obtain new properties from the earlier host by this means of transfer called <u>transduction</u>.

(4) Conjugation. A primitive form of sexuality exists in bacteria. Donor cells possess a <u>fertility factor</u> (F factor) that is absent in recipient cells. During <u>conjugation</u>, donor and recipient cells attach, and the donor chromosome begins to move into the recipient cell. Usually, connection of the cells is interrupted before the entire donor chromosome passes to the recipient. Next, recombination occurs between the recipient chromosome and the donor chromosome fragment. DNA fragments can exist either integrated into the circular chromosome or as independent genetic elements in the cytoplasm; we'll return to episomes and plasmids in the contexts of genetic engineering and cytoplasmic genes.

C. <u>Chemical Structure of Nucleic Acids</u>.--The largest-known biological molecules are nucleic acids, which are polymers of nucleotides. You already are acquainted with <u>nucleotides</u> from earlier in the course; these include the energy currency molecule (adenosine triphosphate) and various coenzymes (NAD, NADP, FMN).

Nucleotides consist of three components: a phosphate group, a five-carbon sugar (a <u>pentose</u>), and a nitrogenous base. These components, in both separate and assembled forms, are illustrated in Fig. 16.1. Only five different bases occur in nucleic acids. Two of these are <u>purines</u> (adenine and guanine) and three are <u>pyrimidines</u> (cytosine, uracil, thymine). Pyrimidines are single-ring molecules, whereas purine molecules consist of two joined rings (textbook Figure 16-11).

Research conducted by Watson and Crick demonstrated that DNA polymer molecules are arranged into a <u>double-stranded helix</u>. In this configuration, the phosphate and sugar molecules link alternately to form the side rails of this twisted ladder. Connection of the two strands occurs by base pairing in which a purine and a pyrimidine bond to each other. The pattern of <u>base pairing</u> is very specific. In DNA, adenine bonds with thymine, and guanine bonds with cytosine. Uracil takes the place of thymine in RNA; hence, the <u>complementary base pair</u> in RNA is adenine and uracil.

D. <u>Replication of DNA</u>.--It is apparent, then, that each strand (a molecule of DNA) of the double helix is <u>complementary</u> to the other strand. Each strand serves as a <u>template</u> from which the complementary strand can be constructed. Indeed, in preparation for karyokinesis (the nuclear divisions in mitosis and meiosis), the quantity of DNA in the nucleus doubles so that each DNA molecule is present in duplicate. The manner in which DNA replicates is referred to as <u>semiconservative replication</u>. (How did Meselson and Stahl demonstrate this?) This method of replication involves separation of the double helix into two complementary strands. Two identical new DNA molecules are then formed, each being half old and half newly constructed from the single strand templates.

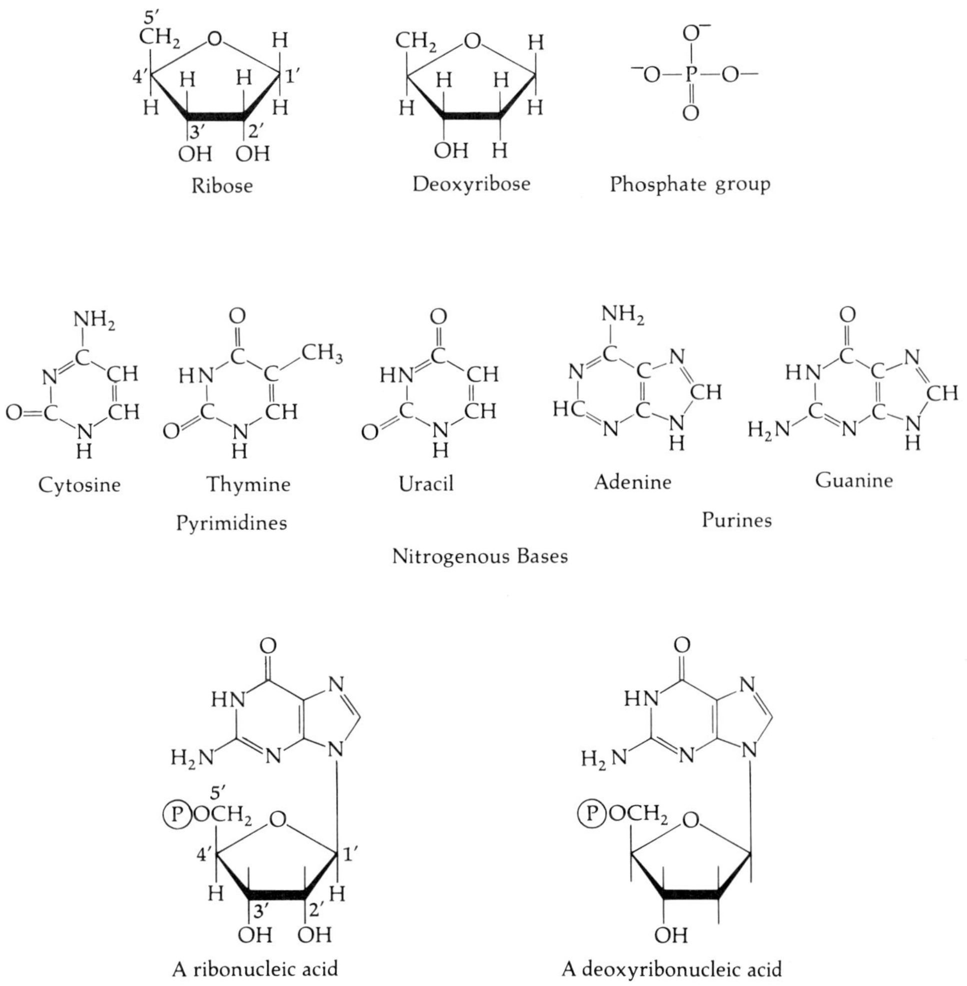

Fig. 16.1: Nucleotides and their components

A few additional points regarding DNA synthesis need to be made here. DNA synthesis requires the presence of the enzyme <u>DNA polymerase</u> (and, as later research demonstrated, an array of other enzymes) as well as an adequate supply of the four nitrogenous bases. Building of the new strand proceeds in an organized, directional manner. <u>Directionality</u> is indicated by numbering the carbon atoms of the sugar molecule and noting which carbons are involved in the <u>phosphodiester bonds</u> that join consecutive nucleotides. Bonding of the sugar-phosphate backbones involves placing a phosphate group between the 3' carbon of one pentose and the 5' carbon of the next pentose. The 5' (pronounced 5-prime) carbon of the pentose is the only carbon that is not a member of the ring; the carbon numbering scheme is shown in Figure 16.1 above. The new DNA strand is built in the 5' to 3' direction.

The two complementary strands of which DNA is made attach to each other in opposite directions. That is, the 5' end of one strand matches to the 3' end of the other strand. As these strands unravel in preparation for replication, one strand (the <u>leading strand</u>) has its 3' end free and the other (the <u>lagging strand</u>) has its 5' end free. Replication proceeds smoothly and continuously along the leading strand, but assembly from the lagging strand occurs in discontinuous segments called <u>Okazaki fragments</u>. See textbook Figure 16-20 for an illustration of the differences in assembly from the leading and lagging strands.

E. <u>Gene Function</u>.--Up to this point, we have summarized information indicating (1) that genes carry hereditary information and (2) that genes are made of DNA. But, <u>how</u> does DNA dictate how the genotype is expressed?

Although suspected as early as around 1908 (by Garrod), it was many decades later until scientists demonstrated that DNA exerts its effects by <u>regulating the synthesis of proteins</u>. Recall that proteins (in the form of enzymes) play a central role in overall metabolism . . . by way of the role of enzymes that regulate virtually every biochemical pathway in an organism.

Experimentation by Beadle and Tatum with the bread mold, *Neurospora crassa*, led to the "<u>one gene, one enzyme</u>" hypothesis. These researchers induced gene mutations in *Neurospora* by irradiating them with X-rays. The mutations they studied concerned nutritional needs of the fungus; nutritional deficiencies resulting from mutations were easily detected by growing fungi on culture media with various combinations of amino acids needed to complete certain pathways. Beadle and Tatum determined that mutations of certain genes resulted an the inability of the fungus to carry certain biochemical pathways to completion because the fungus lacked the genetic instructions required to synthesize specific enzymes needed in that pathway. Hence, they deduced that each gene coded for an enzyme.

Subsequent findings required that the "one gene, one enzyme" label be re-phrased to read "<u>one gene, one protein</u>." This change was required when researchers learned that synthesis of all kinds of proteins, not just enzymes, was encoded by DNA. One further modification in the phrase must be made for the more complex proteins that are made of two or more polypeptides; because each polypeptide is encoded by a separate gene, a more correct name is the "<u>one gene, one polypeptide</u>" hypothesis. Study of the human <u>hemoglobin</u> molecule (a tetramer) supported the "one gene, one polypeptide" hypothesis. Hemoglobin research also led us one step further in understanding how genes determine protein structure: a gene dictates the <u>sequence of amino acids</u> in the polypeptide encoded by that gene. In the next chapter, we'll examine just how DNA indicates amino-acid sequence.

TERMS TO UNDERSTAND: Define and identify the following terms:

gene

deoxyribonucleic acid

ribonucleic acid

transformation

viral replication

transduction

conjugation

transforming principle

bacteriophages

virulent phages

temperate phages

prophage

lysogenic

fertility (F) factor

Hfr cells

episome

plasmid

retrovirus

nucleotide

polynucleotide

purines

pyrimidine

double-stranded helix

base pairing

complementary strand

satellite DNA

DNA polymerase

ribonucleotide reductase

RNA polymerase

template

semiconservative replication

phosphodiester bonds

5' carbon

3' carbon

Okazaki fragments

DNA ligase

alkaptonuria

one gene, one enzyme hypothesis

sickle-cell anemia

one gene, one polypeptide hypothesis

CHART EXERCISE

Complete the chart below (1) to compare and contrast the chemical characterisitcs of deoxyribonucleic acids and ribonucleic acids, and (2) to indicate the taxonomic distribution of these nucleic acids.

CHARACTERISTIC	DNA	RNA
Identity of Pentose		
Phosphate Group		
Nitrogenous Bases		
Taxonomic Distribution		

PEOPLE TO KNOW: What is the significant biological contribution of each of these individuals as indicated in Chapter 16?

Friedrich Miescher

Fred Griffith

O. T. Avery, C. M. McLeod, and M. McCarty

A. D. Hershey and M. Chase

N. D. Zinder and J. Lederberg

R. Axel, S. Silverstein, and M. Wigler

E. Tatum and J. Lederberg

J. D. Watson and F. H. C. Crick

Arthur Kornberg

M. Meselson and F. Stahl

George Beadle and Edward Tatum

Linus Pauling and Harvey Itano

DISCUSSION QUESTIONS

1. Compare and contrast the chemical composition of DNA and RNA. Can you suggest an adaptively valuable reason for the differences in the nitrogenous bases found in these two nucleic acids?

2. Discuss six properties of genes that were known from classical genetics before our recent advances in molecular genetics.

3. How has the technique of X-ray diffraction been important in discovery of the chemical and physical nature of the gene?

4. In what groups of organisms are genes not made of DNA? For those species, what is the chemical composition of genes?

5. Distinguish between the processes of <u>transformation</u> and <u>transduction</u>. What knowledge did each of these processes contribute towards an understanding of the nature of the gene?

6. Suppose that you are working in a laboratory that performs DNA analyses. Your client wants to know the relative amounts of each of the four nitrogenous bases in the DNA sample from a speckled kingsnake. You assume that you must chemically assay the amounts of each base in the sample to provide the information that your client needs. However, the procedure manual states that you need to conduct the chemical assay for only one of the bases. Being new on the job, you ask you supervisor how you can avoid actual lab work in determining the amounts of the other three bases. Discuss the rationale of your supervisor's explanation.

7. Discuss the roles of at least five of the enzymes that are active in DNA replication.

8. Complementary strands of DNA are oriented in opposite directions. What are the implications of this opposite orientation for the replication of that DNA molecule?

9. Discuss the roles of (a) *Neurospora* and (b) hemoglobin in the achievement of our current understanding of molecular genetics.

TESTING YOUR UNDERSTANDING

For each of the test items below, three of the lettered alternatives are true and the other is false. Determine which alternative is false and write its letter in the blank to the left of the question number. On the blank line below alternative D, write the corrected version of the false statement.

_____ 1. A. The alleles of genes produce distinct phenotypic effects.
B. A gene is positioned at a specific locus on a chromosome.
C. Genes simultaneously exhibit a substantial degree of stability from one generation to the next as well as a variable degree of mutability.
D. The location of a gene corresponds to the position of a cross-band on a chromosome.

Correction: _____

_____ 2. A. The genes of viruses and of some bacteria are made of ribonucleic acid.
B. DNA is acidic and is rich in phosphorus.
C. Studies of the transformation of nonvirulent bacteria into virulent bacteria provided evidence that DNA was the identity of the "transforming principle."
D. The life cycle of virulent bacteriophages includes lysis of the host cell after a brief infection period.

Correction: _____

_____ 3. A. The combination of temperate-phage material with host-cell DNA produces a prophage.
   B. Infection of a host cell by a virulent phage involves injection of phage protein into the host cell.
   C. Study of virulent phage life cycles by Hershey and Chase helped to establish that genes are made of DNA.
   D. Infection of a host cell by a temperate phage results in recombination of phage genes with host cell genes.

   Correction: _____

_____ 4. A. Conjugation usually results in the production of bacterial cells that are diploid.
   B. In transduction, genes from one bacterial cell are transferred to another bacterial cell by a temperate phage.
   C. The principle of transduction provides the basis for the emerging field of genetic engineering.
   D. Bacterial donor cells are recognized by the presence of a fertility factor that either is integrated into the circular chromosome or is a separate cytoplasmic chromosome.

   Correction: _____

_____ 5. A. An episome is a genetic unit that can exist as an independent part of the main chromosome.
   B. An episome is a genetic unit that can exist as a free element in the cytoplasm.
   C. Plasmids are genetic units that do not integrate into the main chromosomal ring.
   D. Retroviruses possess DNA arranged into plasmids and episomes.

   Correction: _____

_____ 6. A. Molecules of purines, such as guanine, consist of a single-ring structure.
   B. One of the nitrogenous bases found in RNA is uracil.
   C. Uracil is a pyrimidine.
   D. A nucleotide consists of a phosphate group, a nitrogenous base, and a pentose.

   Correction: _____

_____ 7. A. Complementary base pairs consist of one pyrimidine molecule bound to a purine molecule.
   B. Nucleotides are involved in the transmission of the genetic code and, as coenzymes, in the process of electron transfer.
   C. The "side rails" of a double-stranded DNA helix consist of alternating pentose and phosphate molecules.
   D. In both DNA and RNA, adenine pairs with thymine.

   Correction: _____

_____ 8. A. One of the implications of complementary base pairing is that a single strand of DNA can serve as a pattern for production of an identical strand of DNA.
B. Bonding of molecules of cytosine and guanine involves three hydrogen bonds.
C. The number of adenine molecules in a double-stranded DNA molecule usually is not the same as the number of cytosine molecules.
D. Satellite DNA consists almost entirely of regions of the molecule containing only adenine and thymine.

Correction: _____

_____ 9. A. Polymerase is an enzyme that is required to assemble polynucleotides from nucleotides.
B. Replication of DNA requires an unwinding enzyme called helicase.
C. Okazaki fragments are joined together by an enzyme called DNA gyrase.
D. Assembly of a new DNA strand proceeds directionally from the 5' end towards the 3' end.

Correction: _____

_____ 10. A. The presence of unequal amounts of guanine and cytosine in a DNA molecule suggests that the DNA molecule is single-stranded.
B. Single-stranded DNA is known to occur in a few kinds of viruses.
C. In situations where DNA is normally single-stranded, there is a temporarily double-stranded form called the replicative form.
D. Adenine and uracil of a double-stranded RNA molecule are bonded by phosphodiester bonds.

Correction: _____

_____ 11. A. Each gene codes for one polypeptide.
B. The "one gene, one enzyme" hypothesis derived from study of fruit flies.
C. Human hemoglobin is a tetramer whose polypeptide units are governed by two genes.
D. Because *Neurospora* is haploid during much of its life cycle, this fungus is an excellent model species for study of gene mutations.

Correction: _____

_____ 12. A. Beadle and Tatum proposed the "one gene, one enzyme" hypothesis.
B. Meselson and Stahl demonstrated that replication of DNA is semi-conservative.
C. Using X-ray diffraction techniques, Watson and Crick deduced that the DNA molecule is a double-stranded helix.
D. Griffith was the first person to isolate DNA.

Correction: _____

CHAPTER 17--MECHANISMS OF GENE ACTIVITY

CHAPTER OVERVIEW

In this chapter, we will examine the series of events by which the DNA of a structural gene is interpreted and is used to produce a polypeptide. We will see how the four-nucleotide vocabulary of DNA is read into the twenty amino-acid vocabulary of polypeptides. The major steps in the process are transcription of the DNA message into messenger RNA in the nucleus and translation of the messenger RNA into an amino-acid sequence in association with ribosomes in the cytoplasm.

Having addressed the basic process by which a gene codes for a polypeptide, we'll next examine mechanisms of mutation at yet another level--the molecular level. Gene mutations can occur through errors in coding by nucleotides and by the often-related errors of misreading a nucleotide sequence. Interestingly, protective mechanisms for automatic repair of gene mutations exist. We close the chapter with discussion of new findings in molecular genetics . . . these topics pertain to the potential importance of sequences of DNA that are repetitive, intervening, and movable.

TOPIC SUMMARIES

A. <u>DNA as a Genetic Code</u>.--The instructions for assembly of all proteins are written into a genetic code; this code takes the form of <u>specific sequences of nucleotides</u> in DNA or, in some situations, in RNA. A total of only five nucleotides comprises this code. Three nucleotides (adenine, guanine, and cytosine) are present in both DNA and RNA; thymine of DNA is replaced by uracil of RNA. In this chapter, we will examine how a vocabulary of only four "words" (nucleotides) can code for the 20 kinds of amino acids that form the vast diversity of proteins present in living organisms.

B. <u>The Central Dogma</u>.--The flow of genetic information from DNA in the nucleus of a eukaryotic cell into the final protein product involves three key events. These events are often referred to as the "Central Dogma."

(1) Just before mitosis, the DNA (in the form of chromatin) of a nucleus <u>replicates</u> itself.

(2) <u>Transcription</u> of DNA into a complementary strands of RNA occurs next. This RNA is called messenger RNA (mRNA) because it leaves the nucleus and, thereby, carries with it the genetic message to the sites of protein synthesis in the cytoplasm.

(3) The final step involves <u>translation</u> of the RNA-encoded message into an amino-acid sequence that is specific for the polypeptide chains of the cell's proteins. Much of the rest of this chapter examines steps 2 and 3 in greater detail.

C. <u>Transcription</u>.--Nuclear DNA directs the synthesis of several kinds of RNA, including mRNA, tRNA, rRNA, and hnRNA. At this point, we'll concentrate on <u>messenger RNA</u>. Regions along template DNA which specify mRNA structure (and, hence, protein structure) are called <u>structural genes</u>. The enzyme <u>RNA polymerase</u> catalyzes the synthesis of mRNA when a template of DNA is present. [Note that only one of the complementary strands of DNA serves as a template; if both strands for a given gene were transcribed, then two different mRNA's (and ultimately two different proteins) would be synthesized.] Transcription involves three steps:

(1) Initiation step. Initiation of transcription at the proper location along a strand of DNA requires that RNA-polymerase molecules be able to recognize the beginning of a structural gene. An RNA-polymerase molecule binds to the promoter of that gene. The promoter sequence is located just upstream of the initiation site. The initiation site is marked by the first nucleotide to be transcribed. Within the promoter sequence is a string of six bases (known as the Pribnow box) whose sequence is very similar in all structural genes in all species that have been studied. Even further upstream in prokaryotes is the -35 region whose function is to aid in tight binding between DNA and polymerase molecules. A more complex situation prevails in eukaryotes: DNA sequences required for transcription occur in at least three regions near the RNA initiation site, including the TATA box and often the CAAT box.

(2) Elongation step. In this process, molecules of RNA polymerase meet the unwinding DNA molecule, then position each ribonucleotide, and finally catalyze the polymerization.

(3) Termination step. The elongation process continues until RNA polymerase encounters the stop signal for that gene. The product of elongation is the primary RNA transcript, an inactive form of RNA that migrates from the nucleus into the cytoplasm. Before translation of this inactive transcript can occur, it must undergo a process of maturation within the nucleus (see section H below).

D. Translation.--Assembly of polypeptides requires interaction of mRNA, "adaptor molecules" of tRNA, and amino acids associated with ribosomes located in the cytoplasm.

Amino-acid molecules do not automatically associate with mRNA, but they are placed on the relatively small molecules of transfer RNA (tRNA) by specific charging enzymes. Each amino acid associates specifically with one (or sometimes more) particular tRNA molecules. Ribosomes then position these amino acid-tRNA complexes appropriately along an mRNA strand. Despite their important differences, all tRNA molecules share at least four features:

(1) Each tRNA molecule has loops, stems, and arms that impart a characteristic cloverleaf shape to the molecule. Figure 17.1 labels several important features of tRNA molecules. As we'll see below, at least one of these loops (anticodon loop) has great functional importance.

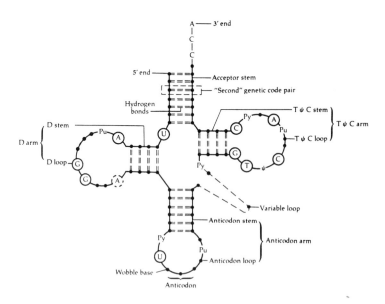

Fig. 17.1: Structure of a molecule of transfer RNA

(2) tRNA's differ from the other nucleic acids we've examined so far in the presence of several "unusual" nitrogenous bases. Most of these unusual bases differ from the "regular" ones by the attachment of one or more methyl groups to a regular base. The function of these bases is unknown, although they might well be involved in the attachment of tRNA to ribosomes.

(3) Each tRNA molecule ends in the same three-base sequence: CCA (cytosine, cytosine, adenine). This triplet identifies the acceptor end of the tRNA molecule, and it is to these three bases that the appropriate activated amino acid molecule attaches. The association of the amino acid and tRNA molecules is called the amino-acid tRNA complex.

(4) Each tRNA contains a critically important nucleotide triplet, an anticodon, that is located in the anticodon loop. The three bases here are complementary to a corresponding triplet of nitrogenous bases (a codon) on the mRNA molecule that codes (in most cases) for a particular amino acid. For example, the tRNA anticodon of UCG allows the tRNA molecule to match the AGC codon of mRNA; the codon AGC is a genetic code word for the amino acid serine.

Refer to textbook Table 17-1 for the complete genetic code. Notice that the first two of the three bases in a codon generally are the most important in indicating an amino acid; some variation in the third base is tolerated. Another interesting feature of the code is that it is degenerate: a particular amino acid can be indicated by more that one codon . . . that is, this language contains synonyms. However, there is no ambiguity; any particular codon can be interpreted in only way. Although most codons code for an amino acid, three triplets (UAA, UAG, UGA) serve as stop signals which indicate the end of a polypeptide chain. The genetic code appears to be universal, i.e., the same code applies in all organisms.

Getting back to the process of translation . . . Appropriate amino-acid tRNA complexes move to a site on the mRNA molecule where a ribosome is positioned. The anticodons of these complexes temporarily bind to the appropriate codons of the mRNA. The amino acids then are transferred to a growing polypeptide chain, and the now "empty" tRNA's are released into the cytoplasm to locate and bring back another appropriate amino-acid molecule.

E. More on mRNA Structure.--Not all segments of mRNA can be translated into polypeptides. For example, the mRNA's of most eukaryotes end in a long stretch of adenines that do not code for polypeptides; the importance of this poly(A) tail is not yet known. Another such region includes the sequence AAUAA plus a terminator codon which ends translation of a eukaryotic gene. The initiator codon is the first codon of a gene to be translated; in prokaryotic and eukaryotic species, the initiator codon specifies the amino acid methionine. The 5' end of mRNA molecules has a 5' cap followed by a leader of up to 50 untranslatable nucleotides; possibly this "leader" somehow indicates the start end of the molecule.

F. The Molecular Basis of Gene Mutation.--We've spoken in earlier chapters of the various levels (e.g., changes in chromosome structure and numbers) at which genetic mutations can occur. By solving the genetic code, we are now able to examine mutations at the molecular level, that is, those that are not visible in the numbers or morphology of chromosomes.

Molecular-level mutations represent errors in coding and/or in reading of the code. Coding errors include substitutions in nitrogenous bases and insertions or deletions of one or more base pairs. When the altered sequence is read, changes in bases can (although it does not always . . . due to degeneracy of the code) cause synthesis of the "wrong" polypeptide (due to amino-acid substitutions, deletions, or additions) or an incomplete polypeptide (if the change results in a stop signal). The condition of sickle-cell anemia apparently resulted from the mutation of only one base in the beta chain of hemoglobin. Other errors might involve incorrect reading (via a shifted reading frame) of an otherwise correct sequence.

Although mutations occur continuously, they are not always expressed or perpetuated between generations . . . in part, because of the existence of some self-repair mechanisms. Cells can produce a battery of enzymes that are parts of a DNA repair system. Some of these enzymes catalyze the removal of damaged base pairs (a process called excision repair). An array of enzymes (called glycosylases) is involved in the removal of incorrect individual bases. These two types of repair occur before replication of the DNA molecule. Sometimes, however, damaged DNA is replicated with the result of "post-replication gaps" in the daughter strands of the DNA and a damaged parental strand. Post-replication repair entails (1) excision of the damaged region and (2) a recombination event with the sister chromosome which (hopefully) carries the correct information. This supplies a correct template for synthetic repair of the damaged strand.

G. Repeated DNA Sequences.--Up to 90% of the DNA in some eukaryotic cells consists of multiple repetitive copies of the same or similar base sequences. Most genes that do code for proteins are located in simple sequence DNA, whereas moderately and highly repetitive DNA generally does not include structural genes. The significance of repetitive DNA is not yet understood. A suggested role for duplication of structural genes, however, includes production of large quantities of the products of those genes (dosage repetition). Because not all repeated DNA sequences include identical copies of genes, variant repetition results in production of arrays of similar (but not identical) proteins.

H. Intervening DNA Sequences.--Some structural genes are not colinear . . . that is, the coding sequences (exons) are interrupted by non-coding stretches called intervening sequences or introns. Although common in eukaryotes, introns are not widely known in prokaryotes. Interestingly, both introns and exons are transcribed into the primary transcript (or pre-mRNA). Maturation of pre-mRNA into mature mRNA requires excision of introns and subsequent joining of the exons. The importance of introns is not yet known.

I. Movable DNA Sequences.--For decades, geneticists have assumed that chromosomes are stable structures. Hence, barring mutation, a particular gene in a particular species always has a particular location on a particular chromosome. One of the newest frontiers in biology pertains to the mobility of structural genes (or parts thereof) within the genome. These movable elements are called transposons. Gene mobility of this sort offers yet another mechanism for introducing genetic variability into the genome. Refer to your textbook for a few general characteristics of transposons. Much remains to be learned about transposons!

TERMS TO UNDERSTAND: Define and identify the following terms:

transcription

messenger RNA (mRNA)

translation

RNA polymerase

structural genes

transfer RNA (tRNA)

ribosomal RNA (rRNA)

heterogeneous nuclear RNA (hnRNA)

promoter

initiation site

Prinbow box

-35 region

stop signal

primary RNA transcript

adaptor molecules

anticodon loop

amino-acid tRNA complex

codon

anticodon

polysome

poly(A) tail

terminator codon

5' cap

overlapping genes

reading frame

insertions

deletions

missense mutation

nonsense mutation

pyrimidine dimer

excision repair

glycosylases

post-replication

repetitive copies of DNA

satellite DNA

dosage repetition

gene amplification

variant repetition

intervening sequences

exons

introns

primary transcript (pre-mRNA)

maturation

biochemical microheterogeneity

transposable elements

mutation by insertion

transposons

CHART EXERCISE

Many different enzymes are involved in the processes of DNA replication, transcription, and translation. Complete the chart below to indicate the cellular location and functions of each of the listed enzymes.

ENZYME	CELLULAR LOCATION	FUNCTIONS
RNA polymerase		
DNA polymerase		
glycosylases		
DNA ligase		

PEOPLE TO KNOW: What is the significant biological contribution of each of these individuals as indicated in Chapter 17?

Robert W. Holley

Alexander Rich

M. W. Nirenberg and J. H. Matthaei

Barbara McClintock

Sidney Brenner

DISCUSSION QUESTIONS

1. What are the various kinds of RNA that occur in eukaryotic cells? Where is each of these types of RNA found in the cell? Briefly discuss the function of each type of RNA.

2. The unlabeled diagram below depicts the translation of a molecule of messenger RNA. Label as many of the components of this process as you can.

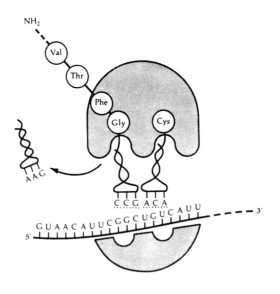

3. Discuss the similarities and differences of transcription and translation processes in eukaryotes, prokaryotes, and viruses.

4. A shift in the reading frame can cause a gene to be read and translated erroneously. What amino-acid sequence is coded by the following sequence of nucleotides in a messenger RNA molecule?

   CUUCGGUCUUUAAUCAUAGUACUCGGGGACGCCUAUUCGAUAACCUAA

   ____ ____ ____ ____ ____ ____ ____ ____ ____ ____ ____ ____ ____ ____ ____ ____

   What amino acid sequence is read from that same nucleotide sequence if the reading frame is shifted one nucleotide to the right?

   ____ ____ ____ ____ ____ ____ ____ ____ ____ ____ ____ ____ ____ ____ ____ ____

   Does this frame shift result in production of a polypeptide different that if the sequence is read properly?

5. Distinguish between "missense" and "nonsense" mutations. Use examples.

6. Discuss three ways by which damage to a DNA molecule can be repaired.

7. Suggest an experimental design by which genetics researchers could make progress towards understanding the importance of movable DNA sequences.

TESTING YOUR UNDERSTANDING

For each of the test items below, three of the lettered alternatives are true and the other is false. Determine which alternative is false and write its letter in the blank to the left of the question number. On the blank line below alternative D, write the corrected version of the false statement.

_____ 1. A. Translation occurs in the nucleus.
     B. Replication of DNA occurs in the nucleus.
     C. Transcription occurs in the nucleus.
     D. Production of messenger RNA occurs in the nucleus.

   Correction: _____

_____ 2. A. Transcription is the process by which messenger RNA is produced from the DNA template.
     B. Transcription is the process by which a polypeptide is constructed according to instructions carried on molecules of transfer RNA.
     C. Translation occurs in association with ribosomes.
     D. The "central dogma" refers to the sequence of events by which genetic information carried in DNA directs synthesis of polypeptides.

   Correction: _____

_____ 3. A. A messenger-RNA molecule can be used as a template to synthesize the complementary RNA molecule.
     B. The reactions catalyzed by molecules of RNA polymerase and DNA polymerase occur in the nucleus.
     C. The primary RNA transcript must undergo a process of "maturation" before it can function in protein synthesis.
     D. An anticodon loop is a feature of transfer-RNA molecules, but not of messenger-RNA molecules.

   Correction: _____

_____ 4. A. In the genetic code, a particular amino acid may be coded for by one or more codons.
     B. A codon is a sequence of three consecutive nucleotides in a messenger-RNA molecule.
     C. Attachment of an amino acid molecule to a transfer-RNA molecule requires activation of the amino-acid molecule.
     D. Both strands of an RNA double helix molecule can serve as templates for synthesis of other nucleic acids.

   Correction: _____

_____ 5. A. All transfer-RNA molecules end with the base sequence CCA.
     B. Activated amino-acid molecules attach to the anticodon of molecules of transfer RNA.
     C. Activation of amino-acid molecules consumes energy in the form of ATP.
     D. The "adaptor molecules" that carry amino acids to messenger RNA are actually molecules of transfer RNA.

   Correction: _____

_____ 6. A. Binding of RNA polymerase with a DNA molecule occurs at specific regions of the DNA molecule called promoters.
B. In a structural gene, the first DNA base to be copied into RNA is located in the Pribnow box.
C. Each gene has a promoter sequence.
D. Frequently, the initiation site of a structural gene immediately follows the promoter.

Correction: _____

_____ 7. A. Within a structural gene, each codon codes for a specific polypeptide.
B. UAG is the codon of messenger RNA that corresponds with the DNA base sequence of AUC.
C. A strand of messenger RNA to which several ribosomes are bound is a polysome.
D. The initiator codon is located at the 5' end of a structural gene, and the terminator codon is located at the 3' end.

Correction: _____

_____ 8. A. Not all structural genes are colinear; rather, the DNA sequences for some genes overlap with each other.
B. Within a codon, mutation of the third base is the least likely to cause an error.
C. Mutations that cause the wrong amino acid to be coded are called missense mutations.
D. Missense mutations can be detected from the production of incomplete polypeptide chains.

Correction: _____

_____ 9. A. Sickle-cell anemia apparently resulted from the alteration of only one base in the DNA sequence that codes for one of the hemoglobin polypeptides.
B. Exposure of DNA to ultraviolet radiation can cause a lesion known as the pyrimidine dimer.
C. Glycosylases repair mutated DNA molecules by catalyzing reactions in which incorrect nitrogenous bases are removed.
D. Post-replication repair of DNA requires use of glycosylases but not of ligases.

Correction: _____

_____ 10. A. Satellite DNA consists of multiple repetitive copies of the same or similar base sequences.
B. Most of the functional genes that are read into polypeptides are classified as repetitive DNA.
C. Gene amplification occurs primarily in genes whose products are needed in larger amounts than could be supplied by reading of only one copy of that gene.
D. Antibody proteins are encoded by genes that exhibit variant repetition.

Correction: _____

_____ 11. A. Exons are the noncoding sequences of DNA found within a gene.
B. Introns and exons are known in both prokaryotes and eukaryotes.
C. Because of intervening DNA, a gene for a particular polypeptide can include more bases than would otherwise be necessary to synthesize that polyepeptide.
D. Maturation of the primary messenger RNA transcript involves excision of the noncoding bases from the RNA molecule.

Correction: _____

_____ 12. A. Heterogeneous nuclear RNA represents a mixture of primary transcript-RNA molecules which are undergoing processing and splicing.
B. The splicing of primary transcript RNA begins at the 3' end and proceeds continuously to the 5' end.
C. For eukaryotes, the majority of nuclear DNA does not code for proteins.
D. Transposons are sequences of DNA which do not occupy a fixed position in the genome.

Correction: _____

CHAPTER 18--REGULATION OF GENE ACTIVITY

CHAPTER OVERVIEW

This chapter addresses many issues related to the following observation: Many multicellular organisms are made of cells (and tissues) of different types. Yet, each cell of each tissue type (e.g., muscle, nerve, skin, parenchyma) of an organism contains an identical, full chromosomal complement. Clearly, not every gene is expressed in every cell. How, then, are genes regulated?

The mechanisms of gene regulation operating for prokaryotes are quite different (and generally simpler) from those controlling eukaryotes. We'll examine the similarities and differences. We'll develop general models of a structural gene and its associated regulators for prokaryotes and for eukaryotes. We'll also see some of the practical applications that have come from a better understanding of gene regulation; genetic engineering is a primary example.

We'll look, too, at inheritance that does not involve nuclear chromosomes, but chromosomes of certain organelles. Such non-nuclear inheritance offers evidence for the evolution of eukaryotes from prokaryotes.

TOPIC SUMMARIES

A. <u>Gene Regulation in Prokaryotes</u>.--Although <u>structural genes</u> (genes that code for proteins) are continuously present within a prokaryotic cell, these genes are not continuously active in producing proteins. The activity of such structural genes is regulated by <u>regulatory genes</u>. Gene activity (that is, protein production) is <u>induced</u> by the presence of an <u>inducer</u>. Induction is an increase in the rate of synthesis of one (or several) enzymes in response to the presence of an inducer. Chemically, an inducer resembles molecules of the substrate of the enzyme.

The protein-producing activity of a structural gene slows by a process called <u>repression</u>. The <u>repressor</u> substance generally is a product (or intermediate) of the protein-producing activity of the structural gene. Hence, the presence of a large concentration of an enzyme tends to inhibit the reaction sequence that produces that enzyme. As the concentration of the enzyme decreases, the reaction sequence becomes <u>derepressed</u> and structural gene activity increases.

B. <u>How Inducers and Repressors Regulate Gene Activity: The Operon</u>.--The familiar bacterium *Escherichia coli* metabolizes lactose as a source of carbon and energy. Interestingly (and economically), the structural genes (called z, y, and a) for three enzymes early in the lactose metabolism pathway follow one another on one strip of messenger RNA. Transcription of this entire polycistronic stretch of DNA is controlled by a single regulatory region called the <u>operator</u>. The location of the operator is upstream of genes z, y, and a. A second regulatory gene (the I gene) and its operator are located farther upstream on the chromosome and codes for production of repressor molecules. Fig. 18.1 shows the relative locations of these structural and regulatory genes.

When inducer concentration is very low, the corresponding structural gene can be thought of as being "turned off." In this situation, transcription does not occur because repressor molecules have bound to the operator gene; such binding of repressor and operator prevents transcription of that structural gene sequence. However, as inducer concentration increases, the rate of transcription increases: inducer molecules bind with and inactivate repressor molecules, thus allowing transcription to proceed unopposed.

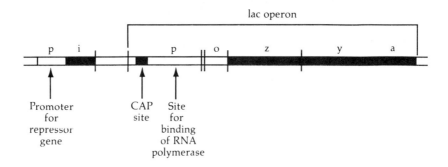

Fig. 18.1: Model of the *lac* operon in the bacterium, *E. coli*

Think back to the previous chapter, Chapter 17. Recall that transcription of a DNA sequence begins at the promoter, a site at which RNA polymerase binds. The lac operon we have just discussed has two promoter sites, one for the regulatory I gene and one for the sequence of operator and associated structural genes.

C. <u>Gene Regulation in Eukaryotes</u>.--The ways in which genes are regulated in eukaryotes are greatly different than for prokaryotes. These differences relate mainly to the much longer life span of most eukaryotes (years or centuries compared to minutes or hours) and to the multicellular nature of most eukaryotes. Associated with multicellularity is the presence of a diversity of kinds of cells; each cell of an organism has the identical genetic information, but different kinds of cells (e.g., muscle, bone, skin) are produced by and are operated by different subsets of the genome.

Gene regulation in eukaryotes is one of the "cutting edge" fields in biology . . . that is, we are only now beginning to understand the mechanisms involved. These mechanisms seem to center on <u>selective (or differential) gene functioning</u>. These mechanisms target different levels of formation and utilization of messenger RNA, including (1) differential transcription, (2) differential processing of mRNA transcripts, and (3) differential translation of mRNA into protein.

D. <u>Signals for Gene Control in Eukaryotes</u>.--Chemicals of various sorts (hormones and various growth factors) influence transcriptional activity of mRNA. The concentrations of such chemicals determine when a gene is transcribed, whether the rate of transcription is increased or decreased, and the quantity of mRNA that is produced. Studies of *Drosophila* have shown that <u>puffing</u> of certain regions of chromosomes can be caused by exposure to hormones; puffed regions are sites of active mRNA synthesis. We'll examine hormones and other chemical messengers in more detail in Chapter 28. As indicated in Chapter 7, cell-to-cell contact also can regulate gene transcription.

E. <u>Steps at Which Control of Eukaryotic Genes is Exerted</u>.--Regulation of genes can occur at any of a number of steps beginning with transcription from DNA in the nucleus to beginning of protein synthesis in the cytoplasm. To this point, most of the examples of such regulation occur in the nucleus, that is, early in the sequence of steps from transcription to translation. Regulation in the nucleus rather than in the cytoplasm clearly is more economical in terms of decreasing the amounts of resources used in a non-productive way. Four nuclear and cytoplasmic junctures where such regulation occurs are listed below:

(1) Control of mRNA synthesis. Transcription of structural genes is regulated by a number of short sequences of DNA, most of which are located "upstream" from the structural

gene. Three different types of these transcriptional signals are known: The sequence TATA (called the "TATA box") fixes the start site of transcription by positioning RNA polymerase. The "cat box" (sequence CCAAT) influences rates of transcription. Other segments of DNA (called enhancer elements) which are located even farther upstream affect the efficiency of transcription. These sequences are evolutionarily conservative, that is, they occur in the same (or slightly modified form) ahead of many structural genes in many eukaryotic species.

(2) Processing of primary RNA transcript. At least a few, and perhaps a great many, genes are controlled by differential processing of the primary RNA transcript. Gene regulation, therefore, seems to be one of the major activities involved in the maturation process.

(3) Stability of mRNA in the cytoplasm. Conditions in the cytoplasm can markedly affect the rate of transport of mRNA and the life-span of mRNA. Both of these factors can greatly affect the amount of protein that can be translated from a molecule of mRNA.

(4) Regulation of translation. Only a few examples of regulation of any of the steps of translation are known. Such regulation occurs during the initiation of protein synthesis.

F. Molecular Means of Controlling Transcription.--One of the primary means of regulating transcription involves the binding of various molecules to specific sites on the DNA molecule. Hormone molecules and hormone-receptor complexes can bind directly to specific sites on the DNA molecule. Such binding affects transcriptional activity.

Changes in chromatin structure likely accompany changes in transcriptional activity. Some evidence suggests that chromatin must unwind before genes in that region can become active. DNase I, a nuclease, is an enzyme that can cut either DNA strand both in active regions and less frequently in inactive regions. Interestingly, the CAAT box (referred to above) is located within such regions of hypersensitivity. It is in this context that the potential importance of nucleosomes and of chromosomal coiling and supercoiling is apparent: Coiling might bring regulatory "boxes" and their structural genes close together, even though they are not positioned one right after another when the DNA is stretched out. Likewise, nucleosomes might serve to "hide" or to protect certain regions of DNA from polymerase molecules while exposing other segments to interaction with polymerase. Heterochromatin (regions of highly condensed DNA) probably represents DNA that is not accessible for transcription. Conversion of normal DNA into other forms, such as Z-DNA, likewise might affect the ability of stretches of a chromosome to be transcribed.

One of the nitrogenous bases in DNA, cytosine, is often present in its methylated form. The amount of 5-methylcytosine in a DNA molecule varies in accordance with the transcriptional status of the DNA molecule. We do not yet know, however, if this correlation has a functional basis; whether such methylation prevents transcription or merely results from it is uncertain. Another point of confusion is that the reverse situation holds in plants: active transcription occurs in many plants having very high levels of methylation.

The proteins (histones) that form complexes with DNA molecules also tend to inhibit transcription. This inhibition apparently is non-specific. That is, the histones do not bind, for example, to specific base sequences. Rather, they simply are attracted due to opposite electrical charges.

G. Variations on the Theme--When Genes Are Made of RNA.--RNA viruses do not contain DNA. Rather, their RNA (as a + strand) functions as a messenger RNA molecule within a host cell. Replication of the + strand requires formation of the complementary strand, the - strand. The - strands then serve as templates for production of additional + strands. A different kind of RNA-polymerase molecule operates here than we have seen in organisms in which DNA carries the genetic code. In RNA viruses, an RNA-dependent RNA polymerase (also known as RNA synthetase or RNA replicase) catalyzes RNA replication.

In such RNA viruses, the entire genome is extremely small, consisting of only three structural genes in the bacteriophage MS2. When the RNA of such viruses is allowed to fold and coil, the molecule takes on a form that is remarkably similar to the cloverleaf form of transfer RNA . . . perhaps this similarity will shed some light on the evolutionary relationships of prokaryotes and eukaryotes?

Retroviruses offer another example in which genetic information flows from RNA to DNA, rather than in the more prevalent DNA-to-RNA direction. An RNA-dependent DNA polymerase called reverse transcriptase synthesizes the complementary sequence of DNA from an RNA template. This complementary DNA (also called copy DNA or cDNA) is flanked by DNA sequences known as long terminal repeats (LTR's) and is incorporated into the host cell's DNA where it acts as a provirus. The proviral DNA is either transcribed into new viral RNA or it is replicated as would be a normal gene of the host. In the latter case, the proviral DNA can trigger malignant cell behavior. An interesting related point is that proviral DNA sequences are similar to those in some transposons of both prokaryotes and eukaryotes! What are the potential implications of these similarities?

H. Genes Not Located In Nuclei.--In eukaryotic cells, not all of the cell's genetic material is located in the nucleus. Rather, other organelles such as mitochondria and chloroplasts contain circular strands of DNA. The presence of DNA in these organelles is evidence that mitochondria and chloroplasts probably originated from prokaryotic cells that were internal parasites of eukaryotic cells. Early in this scenario, the endoparasitic cells probably contained full genetic information arranged on a circular chromosome. Now, these organelles contain only small amounts of DNA, usually not arranged into a chromosome. The other ingredients needed for protein synthesis (ribosomes, tRNA, enzymes) also are contained within these organelles. This mitochondrial DNA (called mtDNA) is not extensive enough to code for all the proteins that compose the organelle. Rather, some proteins of the organelles are encoded by nuclear DNA. Hence, an intricate dependency of host and parasite has evolved.

I. Inheritance by Non-Nuclear Genes.--As geneticists conducted breeding experiments in eukaryotes, they noticed that some traits were passed on in ways not in accord with the rules of Mendelian inheritance. It turns out that non-nuclear, cytoplasmic genes also are passed from parents to offspring . . . by the process of cytoplasmic inheritance. This mode of inheritance also is called maternal inheritance because the bulk of the cytoplasm in a newly-formed embryo was contained in the female gamete, the ovum. Male gametes (sperm) generally are quite small in comparison; sperm can be thought of as merely a nucleus with only those few cytoplasmic structures required to ensure delivery to the ovum. Hence, traits carried on cytoplasmic genes are inherited from the mother even though these traits are expressed in both sexes of the progeny. A widely-studied example of maternal inheritance is variegation, the presence of pale blotches in plant leaves.

J. Cell Hybridization as a Means of Altering Heredity.--Most animal cells possess only one nucleus. However, cells that contain more than one nucleus from more than one kind of cell (heterokaryons) can be produced by the process of fusion. During fusion, a process that is mediated by certain viruses, the cell membranes of adjacent cells dissolve at the points where they come into contact. The cytoplasms of these cells then come together. The nuclei carry out mitosis independently, dividing at their own rates. Nuclei from the different source organisms occasionally divide at the same time; in such situations, nuclear contents may integrate into a single nuclear mass. These composite cells still are capable of carrying out normal functions.

The applications of cell hybridization include <u>gene mapping</u> and preparation of <u>monoclonal antibodies</u> (see Box 23-2 in the textbook).

K. <u>Chromosome-Mediated Gene Transfer</u>.--This technique is yet another way of transferring genes from one genome to another. In this method, <u>fragments of chromosomes</u> are the entities being transferred. This compares to cell hybridization in which the <u>entire chromosomal complement</u> is transferred and to bacterial transformation (also named "DNA-mediated gene transfer") in which <u>small DNA fragments</u> are transferred.

L. <u>Genetic Engineering</u>.--Also known as <u>recombinant DNA technology</u>, genetic engineering applies the principles of transformation (see textbook chapter 16) for purposes including transplanting individual genes from one species to another, production of individual genes in quantity, and manufacture of precious proteins in large quantity. In this process segments of DNA (either natural or artificial) are fashioned into <u>plasmids</u> (small circular DNA molecules not associated with the main chromosome). The plasmids are inserted into the bacterial cells (How?). Both plasmids and chromosomes replicate as part of the cell-division process. Another way of introducing DNA is by using certain <u>DNA tumor viruses</u> as vehicles. Growth of cultures of bacteria containing such a modified genome is called <u>cloning</u>.

M. <u>Definition of a Gene . . . Reconsidered</u>.--Early in our introduction to genetics topics, we used a rather primitive, operational definition for <u>gene</u>, a definition that describes what a gene does (i.e., encodes genetic information that determines how a trait can be expressed) rather than what it is. We then defined "gene" chemically, by describing its chemical components. Then we examined the functional activities of genes, and in doing so, we found that different kinds of genes have different functions: A <u>cistron</u> determines a polypeptide chain. A <u>muton</u> is a smaller unit that can be mutated; this usually is a single base pair. A <u>recon</u> is the smallest unit that can recombine genetically. Even more labels or perspectives are likely to arise as geneticists continue to unravel the secrets of the genome.

TERMS TO UNDERSTAND: Define and identify the following terms:

lac system

beta-galactoside

permease

beta-galactosidase

transacetylase

structural gene

regulatory gene

induction

inducer

repression

derepression

cistron

polycistronic

operator

I gene

repressor

promoter sites

CAP site

constitutive enzyme synthesis

operon

corepressors

repressible pathways

selective gene function

transcriptional signals

TATA box

CAAT box

enhancer elements

protein kinase

Z-DNA

DNA-dependent RNA polymerase

RNA-dependent RNA polymerase

RNA synthetase

reverse transcriptase

retroviruses

complementary DNA

long terminal repeats

extrachromosomal heredity

mitochondrial DNA

variegation

maternal inheritance

cell hybridization

heterokaryon

monoclonal antibodies

gene mapping

chromosome-mediated gene transfer

parasexuality

recombinant DNA technology

cloning

plasmid

restriction enzymes

EcoR1

muton

recon

PEOPLE TO KNOW: What is the significant biological contribution of each of these individuals as indicated in Chapter 18?

Francois Jacob and Jacques Monod

Howard Temin

A. H. Sturtevant

Henry Harris and John Watkins

Har Gobind Khorana

Seymour Benzer

Alexander Rich

DISCUSSION QUESTIONS

1. Most mechanisms of gene regulation operate at the level of transcription rather than at the level of translation. Can you offer reasons to explain this observation?

2. What is the "central dogma?" How are the RNA viruses in contradiction with this principle?

3. Compare and contrast the processes of induction and derepression in prokaryotic cells.

4. Provide several examples of chemical signals that control gene activity in eukaryotes.

5. Transcription of a eukaryotic gene is controlled by several "transcriptional signals." In the box below, indicate and label the relative positions of these signals. Then, briefly describe the base sequence and the function of each signal.

```
┌─────────────────────────────────────┐
│                                     │
└─────────────────────────────────────┘
```

6. What are four primary means of molecular control of eukaryotic genes? Give an example of each.

7. Distinguish between these types of DNA: cDNA, Z-DNA, mtDNA, and B-DNA. In what situation(s) does each occur? What is the function of each type?

8. What are some of the practical applications of genetics techniques based on the principle of transformation?

9. In the unit in your textbook in which we have considered genetics topics, we have used the term "gene" in several different ways. Trace the evolution of the definition for a gene as we have advanced from Mendel's classical experiments through the modern techniques of genetic engineering.

10. What is meant by the term "extra-nuclear heredity?" How does this mode of inheritance differ from Mendelian inheritance? What are the implications of extra-nuclear heredity in certain organelles for the evolution of eukaryotes from prokaryotes?

TESTING YOUR UNDERSTANDING

For each of the test items below, three of the lettered alternatives are true and the other is false. Determine which alternative is false and write its letter in the blank to the left of the question number. On the blank line below alternative D, write the corrected version of the false statement.

_____ 1. A. The lac-operon system demonstrates how gene activity is regulated in prokaryotes and in eukaryotes.
B. The structure of a protein is indicated by structural genes.
C. The rate of protein synthesis is determined by regulatory genes.
D. An operon contains both structural and regulatory genes.

Correction: _____

_____ 2. A. Repression is the inhibitory effect of the product of a reaction sequence on the synthesis of an enzyme early in that reaction sequence.
B. Presence of product molecule in high concentration can repress enzyme synthesis.
C. Inducer molecules cause the cell to increase the rate of production of an enzyme.
D. The inducer molecule for a particular pathway usually is the substrate for the enzyme whose rate of production is increased.

Correction: _____

_____ 3. A. In a situation where the rate of enzyme synthesis is regulated by derepression, the rate of enzyme synthesis increases when the amount of product is low.
B. The regulatory processes of repression and induction can both operate in a single operon.
C. The I gene encodes the inducer molecule.
D. The term "cistron" refers to the section of an operon that encodes a structural gene.

Correction: _____

_____ 4. A. The operator of the lac operon is located immediately upstream of the structural genes it regulates.
B. The I gene encodes molecules that bind with the operator to suppress synthesis of mRNA.
C. The presence of both inducer and repressor molecules can result in synthesis of mRNA for the structural genes.
D. The CAP site is a regulatory site to which RNA polymerase binds.

Correction: _____

_____ 5. A. Hormones are one of the chemical means of regulating transcription.
B. Transcriptional signals are short segments of DNA that influence the activity of structural genes.
C. Most transcriptional signals are located downstream from the structural genes whose activity they regulate.
D. The the role of the "TATA box" in regulating transcription is to fix the start site of transcription by positioning molecules of RNA polymerase.

Correction: _____

_____ 6. A. Some of the regulation of gene activity occurs in the cytoplasm.
B. Geneticists have hypothesized that nucleosomes might be involved in the regulation of gene activity as follows: DNA in the vicinity of nucleosomes might not be transcribable due to the inability of RNA polymerase to bind.
C. Heterochromatin and Barr bodies represent DNA that is actively being transcribed.
D. Recent research suggests that DNA methylation may be involved in regulating or blocking transcription of DNA in vertebrate animals.

Correction: _____

_____ 7. A. Some RNA viruses contain a + strand of RNA whereas other RNA viruses contain a - strand.
B. The - strand can function as a mRNA molecule, but a + strand cannot.
C. RNA synthetase is an RNA-dependent RNA polymerase that catalyzes the formation of a complementary RNA strand upon a single-stranded RNA template.
D. Retroviruses utilize reverse transcriptase to copy RNA into DNA before producing new RNA molecules.

Correction: _____

_____ 8. A. The RNA of a retrovirus becomes incorporated into the genome of its host cells as a provirus.
B. Eukaryotic organelles that contain chromosomes include mitochondria and chloroplasts.
C. One of the possible explanations for progeny ratios that do not follow conventional Mendelian ratios is extranuclear inheritance.
D. Prokaryotes probably evolved from a mutualistic association in which one eukaryotic cell lives within another eukaryotic cell.

Correction: _____

_____ 9. A. The design of chromosomes within cytoplasmic organelles is more similar to chromosomes of prokaryotes than to those of eukaryotic nuclei.
B. Because female gametes contribute much greater amounts of cytoplasm to an embryo than do male gametes, traits carried on mitochondrial chromosomes are inherited from mothers rather than from fathers.
C. Variegation of plant leaves is a trait that is inherited by extranuclear inheritance.
D. The chromosomes contained in certain cytoplasmic organelles carry enough genetic information to produce all the molecules needed for the organelle to function independently of the rest of the cell.

Correction: _____

_____ 10. A. A heterokaryon is a multinucleate cell containing nuclei from more than one species.
B. In cells produced by the technique of cell hybridization, the chromosomes from only one of the nuclei are capable of being transcribed into RNA.
C. One of the major applications of the technique of cell hybridization is the production of monoclonal antibodies.
D. Bacterial transformation can also be referred to as DNA-mediated gene transfer.

Correction: _____

_____ 11. A. Although the field of molecular genetics is advancing rapidly, geneticists have not yet been able to successfully transfer genes between prokaryotic and eukaryotic species.
B. The first artificial gene was synthesized by Har Gobind Khorana.
C. A muton is the mutational unit of a DNA molecule.
D. Jacob and Monod are credited with describing the mechanism of gene regulation in prokaryotes.

Correction: _____

CHAPTER 19--DEVELOPMENT OF FORM IN THE ANIMAL BODY

CHAPTER OVERVIEW

In this chapter, we shift from our study of genetics to an overview of the early stages of development and the processes by which embryos develop. Development includes four major processes: determination, differentiation, growth, and morphogenesis. These are gradual and cumulative processes that lead to irreversible structural change in the individual. We should keep in mind that development does not halt at birth or hatching, but instead that development continues over the entire life span of an individual.

Regardless of species, all animals share a basic sequence of developmental stages. Sex cells are produced by parents by gametogenesis. Fertilization marks the beginning of the new individual. Implantation occurs in some species in which the embryo develops within the female reproductive tract. The embryo then mitotically divides and progresses through several cleavage stages, including the blastula. The three primary germ layers are defined during gastrulation. During neurulation, the organs of the nervous system (as well as many other organs) begin to form.

Later in the chapter, we'll see that various developmental features are quite useful in classifying the diversity of animal species present on Earth.

TOPIC SUMMARIES

A. <u>Processes of Development</u>.--The changes that occur as an organism develops (during that organism's <u>ontogeny</u>) involve several different processes. These processes (1) lead irreversibly to structural changes and (2) are gradual and cumulative. Such developmental processes differ from short-term, reversible processes such as muscle contraction and many other physiological activities. Four primary developmental processes occur during ontogeny:

(1) Determination. In general, every cell of an organism possesses the full set of genetic information, and hence could follow any of the developmental pathways encoded by its genome. However, not every cell is <u>totipotent</u>. Rather, different cells from different parts of an embryo (or after certain periods of time) exhibit different degrees of <u>potency</u>. A cell that can follow only one developmental pathway is said to be <u>unipotent</u>. Similarly, a cell having several to many possible options is <u>pluripotent</u>. Once a cell becomes irreversibly committed to a particular fate, it is said to be <u>determined</u>. For example, a cell can be determined to develop into a muscle cell . . . or into a nerve cell. Interestingly, a committed cell does not necessarily show visibly what its fate is to be; rather, the process of <u>differentiation</u> must occur before determination is apparent.

(2) Differentiation. The process of <u>differentiation</u> produces specialized structural or biochemical properties in a cell. The presence of specific gene products within a cell is an indication that differentiation has occurred. Cells in the process of differentiating into muscle cells, for example, contain large amounts of actin and myosin, proteins involved in the contractile apparatus of muscle tissue. Once a cell has differentiated, it generally does not (indeed, can not) change into some other type of cell. An exception, however, to this rule of irreversibility seems to exist: in the case of injury or of some other sort of damage, some cell types can dedifferentiate enough to enable regeneration of new cells or structures. Examples include regeneration of limbs by salamanders and production of an entire new organism from but one cell of a *Hydra*.

(3) Growth. Additionally, development involves irreversible increase in size of the organism. Growth can occur by either (or both) increase of the number of cells in the organism (i.e., by hyperplasia) or by increase in the size of existing cells (by hypertrophy). From our own observations, we know that a characteristic size can be associated with many species. Most of the animal species with which we are familiar grow to a certain size whereupon growth slows or stops; this observation suggests that growth is a regulated process. The main regulatory substances are hormones and growth factors. Unregulated and invasive growth of cells is the hallmark of cancer.

(4) Morphogenesis. The development of shape and form (of structures as well as of the entire body) is referred to as morphogenesis. Determination, differentiation, and growth necessarily are precursors of morphogenesis. Differences in morphogenesis between plants and animals pertain largely to the mobility of cells within the body. Because of the rigidity of plant cell walls and of the intercellular materials, little cell movement can occur in plant morphogenesis. Oppositely, programmed movements of cells within animals bodies are very important in development of shape and form in animals.

B. Stages of Animal Development.--Animal species (and other organisms, in general) have life cycles, a series of developmental stages through which individuals of that species cycle. A general animal life cycle turns from fertilized egg to embryo to larva to adult, which produces sex cells that re-enter the cycle (Fig. 19.1). In this chapter, we'll focus on the early segments of the life cycle: production of gametes and embryonic development.

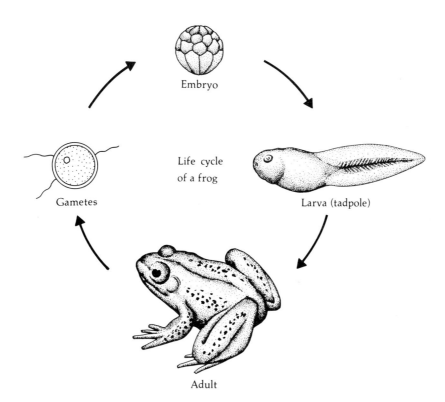

Fig. 19.1: Life cycle of a frog, an amphibian

C. Production of Gametes.--The term gametogenesis refers to the processes by which sex cells (gametes) are produced. Although specific differences exist between the sexes, there is an overriding similarity in gametogenesis for both sexes: The objective of the process is to produce sex cells that have reduced chromosomal (haploid) content and new combinations of genetic traits. You should refer to Chapter 11 to refresh your memory about meiosis.

In males, the process is called spermatogenesis. It occurs in the male gonads, the testes. The general sequence of events proceeds from diploid spermatogonia to haploid spermatocytes which then mature into spermatozoa. Each spermatogonium divides into four spermatozoa, cells specialized for delivery of the male's contribution of half of the genome to the embryo. The anatomical design of a sperm cell reflects this role: The head carries the chromosomal material within a nucleus; associated with the head is the acrosome which contains hydrolytic enzymes needed to allow the sperm to enter the ovum. Movement of the tail imparts motility to the cell. The middle piece, an area of concentration of mitochondria, probably is involved in providing the power needed to operate the tail. A variety of glands add seminal fluid to the sperm to produce semen.

Oogenesis is the corresponding process in females. One major difference from the male process is that only one functional haploid ovum is produced for each diploid cell entering oogenesis; the other three cells produced by meiosis are small polar bodies that generally are not involved in fertilization. Like sperm cells, mature ova contain half the chromosomal complement. Compared to a sperm cell, however, an ovum contains a greater (often enormous) amount of yolk as a nutrient supply for a developing embryo. Yolk quantities of ova usually are greater in species that lay eggs and in which most embryonic development takes place outside of the mother's body; such species are oviparous. In contrast, viviparous species generally produce ova with little to no yolk; instead, nutrients reach the embryo directly from maternal tissues through a placenta.

D. Fertilization.--Embryonic development begins with the process of fertilization. The events of this process include penetration of the ovum by the sperm, activation of the egg, and syngamy, the fusion of the haploid male and female nuclei into one diploid nucleus. Depending on the species, fertilization may occur by release of sperm and ova outside of the body (external fertilization) or by release of sperm within the body of the female (internal fertilization; in mammals this act is called copulation). The first diploid cell of the embryo is zygote. Interestingly, a few species of vertebrates (notably, some salamanders) do not require fertilization for embryonic development to occur; rather ova develop into haploid adults by the process of parthenogenesis.

E. Implantation.--In viviparous species, such as most mammals and a few other vertebrates, the embryo next implants into the maternal tissues of the uterus or some other corresponding segment of the female reproductive tract. Implantation begins the process of formation of the placenta, an intimate association between maternal and embryonic tissues. Virtually all materials needed by the embryo (e.g., oxygen, water, nutrients) and all wastes produced by the embryo (e.g., carbon dioxide, nitrogenous wastes) pass across this thin tissue barrier.

F. Cleavage Stages.--Mitotic divisions of the zygote begin soon (for some species, within only a few hours) after fertilization, indeed often before implantation occurs. The first several divisions may not produce a larger embryo; rather, the existing mass is divided into more cells. The morula is a late cleavage stage in which the cells (approximately 32 in number) are arranged in a solid mass much like a mulberry. The process of blastulation converts the morula into a hollow mass of cells; the space in the blastula is the blastocoel. Each cell of a blastula is a blastomere.

The particular pattern of cleavage corresponds to the taxonomic group of the species and to the amount of yolk in the embryo. The embryos of many invertebrate species divide spirally, whereas most vertebrates undergo radial cleavage. If all cells of an embryo are of equal size and all cells of the embryo divide, then the cleavage pattern is termed equal holoblastic;

examples include sea urchins and many other invertebrates. In <u>unequal holoblastic</u> cleavage, the entire embryo divides but the yolk-laden cells of the <u>vegetal pole</u> are larger and divide more slowly than the smaller cells of the <u>animal pole</u>; amphibians such as frogs are a common example. If only part of the embryo (the <u>blastodisc</u> of a bird embryo, for example) experiences cell division, then cleavage is <u>meroblastic</u>. Fig. 19.2 shows examples of equal and unequal holoblastic and of meroblastic cleavage:

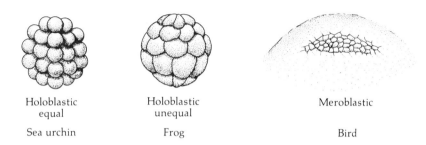

Fig. 19.2: Three patterns of cleavage among animal embryos

During various cleavage stages (the specific stage depending on the species), the orientation of the embryo is established; hence, left and right, top (<u>dorsal</u>) and down (<u>ventral</u>), front (<u>anterior</u>) and rear (<u>posterior</u>) are determined early in development.

G. <u>Gastrulation</u>.--During <u>gastrulation</u>, the general body plan of the embryo emerges. This involves the differentiation of three distinct <u>germ layers</u>: endoderm, mesoderm, and ectoderm. The blastula is hollow sphere bounded by a single layer of cells. The beginning of gastrulation is marked by the movement of some of these surface cells into the interior of the embryo, thus forming two cell layers: the inner layer is <u>endoderm</u>, the outer cells, <u>ectoderm</u>. This inward migration of cells through the blastopore is called <u>invagination</u>. Inward movement of cells eventually obliterates the blastocoel, but simultaneously produces a new space, the <u>archenteron,</u> that will become the gut of the developing individual. The third germ layer, the <u>mesoderm</u>, forms in the region between the other two layers of cells. Another group of cells, the <u>chordamesoderm</u>, that migrated around the <u>dorsal lip</u> of the blastopore turns out to have a very important role in organizing other cells as the embryo develops . . . more about this in Chapter 20. The <u>primitive streak</u> is a modified version of a blastopore that usually is found in vertebrates whose embryos possess huge amounts of yolk.

Another major event during gastrulation is the <u>formation of the coelom</u>, the body cavity within which most visceral organs are located. Most animals possess <u>bilateral symmetry</u>, i.e., the body is built with mirror-image left and right sides. Such animals are subdivided into groups on the basis of the presence or absence (<u>acoelomate</u> animals) of the coelom and the method of coelom formation. A true coelom, formed from openings in the mesoderm, is characteristic of <u>coelomate</u> animals (e.g., mammals and other vertebrates). The body cavities of <u>pseudocoelomates</u> do not develop from within mesoderm. We'll talk more of pseudocoelomates and the coelomates in our study of animal classification in Chapter 22.

Because of its different developmental fates, the blastopore is a feature that is useful in the classification of coelomate animals. In the <u>protostomes</u>, the blastopore develops into the mouth of the individual, whereas the anus forms from or near the blastopore in the <u>deuterostomes</u>.

H. <u>Neurulation and Organogenesis</u>.--Having formed the three germ layers during gastrulation, the development of the embryo continues with the formation of various organs, including those of the nervous system. <u>Neurulation</u>, a developmental feature only of the <u>chordates</u> (e.g., fishes, amphibians, reptiles, birds, mammals, and a few others), is marked by the appearance of the notochord and the neural tube. The notochord develops into various structures related to the vertebral column. The neural tube becomes the brain and spinal cord. An embryo with a closed neural tube has reached the neurula stage.

TERMS TO UNDERSTAND: Define and identify the following terms:

ontogeny

determination

potency

differentiation

growth

hypertrophy

epiphyses

hormones

growth factors

cancer

metastasis

morphogenesis

spermatozoa

ova

spermatogenesis

spermatogonium

spermatocytes

testis

scrotum

oogenesis

oogonium

yolk

polar body

oviparous

viviparous

germinal disc

placenta

seminal vesicle

prostate

Cowper's gland

semen

fertilization

copulation

uterus

Fallopian tubes

granulosa cells

parthenogenesis

syngamy

species selectivity

in-vitro fertilization

sperm capacitation

implantation

cleavage

morula

blastula

blastocoel

larva

gastrulation

invagination

archenteron

blastopore

germ layers

chordamesoderm

primitive streak

coelom

bilateral vs. radial symmetry

acoelomates

pseudocoelomates

coelomates

protostomes

deuterostomes

neurulation

notochord

neural tube

oncogene

MATCHING

For each of the structures or tissues listed below, indicate the germ layer from which it developed. Place the appropriate letter in the blank to the left of the structure: A for endoderm, B for mesoderm, C for ectoderm.

_____ 1. lining of stomach   _____ 2. gonads

_____ 3. epidermis   _____ 4. connective tissue

_____ 5. blood   _____ 6. muscle

_____ 7. bone   _____ 8. trachea

_____ 9. receptor cells in sense organs   _____ 10. urethra

CHART EXERCISE

For the organisms listed as column headings, complete the chart to describe various characteristics of their embryos. Indicate (1) whether cleavage pattern is spiral or radial, (2) whether all blastomeres are approximately equal or unequal in size, (3) how yolk is distributed within the embryo, and (4) what parts of the embryo are involved in cell division.

FEATURE	FROG	BIRD	MAMMAL
Cleavage pattern			
Blastomere size			
Yolk distribution			
Parts of embryo involved in cell division			

DISCUSSION QUESTIONS

1. Distinguish between the terms <u>embryology</u> and <u>developmental biology</u>.

2. Describe the structure and associated functions of the various components of the gametes of animals.

3. Discuss the classification of twinning in mammals.

4. How does <u>differentiation</u> differ from <u>determination</u>? In the sequence of embryological development, which of these two processes occurs first?

5. Discuss cancer in terms of the developmental process of growth.

6. Which of the following developmental strategies is more effective in the survival of an embryo . . . oviparity or viviparity? How?

7. How does embryological development in plants differ from that of animals?

TESTING YOUR UNDERSTANDING

For each of the test items below, three of the lettered alternatives are true and the other is false. Determine which alternative is false and write its letter in the blank to the left of the question number. On the blank line below alternative D, write the corrected version of the false statement.

_____ 1. A. Development of a totipotent cell can progress along only one pathway.
B. A determined cell is irreversibly committed to a particular developmental fate.
C. The process of differentiation follows determination.
D. Differentiation is the actual appearance of specialized characteristics.

Correction: _____

_____ 2. A. The accumulation of specific gene products is evidence that a cell is both determined and differentiated.
B. Regeneration of an organ can be thought of as a form of dedifferentiation.
C. Growth, the irreversible increase in size, is a critical component of development.
D. The increase in size of an organ due to enlargement of its cells is called hyperplasia.

Correction: _____

_____ 3. A. Although growth stops at a certain point in the development of mammals, growth continues throughout the life of many other kinds of organisms.
B. Regulation of growth involves chemicals such as hormones and growth factors.
C. The spread of cancerous cells throughout the body is called epiphysis.
D. A trait shared by all forms of cancer is uncontrolled growth of cells.

Correction: _____

_____ 4. A. Morphogenesis in plants generally does not involve movements of cells from one area of the body to another area because the rigidity of the cell walls restricts cell movements.
B. Implantation is one of the events that occurs during early development of most species of vertebrates.
C. The earliest developmental stage of an animal embryo in which three germ layers are present is the gastrula.
D. The morula stage occurs before the blastula stage.

Correction: _____

_____ 5. A. All of the following are related to the process of spermatogenesis: testes, polar bodies, acrosome.
B. A mature sperm cell is little more than a nucleus enclosed in a specialized delivery apparatus.
C. Housing of testes in a scrotum correlates with the greater efficiency of the process of spermatogenesis at temperatures lower than body temperature.
D. The process of fertilization both restores the diploid chromosome number and activates the ovum.

Correction: _____

_____ 6. A. For each cell entering the process, oogenesis produces only one functional daughter cell whereas spermatogenesis produces four functional cells.
B. Penetration of the ovum by a sperm involves digestion of the egg envelope by the enzyme hyaluronidase.
C. Parthenogenesis is a type of development which does not require a genetic contribution from a female parent.
D. The process of syngamy is triggered by activation of the egg.

Correction: _____

_____ 7. A. The process of implantation results in the establishment of an intimate association between the embryo and the adjacent maternal tissues.
B. Materials exchanged across a placenta include water, wastes, blood cells, and respiratory gases.
C. The division of the zygote into blastomeres occurs via the process of mitosis.
D. In an amphibian embryo, blastomeres of the vegetal hemisphere are larger than those of the animal hemisphere.

Correction: _____

_____ 8. A. Embryos containing large amounts of yolk undergo meroblastic cleavage.
B. Embryos of invertebrates such as sea urchins undego equal holoblastic cleavage.
C. The primitive streak of a bird embryo represents a modified blastopore.
D. The blastocoel of the blastula develops into the archenteron of the gastrula.

Correction: _____

_____ 9. A. The three primary germ layers formed during gastrulation are endoderm, ectoderm, and chordamesoderm.
B. Movement of surface cells into the interior of an early embryo involves a process called invagination.
C. In protostomes, the blastopore eventually develops into the animal's mouth.
D. Chordates are properly classified as deuterostomes.

Correction: _____

_____ 10. A. Endoderm develops into the lining of the digestive tract.
B. The nervous system develops from ectoderm.
C. The outer layer of mammalian skin develops from ectoderm.
D. Muscle and bone tissue develops from ectoderm.

Correction: _____

_____ 11. A. During neurulation, neural folds develop before the neural tube develops.
B. Neural crest cells differentiate into a variety of structures including sensory nerve cells and cartilage.
C. One of the distinguishing features of echinoderm larvae is a notochord.
D. Most species of animals exhibit bilateral symmetry.

Correction: _____

_____ 12. A. The body cavity of coelomate animals develops from space that forms within the mesoderm.
B. Pseudocoelomate animals have a body cavity that forms from mesoderm.
C. The coelom is the body cavity within which most internal organs are located.
D. Some deuterostomes can be properly classified as coelomates.

Correction: _____

CHAPTER 20--DEVELOPMENTAL MECHANISMS

CHAPTER OVERVIEW

In Chapter 19, we described the general events associated with the early development of animal embryos. Here, we examine various of the known and suspected mechanisms related to these developmental changes. Studies of molecular mechanisms center on how different subsets of the genome are allowed to be expressed in some types of cells while their expression is suppressed in other cell types. Also in the context of molecular mechanisms, we are introduced to various methods and model species being used in efforts to understand development.

Another series of developmental mechanisms seems to involve various means of communication between cells. Such communication can direct movements of cells within the embryo as well as induce groups of cells to follow particular developmental pathways. Communication may involve direct contact of cells with each other and/or production and recognition of chemical signals.

A final topic joins development of individuals (ontogeny) with evolutionary development of species lineages (phylogeny). The meaning of the classical statement, "ontogeny recapitulates phylogeny," is examined. We end the chapter with a look at the evolutionary results of asynchronous development of reproductive and somatic tissues.

TOPIC SUMMARIES

A. <u>The Role and Importance of the Nucleus in Development</u>.--From our studies of genetics in earlier chapters, it should come as no surprise that the nucleus of a cell plays a critical role in all aspects of metabolism, including development. This role relates to the presence of chromosomes and, hence, most of the genome, in the nucleus.

Scientists of the past did not, however, have the knowledge of genetics that even we in an introductory biology class possess. Hans Spemann, for instance, experimentally determined that a newt (salamander) embryo would not develop normally unless a nucleus was present in all cells of the embryo. Briggs and King showed also that totipotency of nuclei lasted at least into late blastula stage. Gurdon demonstrated that a nucleus from an adult frog cell still carried all of the genetic information for the development of a complete individual.

Such experiments in nuclear transplantation, however, did not explain <u>how</u> the nucleus exerted regulatory effects. This understanding has come from experiments in molecular biology and biochemistry.

B. <u>Gene Regulation vs. Differential Loss and Retention of Genes</u>.--Once it was realized that chromosomes (and the genes contained thereon) were instrumental in development, the question became: Does differentiation result from <u>loss</u> of those genes that are not needed for a particular cell type, or is the entire genome retained in each cell and <u>regulated</u> so that only a subset of the genes is expressed for a particular cell type? Gurdon's work on *Xenopus* (see your textbook for a description of the experiment) showed that all genes are retained in all somatic cell types, and that genes can be turned on and off (i.e., <u>regulated</u>). Research by Gross on sea urchin eggs and by Brown on *Xenopus* suggest that gene regulation involves production of certain <u>histones</u> (a group of proteins) that complex with DNA to inactivate selected genes. Such histones and other types of molecules (e.g., certain maternal mRNAs, transforming growth factor-beta, transcription factor IIIA) that influeuce the developmental fates of embryonic cells are called <u>cytoplasmic determinants</u>.

Current research into understanding the mechanisms of development involves culturing of cells of various types (e.g., myoblasts as precursors of muscle tissue). A primary research interest is to identify the cellular events that initiate the steps of differentiation of a cell culture. Commitment to a certain developmental fate apparently involves a single, unknown, cellular event.

C. Models and Methods for Tracing Genes During Development.--Progress toward understanding gene regulation during development involves experimental manipulation of living organisms. Three of the favorite animal models are mice, a nematode, and fruit flies.

1. Transgenic mice. One means of introducing genes of one strain of mice, *Mus musculus,* into another strain is to "build" early embryos consisting of a few cells from each strain, and to then implant them into the uteri of females. Such combinations are referred to as allophenic mice, and, surprisingly to early researchers, they develop into normal mice that are mosaics of cells from the two original embryos. Several findings came from experiments on allophenic mice: Up to about the 32-cell stage, each cell of the embryo retained totipotency. Study of melanocytes (hair pigment-producing cells) indicated that only 34 embryonic cells differentiated to produce all of the hair pigment cells. The implication is that commitment to a particular developmental path occurs early and that only a few early cells need to commit to any particular developmental fate.

The next step in producing transgenic mice was to introduce pure DNA into ova, rather than by transferring DNA contained within an entire cell. Foreign DNA injected into eggs inserts itself into the genome of the recipient cell, and thereby, causes mutations. The technique, called insertional mutagenesis, has great practical and scientific application.

2. A nematode, *Caenorhabditis elegans*. For reasons that your textbook outlines, *C. elegans* is a perfect model for study of development mechanisms. The complete developmental history of each of this nematode's 959 cells is known! Thus, the effects of manipulation of any cell during development can be observed and described. Understanding the role of cell death in development is among the most promising of the many areas to which *C. elegans* will contribute.

3. *Drosophila,* an old friend from genetics. Homeotic genes are mutant genes that cause cells to switch from one developmental fate to another. Such genes can affect the segmentation of the body plan of fruit flies and other segmented organisms. Insect bodies are divided into three regions (head, thorax, abdomen) which are made of segments. Each segment is a separate development parcel arising from only a few founding cells, and each segment consists of an anterior and a posterior compartment.

In fruit flies, homeotic genes occur in at least two clusters, the antennapedia complex (influencing determination of head and thorax segments) and the bithorax complex (influencing thorax and abdominal segments). Many (perhaps all) homeotic genes contain a common sequence of DNA, the homeo box. This sequence also has been found outside of the antennapedia and bithorax complexes as a regulatory gene whose role is to distinguish the anterior and posterior compartments in each segment. This particular external homeo box is the one responsible for engrailed, a wing trait in which the rear of the wing edge resembles the front margin of the wing. The significance of this trait, discovered in 1929, did not become apparent until 1975 when the connection between engrailed and homeotic boxes was realized. More recent work has found that such homeo boxes occur in many species, including humans.

D. Embryonic Induction.--The developmental fate of a cell depends not only on the cell type, but also on the location of that cell in the embryo. Spemann arrived at this conclusion after numerous transplant experiments in which he excised chordamesoderm tissue from the dorsal lip region and placed it elsewhere in the embryo. He found that the neural tube formed only in association with the chordamesoderm. Hence, neural tube formation is induced by the presence of an organizer tissue. The mechanisms by which organizers induce certain courses of development are not yet understood. Time, however, seems to be a critical factor as well: Only cells from an early embryo can respond to organizers; with aging, embryos lose their ability to respond (i.e., lose their competence).

E. Cellular Strategies.--The first parts of this chapter focussed on developmental mechanisms operating within cells. The next section of this chapter examines strategies that work at the level of entire cells.

(1) Morphogenetic movements. One of the most interesting aspects of development is that cells and groups of cells move en masse from one place in the embryo to another. Sheets of such moving cells are often referred to as mesenchyme. In the previous chapter, we saw the importance of such morphogenetic movements in the early embryo; for example, movements of cells from the embryo surface through the blastopore and into the interior resulted in the formation of the three germ layers.

Cell movements do not occur at random. Rather, a series of mechanisms seems to govern such movements. Cell movement in general requires that they adhere to a surface; this adherence may somehow direct movements of the cells (via contact guidance). Of course, this adherence must be reversible if the cells are to move any significant distance. Contact inhibition of cell movements often occurs in tissue cultures when the filopodia of one cell (a fibroblast) contacts another fibroblast; such contact causes movement in that direction to stop. In other situation, contact of filopodia of migrating cells results in the formation of tight junctions and the cessation of cell movements.

Experiments in disruption of tissues or even of entire organisms show that cells can recognize other cells of their own kind and that cells can move to reassemble an entire organism. In a famous experiment, entire bodies of sponges were separated into individual cells which then moved to reaggregate into entire bodies with all anatomical details correct. Various embryological experiments have shown that specialized cells (for example, skin pigment cells) injected into the body of a developing embryo do not take up residence just anywhere, but move (or home) to locations appropriate for that cell type (to continue our example, pigment cells moving into the skin). Just how cells recognize their neighbors is not yet understood, but cyclic AMP might well be one of the chemical messengers involved.

(2) Cell interactions. Much of what we know about cell interactions has been learned from the technique of tissue culture. Cells of a tissue type (e.g., kidney cells) can be grown in culture, but in the absence of other cell types these cells form merely an amorphous mass. Yet, if kidney connective tissue is added to the kidney cell culture, then the amorphous field of kidney cells differentiates into kidney tubules. This (and other experiments discussed in your textbook) suggests that some sort of chemical "factors" (whose identities are not yet known) produced by one cell type influences the developmental fate of other cell types. The structural protein collagen probably is an important factor in the cellular interactions that direct organogenesis. Direct cell-to-cell contact is not necessary in many cases. How are such interactions related to the principle of self-assembly?

E. Ontogeny and Phylogeny.--Ontogeny is the sequence of events through which an individual passes during its lifetime. Phylogeny refers to the evolutionary history of a lineage of species or other taxa. During the course of early development, individuals pass through stages that, to some degree, resemble stages of evolutionarily less-advanced species. This observation, in various forms, has been stated as the biogenetic law of von Baer and the principle of recapitulation of Haeckel. Be assured, though, that embryos do not pass through adult or any other stages of any other species. Similarity of developmental mechanisms across species does, however, bear witness to the common ancestry of various groups of organisms.

Ontogeny can be thought of as involving simultaneous development, on a synchronous schedule, of two sets of tissues: the gonads and the rest of the body (the somatic features). Heterochrony refers to shifts in the relative timing of gonadal and somatic development. Several versions of heterochronic development produce different developmental products. Major size increases can occur when gonads develop slowly while somatic development progresses normally and continues beyond the time of gonad maturation; the huge antlers of the extinct Irish elk resulted from this process of hypermorphosis. Retention of juvenile features by adults results in one of two forms of paedomorphosis: Neoteny occurs when somatic development slows while gonads develop normally; the mudpuppy salamander, *Necturus*, is a common example. Progenesis occurs when gonads mature early with the result of a sexually mature individual housed in a small-sized juvenile body.

TERMS TO UNDERSTAND: Define and identify the following terms:

hybridization probe

gene transplantation

commitment

foreign genes

transgenic mice

mosaics

allophenic mice

insertional mutagenesis

homeotic genes

segment

compartment

antennapedia complex

bithorax complex

homeo box

macrophage

morphogenetic movements

cell-to-cell communication

cell motility

filopodia

contact inhibition

contact guidance

chemotaxis

cyclic AMP

self-assembly

collagen

tropocollagen

fibrils

organizer

induction

dorsal lip

chordamesoderm

fate maps

competence

biogenetic law

principle of recapitulation

ontogeny

phylogeny

heterochrony

hypermorphosis

paedomorphosis

neoteny

progenesis

chimera

PEOPLE TO KNOW: What is the significant biological contribution of each of these individuals as indicated in Chapter 20?

Hans Spemann

Sidney Brenner

Walter Gehring

P. H. O'Farrell

K. E. von Baer

E. H. Haeckel

C. L. Markert and R. M. Petters

R. Briggs and T. J. King

John Gurdon

Philip Leder

Beatrice Mintz

Johannes Holtfreter

Warren Lewis

DISCUSSION QUESTIONS

1. Briefly summarize the experimental evidence that indicates that the cell nucleus is a critical structure that controls developmental processes.

2. What is the experimental evidence that genetic regulation of development generally does not involve loss of genes, but involves inactivation of genes instead?

3. For what purpose is a DNA- or RNA-hybridization probe used?

4. How has the technique of tissue culture helped to advance our knowledge of the cellular strategies that regulate development?

5. What aspects of the biology of the nematode *Caenorhabditis elegans* make it a well-suited species for the study of developmental mechanisms?

6. What is the suspected role of collagen in the tissue interactions associated with development?

7. What is meant by the term organizer? Provide at least one example of an organizer tissue that operates in vertebrate embryos.

8. State the biogenetic law and the principle of recapitulation. Are these different ways of stating the same concept, or are the concepts different? If different, how are they different?

9. Distinguish between these three versions of heterochrony: hypermorphosis, neoteny, and progenesis.

TESTING YOUR UNDERSTANDING

For each of the test items below, three of the lettered alternatives are true and the other is false. Determine which alternative is false and write its letter in the blank to the left of the question number. On the blank line below alternative D, write the corrected version of the false statement.

_____ 1. A. For most vertebrate species, the nucleus of a cell from a 16-cell embryo is totipotent.
B. Gurdon conducted experiments demonstrating that differentiation does not result in loss of non-expressed genes from differentiated cells.
C. Histones form complexes with DNA to inactivate genes in those portions of the DNA molecule.
D. Chromosome elimination, a mechanism for inactivating genes, operates in most species of eukaryotes.

Correction: _____

_____ 2. A. Commitment refers to a single cellular event that initiates a coordinated differentiation program in all cells of a tissue culture.
B. Commitment entails a change from unipotency to pluripotency.
C. Allophenic mice are produced by combining into one early embryo cells from the early embryos of two or more different genetic strains of mice.
D. Research by Mintz showed that commitment by cells to a particular developmental pathway occurs early in embryonic life.

Correction: _____

_____ 3. A. During embryonic development, only a few cells commit to any particular developmental pathway.
B. Insertional mutagenesis is a means of causing mutations by inserting foreign DNA into the DNA of a host cell.
C. A teratocarcinoma is a tumor that develops from undifferentiated embryonic cells.
D. Teratocarcinoma cells injected into early embryos of mice produce tumors in those embryos.

Correction: _____

____ 4. A. Among the reasons that *Caenorhabditis elegans* is a useful model species for studying developmental mechanisms is that its life cycle is brief and it can be bred in large numbers.
B. Because so much of its genome is known, fruit flies are a good model species to study of development.
C. Study of the nematode *Caenorhabditis elegans* has demonstrated that the process of cell death plays a key role in development.
D. The ability of *Caenorhabditis elegans* to self-fertilize makes this species especially suitable for study of dominant genes.

Correction: _____

____ 5. A. Each segment of an insect's body is a separate development unit derived from only a few founding cells.
B. Body segments are composed of compartments.
C. The developmental fate of cells in a particular compartment is governed by allophenic genes.
D. In *Drosophila*, determination of head segments is directed by genes in the antennapedia complex.

Correction: _____

____ 6. A. Chemotaxis is an oriented movement of an organism under the influence of a chemical agent.
B. For most types of tissues, communication between cells requires that those cells be in physical contact with each other.
C. Myoblasts are cells that differentiate into muscle cells.
D. Collagen is a fibrous protein that is found in connective tissue.

Correction: _____

____ 7. A. Differentiation of some kinds of cells requires that the cells be exposed to a conditioned medium.
B. One of the activities of fibroblasts is the synthesis of collagen.
C. Experimental evidence suggests that collagen might have an important role in the assembly of cells into tissues.
D. Fibrils of collagen form via self-assembly of molecules of tropomyosin.

Correction: _____

____ 8. A. The organizer cells at the dorsal lip of the blastopore begin their organizing activities at the morula stage.
B. Chordamesoderm cells differentiate to form mesoderm and notochord.
C. Chordamesoderm cells induce the overlying ectoderm to form structures that will become major parts of the nervous system.
D. Transplanting dorsal lip cells into a different part of an embryo will induce a nervous system to form in the area into which the dorsal lip cells were transplanted.

Correction: _____

_____ 9. A. Fate maps show what organs will eventually be produced by various regions of an early embryo.
   B. The capability of a group of cells to respond to an organizer is called competence.
   C. Phylogeny refers to the developmental stages that an individual passes through from the zygote stage until death of the individual.
   D. Von Baer's biogenetic law states that the embryos of "higher" organisms pass through stages resembling the embryos of "lower" organisms.

   Correction: _____

_____ 10. A. Haeckel proposed the principle of recapitulation, which states that ontogeny recapitulates phylogeny.
   B. The principle of recapitulation more correctly describes embryonic development than does the biogenetic law.
   C. Heterochronic development occurs when somatic traits and gonadal maturation occur out of synchrony with each other.
   D. Neoteny occurs when somatic development lags behind reproductive maturation.

   Correction: _____

CHAPTER 21--HOW ORGANISMS REPRODUCE

CHAPTER OVERVIEW

All species are capable of producing more individuals of the same species. Such reproduction allows growth of populations, provides for genetic diversity, and allows replacement of injured or diseased individuals. Additionally, reproduction is critical for the continuation of a species because no individuals of any species are immortal.

In Chapter 21, we examine the basic life cycle of an organism. We see how, in sexually-reproducing species, this cycle is divided into a haploid phase that follows meiosis and a diploid phase that follows fertilization. We compare various species of eukaryotes to see variations of this basic life cycle. We examine the asexual ways in which many species also are capable of reproducing.

In our study of plant reproduction, we introduce the concept of alternation of generations. In some plant species, the gametophyte phase is prominent whereas the sporophyte predominates in others. We'll look at ways in which plant and animal reproduction are similar and yet different. The chapter ends with treatment of human reproduction and contraception.

TOPIC SUMMARIES

A. <u>Why Reproduce? . . . Why Not Immortality</u>?--The ability of existing individuals to produce new individuals of its kind is a characteristic of all species. Reproduction is necessary to <u>replace</u> ineffective, injured, diseased, and dying individuals. Reproduction (if it exceeds the death rate) is a means by which populations <u>grow</u>. Perhaps the most important value of reproduction (at least for sexually reproducing species) is that the cell-division process of meiosis introduces <u>genetic variability</u> into a population. This genetic variability is the source of genetic diversity that enables a population to cope with unpredictable future changes in the environment. Species that do not adapt genetically probably are destined to extinction.

B. <u>Primary Modes of Reproduction</u>.--Being a species that reproduces sexually, we humans might think that sexual reproduction is the only (or at least the best) way to reproduce. However, probably more species reproduce <u>asexually</u> than <u>sexually</u>. We should not think of one reproductive strategy as being better than another because the reproductive mode(s) used by a particular species is (are) very effective in perpetuating that species in the surroundings to which it is adapted.

(1) Asexual reproduction. In earlier chapters, we have mentioned at least a few of the modes of asexual reproduction. Barring chromosomal mutations, genetic information does not change from one generation to the next.

One means by which unicellular organisms reproduce is <u>fission</u>, the splitting of one cell into two, genetically identical cells. Steps involve replication and partitioning of genetic material (e.g., the single circular chromosome in bacteria) and sharing of the parental cytoplasm by the daughter cells.

<u>Vegetative reproduction</u> occurs when a parent organism produces an offshoot that develops into an independent organism. As the name suggests, a great many plants reproduce vegetatively (by producing bulbs, tubers, corms, runners, etc.). Perhaps surprisingly, a few animals also reproduce vegetatively: polyps that bud off of *Hydra* develop into new organisms.

Parthenogenesis is an asexual mode of reproduction that verges on sexual reproduction. In this mode, ova develop without fertilization. Hence, individuals produced by parthenogenesis are haploid. In many parthenogenetic species, no males have ever been discovered! Examples include rotifers, aphids, some vertebrates (e.g., some fishes, some amphibians, some lizards), and an array of plants (e.g., dandelions).

(2) Sexual reproduction. This approach is more "expensive" to an organism than is asexual reproduction, largely due to the associated production of specialized reproductive structures (e.g., gonads, flowers, duct systems, etc.). However, this investment pays off in terms of increased genetic diversity--the genetic key to coping with a changing environment. This process involves meiosis to produce haploid gametes for each sex; fusion of gametes from opposite (or different) sexes restores diploidy.

Although sexual reproduction is known for at least some species in all five kingdoms of organisms, tremendous variation in the specific structures and events exist. "Sexual" mechanisms in prokaryotes include conjugation and syngamy (refer to Chaper 16). In eukaryotes, more conventional sexual mechanisms prevail; we will define sex as "meiosis followed by nuclear fusion." Why is it difficult to apply this definition of sex to prokaryotes?

C. Reproductive Strategies among Sexual Life Cycles.--In the life cycle of any sexuallyreproducing species, two phases are recognized: Haplophase is that interval after meiosis but before fertilization, i.e., when haploid germ cells are present. The remainder of the cycle is diplophase, extending from fertilization to meiosis, when all cells are diploid. Fig. 21.1 depicts the general scheme from which we can derive the life cycle of all species.

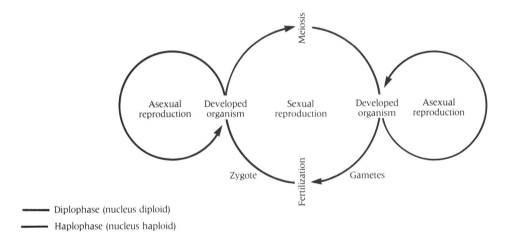

Fig. 21.1: Generalized life cycle

On the basis of the proportion of the life cycle spent in haplophase vs. diplophase, we can define three types of life cycles: Our own life cycle exemplifies gametic meiosis, wherein gametes are the only haploid cells. In the opposite extemre (zygotic meiosis, as in various algae and fungi), the zygote is the only diploid cell; it immediately undergoes meiosis to restore the haploid condition. Occupying the middle ground of this reproductive continuum is sporic meiosis, in which relatively longlived individuals occur in both gametophyte (haploid) and sporophyte (diploid) generations; examples include plants and many algae. In the sections below, we'll examine various examples of such alternation of generations between haplophase and diplophase.

D. <u>Reproduction in Protists</u>.--Protists are the eukaryotic microorganisms, including primarily the algae and protozoans. *Chlamydomonas*, a flagellated, unicellular green alga, spends most of its life cycle in haplophase (textbook Fig. 21-7), with haploid cells mitotically producing more haploid cells. However, two haploid cells occasionally unite to form a one-celled <u>zygote</u> (the sporophyte). The sporophyte does not divide to form a multicellular body, but immediately undergoes meiosis to produce four, flagellated, haploid <u>spores</u>, which can be considered the gametophyte.

Many other algae emphasize diplophase. Textbook Figs. 21-9 and 21-10 depict the life-cycles of sea lettuce *(Ulva)* and a brown alga *(Ectocarpus)*. Both sporophyte and gametophyte are prominent and multicellular. Specialized cells (<u>sporocytes</u>) within the leafy diploid sporophyte undergo meiosis to form haploid spores having four flagella. Each spore swims away to divide mitotically into a leafy haploid gametophyte. This gametophyte mitotically produces haploid gametes which can either fuse to form a diploid zygote or mitotically grow into another gametophyte. Note that still other protists have alternations of generations different than either of those presented here.

E. <u>Reproduction in Plants</u>.--Meiosis occurs in plants, but this process does not lead directly to gametes. Rather, one or several mitotic divisions often occur between meiosis and formation of the gametes. In plants, we will examine variations on this two-phase theme: a diplophase spore-producing phase alternating with a haplophase gamete-producing phase.

(1) Mosses. Textbook Fig. 21-13 shows the life cycle of a moss. The familiar, leafy plant that you recognize as moss is the gametophyte, which actually exists in two versions--male and female. The gametophytes are haploid, and each sex produces specialized structures (<u>antheridia</u> in males, <u>archegonia</u> in females) that produce gametes by mitosis. Male gametes are motile; they swim through a thin film of water to the archegonia of female gametophytes where fertilization occurs. Hence, the zygote (the first cell of the sporophyte) is formed and grows on the female gametophyte. The sporophyte grows to form a stalk that ends in a capsule (<u>sporangium</u>) in which meiosis occurs. Meiosis produces haploid spores that are released from the sporangium to germinate and grow into gametophytes, thus completing the cycle.

(2) Ferns. The familiar (and conspicuous) leafy phase of a fern's life cycle is the sporophyte (textbook Fig. 21-14). Dot-like clusters of <u>sporangia</u> on the underside of the leaf are the sites of meiosis. Haploid spores are forcibly ejected from the sporangia into the air; if they land in a suitably moist location, they germinate into small, inconspicuous gametophytes, each bearing male (<u>antheridia</u>) and female (<u>archegonia</u>) reproductive structures. Sperm swim into archegonia to fertilize the ova; these zygotes grow on the gametophyte to develop into the familiar sporophyte. Note how ferns are much like <u>bryophytes</u> (such as the moss above) in that a moist environment is required for sperm to swim to the site of the female gametes.

(3) Seed plants. These are the most evolutionarily advanced plants. The life cycle of seed plants is a reversed version of those for ferns and bryophytes: In seed plants, the gametophyte grows on the sporophyte. The sporophyte is, far and away, the predominant phase of the life cycle. Although our textbook example of the seed plant life cycle will address flowering plants (the <u>angiosperms</u>), do realize that another major group of seed plants exists: these are the <u>conifers</u> (or <u>gymnosperms</u>), a large group of cone-bearing, non-flowering seed plants.

Flowers are the reproductive structures of the sporophyte of a flowering plant. A flower consists of many features that are specialized to produce gametes (<u>stamens</u> in males, <u>pistils</u> in females) and in many cases to attract pollinators (e.g., colorful <u>sepals</u> and <u>petals</u> attract bees, butterflies, birds, bats, etc.). Many flowers lack showy parts and are wind-pollinated. The <u>anther</u> (male) and the <u>ovary</u> (female) are the sporangia because they are the sites of meiosis. Let's look at the steps that form the gametophytes separately for each sex:

Within the anthers of males, diploid <u>microspore mother cells</u> divide meiotically to produce haploid <u>microspores</u>. Each microspore divides mitotically to form a <u>microgametophyte</u> consisting of two cells: a <u>tube cell</u> and a <u>generative cell</u>. A mitotic division of the generative cell produces two <u>sperm cells</u>. The two sperm cells and the single tube cell constitute the mature male gametophyte, a germinated <u>pollen grain</u>.

The female gametophyte develops within the pistil. Only one <u>megaspore mother cell</u> develops in each sporangium (or ovule). This cell divides by meiosis to form four <u>megaspores</u>, only one of which survives. The surviving megaspore divides mitotically to form an eight-celled gametophyte. The <u>embryo sac</u> consists of seven of these nuclei together with an egg formed from the eighth nucleus.

Fertilization is uniquely <u>double</u> in seed plants! After a pollen grain contacts the <u>stigma</u> of the pistil, a pollen tube grows through the pistil into the ovary. One sperm nucleus fertilizes the ovum to form the zygote. The other sperm nucleus fuses with the <u>fusion nucleus</u> (a specialized pair of nuclei from the embryo sac) to form the <u>endosperm</u>, a triploid tissue that stores energy for use by the developing sporophyte. You are familiar with endosperm in the form of the starch of corn and other grains. A <u>seed</u> consists of a sporangium of the earlier sporophyte generation surrounding the new sporophyte and its supply of endosperm. In flowering plants, seeds develop within ovaries. The ovaries surrounding such seeds often develop into <u>fruits</u>.

Be sure to read in your textbook about some of the marvelous adaptations of seeds, fruits, pollen, and flowers that resolve the dilemmas presented by immobility in plants. Read also the note on embryonic development in plants.

F. <u>Reproduction in Fungi</u>.--Depending on the species, fungi (yeasts, molds, mushrooms, etc.) may reproduce sexually, asexually, or parasexually. Asexual reproduction is often the primary mode for many fungi. We'll examine some of the diversity of form and reproduction of fungi in more detail in Chapter 40.

G. <u>Reproduction in Animals</u>.--Several features mark (some uniquely) animal reproduction. Diplophase is dominant. The gametophyte is absent. For the more advanced animal species (e.g., vertebrates), asexual reproduction does not occur.

(1) Asexual reproduction. Although some animals (mainly invertebrates) can reproduce asexually, no animals rely solely on asexual reproduction. A common example is the <u>budding</u> of new individuals from older individuals, as is *Hydra*. Some flatworms (e.g., <u>planarians</u>) can split their bodies to form new individuals by a process called <u>fission</u>.

A more advanced approach that relies in part on asexual reproduction is seen in <u>coelenterates</u> (a group that includes jellyfish). Here, sexual and asexual generations alternate with each other. The prominent phase of this life cycle is almost plant-like in appearance: The branched body, attached to the substrate, is really a colony of many individuals (<u>polyps</u>); these polyps are sexless. <u>Medusas</u> are specialized parts of these colonies. These medusas are either male or female; they swim away from the colony and release gametes into the water. Fusion of male and female gametes produces a zygote which then develops into an attached colony.

(2) Sexual reproduction. A tremendous array of strategies, structures, and behaviors has evolved in animals to facilitate sexual reproduction. A common theme for all animals is that sexual reproduction always occurs in an "aquatic" situation: This is obvious for aquatic species in which gametes of one or both species may be shed into the water outside of the body (a form of <u>external fertilization</u>). But this is less apparent for land-dwelling species; the "aquatic setting" for these species might be contained <u>within</u> an egg (e.g., bird) or fertilization and

development might occur within the body of the female (e.g., most mammals). The amnion, one of several extraembryonic membranes (what are the names and functions of the other membranes?), surrounds the developing embryo of reptiles, birds, and mammals; the amniotic fluid within this sac is the aquatic setting in which the embryo develops. The appearance of this membrane marked the true independence of higher vertebrates from aquatic environments.

The anatomy of reproductive structures directly reflects reproductive strategies. For species with internal fertilization, a copulatory structure (e.g., penis in many species) delivers sperm into a receptacle (e.g., a vagina) in the female wherein the ova are located. Shell glands may be present in many reptiles and birds, or for species in which the developing embryo is held internally, a uterus is characteristic in females.

Larvae represent a significant stage in the life cycles of most animals (including insects and most other invertebrates, many fishes, and many amphibians). A larva is an early developmental stage whose form is not like that of the adult (compare a caterpiller to a moth, or a tadpole to a from). Larval stages may facilitate dispersal (especially important in species that are sessil as adults) or may be the only phase of the life cycle that is capable of feeding. Be sure to study the textbook examples of metamorphosis in amphibians and insects.

H. Mammalian Reproduction.--We can examine humans as an example of mammalian reproduction. In mammals (with few exceptions), fertilization is internal, early development is within the uterus, nutrients cross from mother to embryo via a placenta, young are born rather than hatched, and nourishment of young is via suckling of milk from specialized mammary glands. We'll next review the anatomy and physiology of human reproduction.

(1) Male reproductive system. Both male and female reproductive systems have corresponding roles. These include gametogenesis, copulation, and production of certain sex hormones. In males, meiosis occurs within the seminiferous tubules of the testes. As sperm cells pass from the testes to the penile orifice, several glands (seminal vesicles, prostate, and Cowper's gland) contribute to the fluid which becomes the semen. Secretions of these glands provide for the sperm a medium having the proper nutrition and acid-base balance. Semen is conducted to the exterior by rhythmic contractions of muscles surrounding the urethra, a tube that in males also carries urine. The penis, the copulatory organ, is an example of a hydrostatic skeleton; the structure is rigid when erectile tissues of the corpora cavernosa fill with blood.

The male hormones are produced primarily by endocrine cells (interstitial cells) located in the spaces between adjacent seminiferous tubules in the testes. Testosterones stimulate the development of many male characters, including facial and pubic hair, deepening of voice, muscular development, and others.

(2) Female reproductive system. Ova are produced in a pair of ovaries located in the abdominopelvic cavity. The schedule (ovarian cycle) on which ova mature and are released is regulated by hormones (e.g., estrogen and progesterone), some of which are produced by the ovaries themselves, and by activity of the pituitary gland and hypothalamus. Progesterone affects the uterus such that its lining (endometrium) is prepared to receive an embryo (implantation) in the event that fertilization occurs. In the absence of fertilization, the endometrium breaks down and sloughs away as menstrual flow. This approximately monthly cycling enables humans to mate throughout the year. Most other mammals, in contrast, experience an estrus cycle in which a female is prepared and receptive for mating only once (or a few times) per year.

The gestation period is the time from fertilization until birth (parturition). The length of gestation varies greatly by species, in humans being about 280 days. Birth is preceeded by labor, a series of muscular contractions that eventually expel the fetus from the uterus. Uterine contractions are, in part, under the control of hormones such as oxytocin. Nourishment of the neonate during early post-natal development is by milk produced by mammary glands of the mammae (breasts). Milk production (lactation) and ejection is stimulated, in part, by hormones such as luteotropic hormone and oxytocin.

I. Contraception.--Contraception is a category of methods of birth control. Contraception involves any of a variety of biochemical, physiological, mechanical, and behavioral means of preventing fertilization of an ovum from occurring. These range from abstinence to surgical procedures which prevent an ovum from entering an oviduct (or sperm from leaving the male system, vasectomy). Contraceptive efficiency varies greatly by method used. Each method also has health risks associated with it. Abortion is not appropriately classified as a contraceptive method because the need to employ abortion occurs only if contraception has failed. Although we'll not examine each method in detail here, you can review contraception by working Chart Exercise A in this chapter of the Study Guide.

TERMS TO UNDERSTAND: Define and identify the following terms:

sexual reproduction

asexual reproduction

fission

vegetative reproduction

sporulation

sporophyte

gametophyte

parthenogenesis

automixis

apomixis

agamospermy

conjugation

hermaphrodite

perfect vs. imperfect flowers

complete vs. incomplete flowers

monoecious vs. dioecious

haplophase

diplophase

gametic meiosis

zygotic meiosis

sporic meiosis

alternation of generations

archegonium

gymnosperms vs. angiosperms

sporangia

antheridium

archegonium

pollen

seed

flower

sepal

petal

stamen

filament

anther

pistil

stigma

style

ovary

ovule

megaspore

microspore

microspore mother cell

tube cell

generative cell

megaspore mother cell

embryo sac

fusion nucleus

endosperm

gametangia

hypha

mycelium

parasexuality

homokaryotic

heterokaryosis

heterokaryotic cell

polyp

medusa

larva

metamorphosis

penis

vagina

metamorphosis

molting

amnion

uterus

mammary glands

androgens

testis

spermatic cord

fallopian tube (oviduct)

ovulation

estrogen

progesterone

ovarian cycle

uterine (menstrual) cycle

endometrium

adenohypophysis

follicle-stimulating hormone

luteinizing hormone

luteotrophic hormone

gestation

parturition

neurohypophysis

lactation

contraception

CHART EXERCISES

A. Compare the commonly-used means of contraception listed below. Fill in the chart to indicate (1) the mechanism by which the method works, (2) the device used (if any), (3) when the practice or device is used, and (4) the percent efficiency for each method.

METHOD	MECHANISM	DEVICE USED	WHEN USED	EFFICIENCY
Coitus interruptus				
Rhythm method				
Diaphragm				
Condom				
Intrauterine device				
Sterilization				
Contraceptive pill				

B. Reproduction in mammals, such as humans, is regulated by an array of hormones. Complete the chart below to indicate (1) the sex(es) in which each hormone is found, (2) the structure(s) that produce the hormone, and (3) the physiological effect(s) of each hormone.

HORMONE	SEX	STRUCTURE	EFFECT
Oxytocin			
Testosterone			
Progesterone			
Estrogen			
Follicle-stimulating hormone			
Luteotropic hormone			
Luteinizing hormone			

DISCUSSION QUESTIONS

1. a. Draw a general diagram that shows the life cycle that is characteristic of any species. Briefly discuss the events of each phase of that life cycle.

b. Draw a general diagram of a life cycle that shows both sexual and asexual modes of reproduction. Briefly discuss the events of each phase of that life cycle.

2. Considering species from each of the five kingdoms, discuss the relative importance of motility of gametes to the reproductive strategies found in each kingdom. Include in your response items such as whether gametes of both sexes are motile and whether or not fertilization occurs in an aquatic environment.

3. Considering species from each of the five kingdoms, discuss the taxonomic occurrence of sexual and asexual reproduction. Are there kingdoms in which only sexual or asexual reproduction occurs?

4. a. What advantage(s), if any, does asexual reproduction have over sexual reproduction?

   b. What advantage(s), if any, does sexual reproduction have over asexual reproduction?

5. A variety of reproductive strategies can be found among plants. Compare the differences in the relative prominence the gametophyte phase between bryophytes and angiosperms.

6. Discuss parthenogenesis and hermaphrodism in terms of a trend of regression of sexuality.

7. What problems associated with reproduction are presented by the immobility of plants? How have these problems been resolved?

8. Summarize the events of early development of a seed plant.

9. How was evolution of the amnion important in the invasion of land by vertebrates?

TESTING YOUR UNDERSTANDING

For each of the test items below, three of the lettered alternatives are true and the other is false. Determine which alternative is false and write its letter in the blank to the left of the question number. On the blank line below alternative D, write the corrected version of the false statement.

_____ 1. A. Reproduction is necessary because individuals are not immortal.
B. Asexual reproduction provides means of increasing genetic diversity whereas sexual reproduction does not.
C. Many species reproduce by both sexual and asexual means.
D. Fission is considered to be an asexual means of reproduction.

Correction: _____

_____ 2. A. Some species of animals can reproduce asexually.
B. A sporophyte produces asexual spores whereas a gametophyte produces sex cells.
C. The predominant phase of the life cycle in angiosperms is the sporophyte.
D. In angiosperms, the sporophyte grows upon the tissues of the gametophyte.

Correction: _____

_____ 3. A. Genetic recombination occurs in mitosis but not in meiosis.
B. Sexual reproduction in bacteria occurs by the process of conjugation.
C. Hermaphrodites are individuals that contain sex organs of both sexes.
D. Hermaphrodism occurs in some plant species and in some animal species.

Correction: _____

_____ 4. A. The haplophase segment of the life cycle occurs after meiosis but before fertilization.
B. In gametic meiosis, the gametes are the only haploid cells; hence, haplophase is a very brief segment of the life cycle.
C. Zygotic meiosis is the type of life cycle that occurs in humans.
D. Diplophase occurs after fertilization but before meiosis in the offspring.

Correction: _____

_____ 5. A. The archegonium is composed of haploid cells.
B. The spores of the green alga *Chlamydomonas* are motile.
C. In sea lettuce *(Ulva)*, both sporophyte and gametophyte grow into prominent, leafy plants.
D. The life cycle of the brown alga *Fucus* includes no sporophytic generation.

Correction: _____

_____ 6. A. The bryophytes are multicellular plants that lack vascular tissue.
B. Tracheophytes are multicellular plants that have vascular tissue.
C. Angiosperms are tracheophytes.
D. Mosses are tracheophytes.

Correction: _____

_____ 7. A. In plants, haploid gametes are directly produced by meiotic divisions.
B. In the plant life cycle, a diplophase spore-producing generation alternates with a haplophase gamete-producing generation.
C. The antheridium is the organ of a gametophyte in which sperm cells are produced.
D. The sperm of bryophytes and of some tracheophytes are flagellated.

Correction: _____

_____ 8. A. In the fern life cycle, the sporangia on the undersurface of the leaves are the sites of meiosis.
B. The mature male gametophyte of a flowering plant is the pollen grain.
C. Each microspore mother cell of a male angiosperm divides meiotically to produce four haploid microspores.
D. Each microspore of a male angiosperm divides to form a microgametophyte consisting of a tube cell and a generative cell.

Correction: _____

_____ 9. A. Only one of the four cells produced by meiotic division of a megaspore mother cell survives as a megaspore.
B. The female sporophyte forms by mitotic division of a megaspore.
C. The eight-celled embryo sac comprises the female gametophyte of a flowering plant.
D. In flowering plants, the mature male gametophyte is motile whereas the mature female gametophyte is not motile.

Correction: _____

_____ 10. A. The double fertilization event of flowering plants produces the zygote and the first cell of endosperm.
B. A seed constitutes an embryonic sporophyte surrounded by the sporangium of the previous sporophytic generation.
C. Fusion nucleus is the term applied to the product of the union of a sperm nucleus and an ovum nucleus.
D. Endosperm tissue is triploid.

Correction: _____

_____ 11. A. The designs of seeds have evolved to facilitate dispersal of seeds from parent plants.
B. All of the following are agents that are responsible for transport of pollen to the pistil: insects, wind, water, and bats.
C. One of the roles of petals and sepals is to attract pollinators.
D. All of the following are components of the female reproductive organ: carpels, filament, stigma, and ovary.

Correction: _____

_____ 12. A. The attached phase of the life cycle of the coelenterate Obelia represents a colony of medusas.
B. In parthenogenetic species, eggs are produced but they develop without fertilization.
C. The reproductive strategy of hermaphroditic self-fertilization is advantageous for species in which an individual is unlikely to find a mate.
D. Fertilization and early development of vertebrates always occurs in an aquatic setting.

Correction: _____

_____ 13. A. The following is the correct sequence of metamorphosis in insects: egg to larva to pupa to imago.
B. Molting is the process of shedding of an exoskeleton.
C. All of the following are membranes associated with the embryos of mammals: allantois, chorion, amnion, and epididymis.
D. The developing embryo of a mammal implants into the endometrium of the uterus.

Correction: _____

_____ 14. A. A young mammal receives nourishment from its mother through both a placenta and from mammary glands.
B. The following is a correct sequence of structures through which reproductive products of a mammalian male pass: seminiferous tubules, epididymis, vas deferentia, and urethra.
C. The following is a correct sequence of structures through which reproductive products of a mammalian female pass: oviducts, uterus, cervix, and vagina.
D. Development of the lining of the mammalian uterus is regulated by hormones such as oxytocin and estrogen.

Correction: _____

_____ 15. A. The most effective method of contraception is complete abstinence from sexual intercourse.
B. The following are mechanical means for preventing sperm from entering the uterus: condom, intrauterine device, and diaphragm.
C. Vasectomy is a surgical procedure in which the vasa deferentia are severed.
D. Use of the "pill" allows artificial duplication of hormonal conditions during which ovulation does not occur.

Correction: _____

CHAPTER 22--BLOOD

CHAPTER OVERVIEW

This chapter begins our study of homeostasis, the maintenance of a steady internal state in the face of changing conditions. We focus here on the role of blood in homeostasis. Although this chapter uses humans in most examples, you should realize that most animal species possess blood of one kind or another. The blood of these other species allows those organisms to face and resolve similar needs and problems that we humans encounter.

Blood is different from most other tissues in that it is fluid; blood cells are not joined to each other as are cells that form most other tissues. Among the many roles of blood is transport of an array of materials, including hormones, respiratory gases, glucose, and many others. Blood also is involved in defending the body immunologically from invasion by microorganisms and other foreign materials. Blood contains factors that prevent bleeding via a complex clotting system. Another major function of blood is the regulation of salt and water distribution, temperature, and acid-base balance within the body. In this chapter, we examine the composition of both the fluid (plasma) and cellular portions of blood and the chemical mechanisms of blood clotting.

TOPIC SUMMARIES

A. Biological Significance of Blood.--All but the smallest animals have an internal fluid called blood (or some other term appropriate for that kind of animal). Why is such a fluid important? Among other functions, blood is the primary medium for transporting materials (e.g., respiratory gases, nutrients, wastes, water, heat, hormones, etc.) within an animal's body. The need for such a transport medium arose as animal bodies increased in size. Small, thin animals have large surface areas compared to the volumes of their bodies. In such animals, the surface area is great enough for needed materials simply to diffuse from the exterior across the skin to the interior as fast as they are needed. However, larger animals (having a smaller ratio of surface area to body volume) would starve or overheat if no special transport medium were available. We'll examine in a later chapter the systems of tubes within which blood flows.

Other functions of blood include defense mechanisms such as a system for clotting of blood and the immune system which protects the body from invasion by other organisms or other foreign materials. Regulation of many aspects of the internal environment (e.g., acid-base balance, temperature, distribution of salts and water) comprises the final major function of blood. We'll look at some of these in more detail below.

B. Physical Characteristics of Blood.--Blood, a type of connective tissue, has the peculiar quality of being a liquid tissue. Being a tissue, blood is made in part of cells; the liquid intercellular material is plasma. In humans, slightly over half of the blood volume is made of three major cell types: erythrocytes (red blood cells), leukocytes (white blood cells), and thrombocytes (platelets). Each of these cell types has specific functions to be examined in more detail below. The plasma consists primarily of water in which is dissolved or suspended many chemicals, including fibrinogen (for clotting), albumins (for maintaining osmotic pressure), globulins (for the immune system), vitamins, glucose, salts, hormones, etc.

C. Formation of Blood Cells.--The embryonic stem cells are a small group of specialized mesodermal cells from which all other blood-producing cells arise. As with most other tissues of the body, there is a need to continuously produce new blood cells throughout the life of the human organism. The process of blood-cell formation is called hemopoiesis. At different ages, different organs are involved in hemopoiesis. Hemopoietic tissues active before birth include the liver, bone marrow, spleen, lymph nodes, and thymus. After birth, only bone marrow continues this activity.

D. <u>Erythrocytes and Oxygen Transport</u>.--RBC's, as they are affectionately called, are a curious type of cell in that mature erythrocytes can<u>not</u> mitotically divide to produce more RBC's. Rather, new RBC's are produced by division of precursor cells of bone marrow. Mature RBC's cannot divide because they lack nuclei. The absence of a nucleus, which imparts a biconcave disc shape to the cell, enhances the ability of these cells to carry out their primary role: transport of oxygen. The enucleate condition allows more efficient packing of <u>hemoglobin</u> molecules into RBC's. What would happen to hemoglobin molecules if they simply were dissolved in plasma?

Hemoglobin (abbreviated Hb) is the primary <u>respiratory pigment</u> in most vertebrate species. Hb is an example of a category of conjugated proteins called <u>chromoproteins</u>. These consist of protein subunits (<u>globins</u>) surrounding a porphyrin derivative (a <u>heme</u> group). The metal in Hb is iron. Different chromoproteins, such as <u>hemocyanin</u> which contains copper instead of iron, are found in some invertebrate species.

The behavior of a respiratory pigment in binding oxygen molecules is depicted by <u>oxygen dissociation curves</u> (Fig. 22.1). The curve shows the percent of pigment molecules that are saturated by oxygen at various levels of oxygen availability. Different kinds of pigments have different strengths of attraction for oxygen. The relative affinity of a pigment for oxygen is generally indicated by the oxygen level at which 50% saturation occurs. Thus, the fetal Hb molecule has a greater affinity for oxygen than does adult Hb.

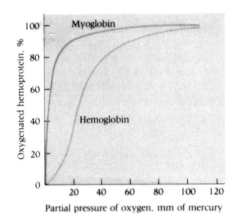

Fig. 22.1: Oxygen dissociation curves for hemoglobin and myoglobin

<u>Myoglobin</u> is another respiratory pigment. It is found in greatest concentrations in muscle tissue, a tissue where oxygen demand is great. Myoglobin can strip oxygen from Hb. In relation to the Hb curve in Fig. 22.1, would the dissociation curve for myoglobin be positioned to the left or to the right?

E. <u>Leukocytes</u>.--Compared to RBC's, white blood cells are very different in structure and function. Leukocytes retain their nuclei. Some types of leukocytes resemble amebae in terms of cell shape. Indeed, their capacity for <u>ameboid movement</u> enables them to protect the body from bacterial invasion or infection. Much of the protective work of leukocytes is done in the tissues, rather than within the bloodstream; ameboid movements allow leukocytes to squeeze between the cells of the capillary walls and pass out into the tissues.

The many types of white blood cells are classified in various schemes on the basis of their structure, source, and function. In one scheme, WBC's containing cytoplasmic granules are called granulocytes; nongranulocytes lack such granules. Staining properties of granules allow further subdivision of granulocytes. In another scheme, the shape of the nucleus distinguishes mononuclear and polymorphonuclear leukocytes. In terms of tissues of origin, leukocytes are called myeloid cells if they come from bone marrow or lymphoid cells if from lymphoid tissue. A final classification reflects the importance of phagocytosis in cell function; phagocytic cells employ phagocytosis whereas nonphagocytes do not.

Localized inflammation is a common response to invasion or injury of the body. Certain types of leukocytes have a direct role in inflammation. Mast cells in damaged tissues release histamine which increases blood flow to the site (via vasodilation) and which causes plasma to leak into the tissues. The gathering of a variety of leukocytes (e.g., macrophages, granulocytes) at the injury site further enhances swelling.

Three cascades, or reaction sequences, are initiated by the inflammation reaction. The first cascade releases kinins, a group of peptides that function much like histamine in increasing vascular permeability and vasodilation. The second cascade activates the complement, a series of 18 proteins with a variety of effects directed towards protection against bacterial infection. The final cascade is a clotting reaction that plugs bleeding blood vessels.

Phagocytosis is the means by which many types of leukocytes function. The first step in the process is reception by the phagocytes of chemical signals indicating the entry of an invader. Leukocytes then move, via chemotaxis, toward the site of invasion. Phagocytic cells then engulf (ingest) and enzymatically digest the foreign material.

F. Platelets.--Also known as thrombocytes, platelets are enucleate fragments of megakaryocytes, large cells found in bone marrow. Platelets play an essential role in hemostasis, the entire process by which bleeding from a damaged blood vessel is stopped. Several special properties enable this role in hemostasis: Platelets adhere to each other and to the walls of injured blood vessels to form a temporary hemostatic plug. They promote activation of clotting proteins in the plasma. They help to maintain the integrity of the endothelium of blood vessels. Platelets produce chemical mediators that initiate repair of blood vessel walls and regulate inflammatory reactions.

G. Plasma, the Noncellular Component of Blood.--The components of plasma are divisible into three categories: Water composes well over 90% of plasma volume; water is the solvent in which other substances are carried (see textbook Table 22-2). Additionally, water is required for metabolic processes in every living cell. An array of proteins is dissolved or otherwise carried in the plasma; we'll examine some of the protein functions below. The third group of plasma materials is a myriad of non-protein molecules and ions being transported to and from the tissues; these include nutrients, salts, metabolites, hormones, and others.

The roles of proteins in defense of the body against invasion already have been mentioned in this chapter and will be studied in greater detail in the next chapter (Immunity). Many of these immunological proteins are grouped as globulins. Certain other globulins work as carriers: They may, for example, bind to metals (iron, copper, etc.) that otherwise would be toxic to the body, or they may attach to lipids to enhance their solubility. Below we'll review the role of proteins in hemostasis. Let's note another critical function of plasma proteins--that of maintaining osmotic balance between blood within the vessels and fluids within the tissues (i.e., external to blood vessels). An insufficient amount of albumins within the plasma can cause an accumulation of fluids within the tissues, a condition known as edema. A consequence of edema is reduced blood volume and, therefore, decreased blood pressure. Albumins constitute up to 60% of total plasma proteins.

H. <u>Blood Clotting</u>.--The purpose of clotting is to stop bleeding. Hemostasis involves two stages: In the <u>primary stage</u>, blood vessels in the immediate vicinity of an injury constrict; such <u>vasoconstriction</u> decreases blood flow to the injured area. Also in the primary stage, <u>platelets</u> are attracted to and adhere to collagen fibers exposed by injury to the blood vessel. These platelets then swell and release various chemicals (e.g., <u>serotonin</u>, ADP, various proteins) which prolong local vasoconstriction and cause additional platelets to aggregate. The result is a <u>temporary plug</u>. The <u>secondary stage</u> of clotting involves formation of a more durable, long lasting <u>clot of fibrin</u> within which erythrocytes are entrapped.

Blood clotting is <u>not</u> a simple process. Rather, nearly 20 different proteins and other agents are involved in a <u>cascading sequence</u> for a clot to form. These agents are labelled as numbered <u>factors</u>, each being activated by a factor earlier in the cascade and each activating a factor later in the cascade. Interestingly, two different mechanisms can activate the clotting reaction. As noted in the preceding paragraph, injury to a blood vessel can trigger clotting; this is the <u>intrinsic mechanism</u>. Alternatively (via the <u>extrinsic mechanism</u>), injury to other tissues results in the release of a glycoprotein called <u>tissue factor</u> (which promotes the clotting reaction) and the production of <u>activated factor X</u>, a member of phase 1 of the cascade (text Fig. 22-21). In either case, however, the final steps of the cascade involve conversion of <u>prothrombin</u> into <u>thrombin</u> which then causes soluble <u>fibrinogen</u> to polymerize into insoluble <u>fibrin</u>. Fibrin forms the protein mesh of the persistent clot.

<u>Fibrinolysis</u> occurs after injured tissues have healed. It is the process by which the clot <u>lyses</u> (dissolves). <u>Plasmin</u>, the primary agent that catalyzes fibrinolysis, is produced by a separate reaction cascade.

It is important to realize that both the clotting and fibrinolysis processes are (of necessity) <u>highly regulated processes</u>. They are regulated in both location and timing. What would happen if clotting occurred in blood vessels where no injury exists? What would happen if clots were not resorbed after injuries were healed? Your textbook offers insight into the need for such orchestration.

TERMS TO UNDERSTAND: Define and identify the following terms:

homeostasis

plasma

erythrocytes

leukocytes

platelets (thrombocytes)

serum

fibrinogen

fibrin

hematocrit

stem cells

hemoglobin

hematopoiesis

methemoglobin

haptoglobin

chromoproteins

hemoproteins

hemocyanin

oxygen dissociation curve

myoglobin

granulocytes

neutrophils

eosinophils

basophils

nongranulocytes

lymphocytes

monocytes

myeloid cells

lymphoid cells

phagocytosis

mast cells

histamine

macrophages

monocytes

reticuloendothelial system

kinins

complement

hemostasis

serotonin

amebocyte

megakaryocyte

endomitosis

release reaction

transport proteins

antibodies

albumin

thrombin

zymogens

fibrinolysis

plasminogen

plasmin

CHART EXERCISE

Many different types of cells compose the cellular fraction of blood. Complete the chart below to indicate for each major cell type (1) whether a nucleus is present, (2) the type of movement of which that cell type is capable, (3) the density in number of cells per mm^3, (4) function(s) of that cell type, and (5) examples of specific kinds of cells classified into that cell type.

CHARACTERISTIC	ERYTHROCYTES	LEUKOCYTES	THROMBOCYTES
Nucleus present?			
Type of motility			
Density			
Function(s)			
Specific examples			

DISCUSSION QUESTIONS

1. In what ways do the size and shape of an organism correspond to the presence and degree of development of blood and a circulatory system? If neither blood nor a circulatory system were present, then how would materials be transported to each living cell of the organism?

2. Compare the respiratory pigments hemoglobin and myoglobin in terms of the following:

   a. their chemical structure,

   b. their functions, and

   c. their relative abilities to transport oxygen. [In answering part c, you might wish to draw oxygen dissociation curves for these pigments.]

3. In many species (such as humans), the sites of hemopoiesis vary according to age of an individual. Briefly summarize below for various ages which tissues are involved in the production of new blood cells.

4. Animals have at least three major means of defending against infection. Briefly discuss these mechanisms below.

5. What is meant by a cascade of reactions? Summarize below at least two such cascade processes related to blood.

6. What is the adaptive value of having the processes of blood clotting and of fibrinolysis being effectively regulated in both time and space?

7. Discuss the role of plasma proteins in maintaining the proper balance of water within blood vessels as compared to within the tissues surrounding these vessels.

TESTING YOUR UNDERSTANDING

For each of the test items below, three of the lettered alternatives are true and the other is false. Determine which alternative is false and write its letter in the blank to the left of the question number. On the blank line below alternative D, write the corrected version of the false statement.

_____ 1. A. Fibrinogen is an insoluble protein involved in blood clotting.
B. Plasminogen is a soluble protein involved in fibrinolysis.
C. Thrombin promotes formation of a blood clot.
D. Hemoglobin is a respiratory pigment involved in transport of oxygen.

Correction: _____

_____ 2. A. A function of blood plasma is the transport of respiratory gases.
B. A function of blood is protection of the body from infection via phagocytosis.
C. Blood functions to transport heat.
D. Various materials transported by blood include hormones and metabolic wastes.

Correction: _____

_____ 3. A. Erythrocytes are enucleate.
B. Platelets lack nuclei.
C. Leukocytes do not possess nuclei.
D. Thrombocytes are involved in clotting of blood.

Correction: _____

_____ 4. A. All blood cells originate from a very small population of multipotent stem cells that form very early in embryonic life.
B. All of the following are hemopoietic tissues or organs: spleen, lymph nodes, thymus.
C. After birth, hemopoiesis occurs primarily in bone marrow.
D. Blood vessels and blood cells develop from endoderm.

Correction: _____

_____ 5. A. Leukocytes are capable of amoeboid movement.
B. Mature erythrocytes divide by mitosis to produce new red blood cells.
C. Thrombocytes are fragments of megakaryocytes.
D. Although hemoglobin in humans is carried in erythrocytes, the hemocyanin of some invertebrates is merely dissolved within the plasma.

Correction: _____

_____ 6. A. The normal erythrocyte density in human blood is about 5 million cells per cubic milliliter.
B. Edema is a condition resulting from a low red-blood-cell count.
C. The active sites of various hemoprotein molecules contain metal atoms such as iron and copper.
D. Myoglobin has a greater affinity for oxygen than does hemoglobin.

Correction: _____

_____ 7. A. Carbon monoxide is a poison because hemoglobin has a greater affinity for carbon monoxide than it does for oxygen.
   B. Fetal hemoglobin has a higher affinity for oxygen than does adult hemoglobin.
   C. Sickle hemoglobin and thalassemia are two of the many conditions based on abnormalities in hemoglobin.
   D. On a graph of percent saturation with oxygen against concentration of oxygen, the oxygen dissociation curve of a pigment with a high affinity for oxygen would be located to the right of the curve for a pigment with low oxygen affinity.

   Correction: _____

_____ 8. A. The reducing power of glycolysis occurring within erythrocytes prevents hemoglobin from being oxidized into methemoglobin.
   B. Leukocytes can be classified in a variety of ways on the basis of characteristics including function, structure, and developmental source.
   C. Myeloid leukocytes develop from bone marrow.
   D. A neutrophil is a type of leukocyte that lacks cytoplasmic granules.

   Correction: _____

_____ 9. A. The density of leukocytes in normal blood is greater than the density for erythrocytes.
   B. Inflammation is a normal reaction to all forms of injury.
   C. Metchnikoff demonstrated the importance of phagocytosis in body defense.
   D. Some leukocytes defend the body via phagocytosis.

   Correction: _____

_____ 10. A. Histamine enhances local blood flow.
   B. Histamine increases the permeability of blood vessels in the vicinity of an injury.
   C. As part of the inflammation response to an injury, blood vessels in the area vasodilate to allow plasma to leak into the tissues; this condition of fluid accumulation in the tissues is called anemia.
   D. The first phase of phagocytosis is chemotaxis.

   Correction: _____

_____ 11. A. The first reaction cascade of inflammation response leads to the release of kinins.
   B. Kinins cause increased vascular permeability and vasodilation.
   C. The second cascade of inflammation activates plasma proteins collectively called complement.
   D. The complement proteins promote the clotting reaction.

   Correction: _____

_____ 12. A. The role of platelets in hemostasis is to form the hemostatic plug.
B. Platelets produce mediator substances, such as opsonins, substances that constrict small blood vessels to slow bleeding.
C. The platelet count in normal blood is greater than the leukocyte count, but lower than the erythrocyte count.
D. Some invertebrate species, such as the horseshoe crab, possess only one type of blood cell.

Correction: _____

_____ 13. A. The three major groups of plasma proteins are albumin, globulins, and fibrinogen.
B. All of the following proteins are types of globulins: antibodies, clotting factors, glycoproteins, and metal-binding proteins.
C. Albumin is the most abundant protein in blood.
D. Lipoproteins serve primarily to maintain proper osmotic pressure and blood volume.

Correction: _____

_____ 14. A. The importance of collagen to hemostasis is that platelets are attracted to collagen and they adhere to collagen.
B. The clotting reaction can be stimulated intrinsically by injury to a blood vessel wall and extrinsically by injury to non-vascular tissues.
C. The primary stage of hemostasis involves production of a meshwork of fibrin strands.
D. Hemophilia is a condition in which clotting time is prolonged.

Correction: _____

_____ 15. A. The conversion of fibrinogen into fibrin is catalyzed by prothrombin.
B. The widespread presence of thrombin in the bloodstream would cause widespread clotting.
C. A value of cascade arrangements for certain processes is that a cascade amplifies small effects.
D. A value of cascade arrangements for certain processes is that it provides many opportunities for the process to be limited and modulated.

Correction: _____

CHAPTER 23--IMMUNITY

CHAPTER OVERVIEW

In this chapter, we examine the immunological system, the system that protects a body from invasion by microorganisms or other foreign materials. The system operates through both cellular and noncellular means. Noncellular strategies involve activity of antibodies, chemicals that enlist other agents to protect the body. The cellular approach involves phagocytosis as well as cell-mediated chemical attack.

The system is complex and highly specific. The system can chemically recognize specific invader cells on the basis of molecules embedded in the cell membrane of the invader. In this chapter, we briefly present information that describes how specific antibodies are encoded into the genome. In doing so, we can explain how a vast array (hundreds of millions) of different kinds of antibodies, each directed against a specific antigen, can be produced. Additionally, we examine the roles of the different types of cells (e.g., T cells, B cells) that are the central players in this system.

Knowledge of the immune system has tremendous practical application, primarily related to medicine. Medical areas related to immunology include tissue transplants, blood transfusions, numerous autoimmune dieseases, and AIDS (acquired immune deficiency syndrome; textbook Box 23-3).

TOPIC SUMMARIES

A. <u>Early Theories of Immunity</u>.--Scientific study and experimentation in <u>immunity</u> began only two centuries ago. Even so, the ancients realized long before that people who had experienced certain diseases generally did not catch those diseases again (or at least did not have severe reoccurrences of those diseases). It is the immunological system that provides such protection against invasion by microorganisms and other foreign materials.

Two theories were proposed to explain how immunity operates. One, the <u>humoral theory of immunity</u>, held that noncellular substances in the blood plasma were the protective agents (called <u>antibodies</u>). Any material that stimulates the body to produce antibodies is called an <u>antigen</u>. The other, the <u>cellular theory of immunity</u>, held that certain blood cells (<u>leukocytes</u>) protected the body via directly attacking invaders. The evidence supported both theories, and modern views of immunity embrace aspects of both of these early theories.

B. <u>Humoral Immunity</u>.--This type of immunity is based on the activities and properties of antigens and antibodies. Such antibodies may be produced by one's own body (<u>active immunity</u>) or they can be transferred (via the placenta or suckling or via injections of <u>transfer serum</u>) from an individual that already is immune (<u>passive immunity</u>). Artificially-induced passive immunity is useful medically because of its immediate effects, but it is temporary. Lasting active immunity is built up over time.

<u>Hypersensitivity</u> is a non-beneficial manifestation of humoral immunity. The first contact an individual has with some foreign substance is harmless, producing no apparent reaction. Subsequent exposure elicits a reaction that sometimes is quite serious. In immediate hypersensitivity (also called <u>antibody-mediated hypersensitivity</u> or <u>anaphylaxis</u>), the antigen-antibody complex stimulates <u>mast cells</u> to release harmful substances, such as <u>histamine</u> (refer to the previous chapter if you don't recall the role of histamine).

(1) Characteristics and types of antibodies. All known antibodies are glycoproteins found in plasma and collectively are called immunoglobulins. Antibody molecules are made of subunits that, on the basis of molecular weight, are called light (L) chains and heavy (H) chains. The ends of the L and H chains are unique for each antibody; these ends confer the specificity of each antibody. The antibodies are classified into five groups (IgG, IgA, IgM, IgD, IgE) according to characteristics of H chains.

Antibodies have several capabilities: They opsonize bacteria. They neutralize bacterial toxins. They enlist agents (such as complement and phagocytes) to kill and lyse bacteria in serum. They can passively induce allergic and anaphylactic reactions.

(2) Characteristics and types of antigens. Although antigens usually are proteins, polysaccharides and DNA can have antigenic effects. As yet, no particular chemical feature has been identified that describes a molecule as an antigen. However, it is known that only a small portion of the molecule (the antigenic determinant, usually only a few amino acids long) is involved in stimulating antibody formation. This is the site to which the antibody molecule binds. Interestingly, most natural antigens possess more than one antigenic determinant.

(3) Antibody production. Antibodies do not appear immediately following initial exposure to an antigen. Rather, there is a latent period, usually a week or more, required for the antibody-synthesizing apparatus to produce antibody molecules (textbook Fig. 23-6). This primary response consists mainly of IgM antibodies and is shortlived. After antibody levels decline to undetectable levels, a second exposure to the antigen causes the secondary (anamnestic) response which is faster and stronger. The antibody produced by the secondary response is an IgG. Because this response is longer-lived than the first, it is referred to as immunological memory. This is the principle used in medical immunizations.

Lymphocytes of two major classes interact to produce antibodies. T cells of the thymus mature by interacting with stromal cells of the thymus. Mature T cells then move into the bloodstream and circulate between lymph nodes and the blood. The three categories of T cells correspond to major functions. T helper cells help other immunological cells to conduct their jobs. T suppressor cells limit or terminate the immune response. T cytolytic cells (also called killer T cells) can recognize antigens of abnormal cells, to which they attach and then destroy.

B cells develop from precursor cells found in lymphoid tissue and in bone marrow. Unlike T cells, B cells have immunoglobulins on their surfaces. The major role of B cells is antibody production. The first step in the process is antigen trapping, in which specific antigens bind with antibody molecules that are attached to the B cell's surface. Next, the B cell transforms into a plasma cell that actively produces and releases antibody molecules.

(4) Theories of immunity. Theories proposed to explain how immunity operates must address how it can distinguish tissues of its own body (self) from foreign tissues (non-self). Such an explanation must examine the great diversity of antibody specificities. The various instructive theories postulated that the body possesses a general antibody-synthesizing apparatus that is capable of producing antibody for any antigen; the antigen serves as an "instructor" that guides the apparatus to produce the appropriate antibody. In contrast, selective theories hold that the body possesses a massive "dictionary" of "words" labelling all possible non-self antigenic determinants. Appropriate antibodies are produced upon exposure to the corresponding antigens.

(5) Mechanisms of immunity. Experimentation ultimately supported the selection theories. B and T cells possess on their surfaces proteins (antigen receptors) that allow them to recognize specific foreign antigens. The mechanisms of recognition differ, however, for B and T cells. B-cell receptors are monospecific immunoglobulins; they can respond only to one

antigen. T cells can recognize only those antigens that are bound to cells. The T-cell receptors are proteins (but are not immunoglobulins) that recognize both the antigen and the MHC (major histocompatibility complex). The MHC apparently is unique to each individual organism; T cells recognize cells of the body to which it belongs on the basis of the MHC, and, hence can distinguish self from non-self.

Following antigen recognition, the selected B cells proliferate into large populations of daughter cells having the same antigen specificity. The selected T cells divide and differentiate into corresponding helper, suppressor, and cytolytic T cells. The complex of B cells having antigens already bound to the receptors with helper T cells causes the release of lymphokines that enable the B cell to become a plasma cell which then produces and releases antibody. Fig. 23.1 depicts this complex.

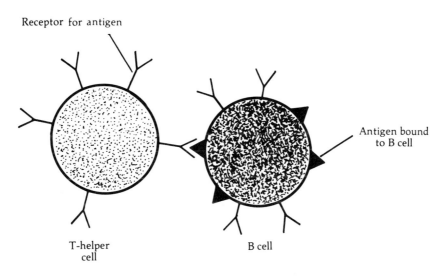

Fig. 23.1: Model of B and T cells joined by antigen

(6) The diversity of antibodies. Once the vast diversity of antibodies was realized (perhaps in the hundreds of millions), researchers searched for the mechanism by which the genome encoded these proteins. Recall that antibody specificity lies in features of the light and heavy chains of the molecules. It turns out that the light and the heavy chains of antibodies of each immunoglobulin class contain two regions: constant (C) regions having the same amino-acid sequence for all antibodies in that class, and variable (V) regions that differ in amino-acid sequence in accordance to antibody specificity.

Studies using recombinant DNA technology showed that V and C regions of light chains are encoded by separate genes on the same chromosome (see textbook Fig. 23-23). A variable region of a light chain consists of the union (via somatic recombinational events) of one of many V segments (and its associated L [leader] segment) with one of several J (joining) segments; the L segment is separated from the the V segment by intervening noninformational DNA. The constant region is encoded by a gene lying several kilobases to the 3' side of the J segment. As yet another source of diversity, single-point mutations occur within the V region. Heavy chains allow for even more combinations than do light chains; heavy chains consist of V, D (diversity), and J segments joined to the constant region (see textbook Fig. 23-24).

C. Cell-Mediated Immunity.--Aspects of immunity that are cell-mediated include aiding in defense against microorganisms, responsibility for cell-mediated allergies, rejection of tissue transplants and graft-versus-host reaction, resistance to malignant tumor growth, and causing many autoimmune diseases.

(1) Lymphokines and monokines. Lymphokines (noted in B4 above) and monokines (produced by monocytes and macrophages) are chemicals that do not directly attack antigens. Rather, they are factors that enhance the weak signals of an antigenic challenge, and, therby stimulate proliferation of appropriate immunological cells. Apparently, a large number of these factors exists. Most are soluble proteins of low molecular weight, and some have been proposed as prospective cures for various cancers.

(2) Tissue transplantation. We already have noted the concept of "self" vs. "non-self." This concept is visibly manifested in <u>tissue transplantation</u>. Not only can the body distinguish its own species from another species, but even from other individuals of its own species. The lone exception is for genetically identical twins. Skin grafts and organ or tissue transplants between individuals of the same species generally fail unless the recipient's immunological efforts to <u>reject</u> the transplant are artificially suppressed. Immunological rejection generally does not occur in transplants between genetically identical twins (<u>isogeneic grafts</u>) or within the same individual (<u>syngeneic grafts</u>).

Studies by Burnet, by Medawar, and by Billingham and Brent indicated that the immunological definition of self comes into existence during later stages of embryological development or even slightly after birth. Tissue transplants during this early period are not rejected, but instead are incorporated and eventually become part of self. The value of such an arrangement is to prevent the body from reacting against its own constituents.

The major histocompatibility complex (MHC; see B5 above) defines self. The MHC is a gene cluster that encodes the cell-surface antigens responsible for graft rejection. Allogeneic graft rejection is mediated by sensitized lymphoid cells. A recipient that has been sensitized to tissues from a donor will rapidly reject (via second-set reaction) any subsequent transplant of any tissue from that donor. The chances of successful transplantation increase with the degree of donor and recipient MHC compatibility.

D. <u>Immunologically Mediated Diseases</u>.--Many diseases are based in imbalances of the immune system. Examples of such autoimmune diseases, in which antibodies (autoantibodies) are produced against self, include rheumatoid arthritis, myasthenia gravis, and lupus erythematosis (textbook Table 23-3). Production of autoantibodies can occur via (a) attachment of some agent (e.g., a drug) to a self cell that then elicits an immune response, (b) reaction to a self antigen that normally is not accessible to the immune system, and (c) general autoimmunity directed against normal body parts throughout the body.

E. <u>Blood Groups and Blood Transfusion</u>.--In earlier chapters, we have commented on the diversity of types of blood, for example, the A, B, O system, the M and N system, and Rh system. Definition of these groups is based on reactions between antigens on the surface of erythrocytes with antibodies in the testing sera. So many different blood group systems exist that they provide a distinct "fingerprint" for each individual.

Understanding of these blood groups has many applications. Because the genes related to some diseases are linked to certain blood-group genes, the presence of a particular blood group can be useful in predicting likelihood of an individual having that disease. Study of the pattern of geographic distribution of blood groups can be very enlightening in studies of evolution of humans (and presumably of other species). Of direct medical importance, understanding of blood-group compatibilities is critical in making safe transfusions.

TERMS TO UNDERSTAND: Define and identify the following terms:

immunology

vaccination

antibody

humoral defenses

cellular defenses

toxin

antitoxin

antigen

hypersensitivity

complement fixation

anaphylaxis

histamine

mast cells

tuberculin reaction

immunoglobulins

antiserum

antigenic determinant

monoclonal antibodies

primary response

secondary (anamnestic) response

immunological memory

T (thymus) lymphocytes

stromal cells

B lymphocytes

bursa of Fabricius

lymphoid follicles

plasma cells

instructive theory

selective theory

antigen receptor

major histocompatibility complex

lymphokines

pre-B lymphocyte

immunocompetent B cell

somatic recombinational event

joining segments

leader segment

single-mutation point

T-cell replacing factor

interleukin 2

rejection

chimera

histocompatibility antigens

autoimmunity

agglutinins

titer

immunoelectrophoresis

PEOPLE TO KNOW: What is the significant biological contribution of each of these individuals as indicated in Chapter 23?

Edward Jenner

Paul Ehrlich

C. F. von Pirquet

Neils K. Jerne

Joshua Lederberg

Sir McFarlane Burnet

W. J. Dreyer and J. C. Bennett

Philip Leder and Susumu Tonegawa

Peter Medawar

Karl Landensteiner and Alexander Wiener

Louis Montagnier

CHART EXERCISE

A variety of kinds of lymphocytes is responsible for immunological protection. Several kinds of lymphocytes are listed in the left column. For each cell type, complete the chart to indicate (1) the chemicals produced (if any), (2) the function(s), and (3) the mechanism of antigenic recognition.

CELL TYPE	CHEMICALS PRODUCED	FUNCTION	MECHANISM
T-helper			
T-suppressor			
T-cytolytic			
B-lymphocyte			
Plasma cell			

DISCUSSION QUESTIONS

1. Distinguish between active immunity and passive immunity.

2. What is the immunological significance of becoming sensitized?

3. Compare and contrast the chemical composition of the light chains and heavy chains of antibody molecules.

4. Explain how the genes encoding the structures of antibody molecules can account for the tremendous number of different kinds of antibody molecules. What is the role of somatic recombination in producing diversity? (In your answer, you will find it useful to draw diagrams showing genes.)

5. Select any of the immunological diseases mentioned in Chapter 23. Go to the library to find the current status of research into the genetics of the disease and information on the genetic basis for that disease. Summarize your findings in the space below.

6. How is the role of an antibody different from the role of a lymphokine?

7. How is study of chimeras important in understanding how the immunological system works?

TESTING YOUR UNDERSTANDING

For each of the test items below, three of the lettered alternatives are true and the other is false. Determine which alternative is false and write its letter in the blank to the left of the question number. On the blank line below alternative D, write the corrected version of the false statement.

_____ 1. A. Immunization against disease can involve injection of attenuated forms of the organisms that cause that disease.
B. Early in the history of the field of immunity, two theories were proposed to explain how immunity worked; later research showed that aspects of both theories were correct.
C. Immunity obtained by inoculation is called active immunity.
D. A non-beneficial form of immediate hypersensitivity is called anaphylaxis.

Correction: _____

_____ 2. A. During anaphylaxis, plasma cells release harmful substances such as histamine.
B. All antibodies belong to a class of plasma globulins known collectively as immunoglobulins.
C. All immunoglobulins contain carbohydrates, and hence can also be called glycoproteins.
D. IgG immunoglobulins possess both light chains and heavy chains, whereas IgM immunoglobulins possess light chains but lack heavy chains.

Correction: _____

_____ 3. A. An antiserum is a serum containing an antibody to a given antigen.
   B. Although antigens usually are proteins, DNA and polysaccharides may be antigenic.
   C. Only a small portion of an antigen's molecular structure, the antigenic determinant, is involved in stimulating antibody formation.
   D. The latent period for a anamnestic response is longer than the latent period for the primary response.

   Correction: _____

_____ 4. A. T lymphocytes originate in the thymus gland of humans.
   B. The maturation of T cells requires interaction with stromal cells of the thymus.
   C. T-helper cells assist B cells to secrete antibody.
   D. B lymphocytes originate by differentiation from T-cytolytic cells.

   Correction: _____

_____ 5. A. T cells produce a group of protein molecules called lymphokines.
   B. Mature T cells reside in lymphoid tissues in clusters called lymphoid follicles.
   C. B cells differentiate into active antibody-forming cells called plasma cells.
   D. The primary function of B cells is to produce antibody.

   Correction: _____

_____ 6. A. The selective theory of antibody diversity holds that an antigen stimulates the synthesis of an antibody whose specificity preexists in the body's genetic information.
   B. The instructive theory states that an antigen impresses its own new pattern upon the body's antibody-synthesizing apparatus.
   C. Research demonstrated that the mechanism leading to antibody diversity is better explained by the instructive theory than by the selective theory.
   D. In the instructive theory, antigenic determinants instruct the antibody-synthesizing apparatus on which antibody to produce.

   Correction: _____

_____ 7. A. On their surfaces, B and T cells have proteins called antigen receptors to allow these cells to recognize specific antigens.
   B. The antigen receptors of B cells are monospecific immunoglobulins.
   C. The antigen receptors of T cells can recognize only antigens that are on a cell.
   D. The antigen receptors of T cells are immunoglobulins.

   Correction: _____

_____ 8. A. The major histocompatibility complex (MHC) is a gene cluster that encodes a protein that indicates "self."
   B. The antigenic receptor of B cells recognizes both an antigen and the protein encoded by the MHC.
   C. Somatic recombination of genes allows the genome to encode the vast diversity antibodies that can be produced in our bodies.
   D. The variable region of a light chain is encoded from three segments: leader, V segment, and J segment.

   Correction: _____

_____ 9. A. Single-point mutations are a genetic source of the diversity of antibodies produced.
   B. The potential diversity of heavy chains is greater than the diversity of light chains.
   C. Monokines are produced by lymphocytes and macrophages.
   D. Lymphokines amplify the weak incoming signals of antigenic challenge and stimulate a surge of cell proliferation.

   Correction: _____

_____ 10. A. Allogeneic grafts survive longer than do syngeneic grafts.
   B. Tissue rejection is based on recognition of non-self major histocompatibility proteins.
   C. Much of our understanding of the mechanisms of self recognition is based on study of chimeras.
   D. The body's definition of self develops during prenatal and early postnatal development.

   Correction: _____

_____ 11. A. In autoimmune diseases, the body perceives itself (or parts of itself) as nonself.
   B. All of the following are examples of autoimmune diseases: rheumatoid arthritis, myasthenia gravis, and hemolytic anemia.
   C. Blood types are based on reactions between antigens found on leukocyte surfaces with antibodies in testing serum.
   D. Blood groups are different than most phenotypic traits in that blood groups are not influenced by the environment; hence, blood groups are useful in studying human evolution.

   Correction: _____

CHAPTER 24--TRANSPORT

CHAPTER OVERVIEW

The movement of materials from one place to another within an organism is a crucial activity. Without transport of materials, it is doubtful that any organism could carry out any of its other activities and functions. As we will see in this chapter, movements occur both within cells and between cells.

Many materials must be transported. We saw in the two previous chapters how the bloodstream of vertebrates transports materials such as water, antibodies, respiratory gases, blood cells, nutrients, hormones, vitamins, and many others. In this chapter we will look more closely at the tubular systems of vertebrates that transport blood. We'll see how vascular plants have independently evolved a different sort of tubular transport system as a response to the same needs for transport that animals have. We'll see, too, how the structural complexity of the transport system correlates with body size and shape.

TOPIC SUMMARIES

A. <u>Transport in Structurally Simple Organisms</u>.--All organisms are made of cells. The organisms having the simplest structures are unicellular. Examples of unicellular organisms include the monerans, protists, some fungi, and some plants. Like multicellular organisms, these organisms import and export materials across their external membranes, and they transport materials within each cell. Various versions of <u>diffusion</u> and <u>active transport</u> usually are adequate means of movement if the distance from source of the material to the point of use of that material is short--on the order of a few micrometers or angstroms. Once the material is within the cell, diffusion still is rapid enough for transport of some substances. <u>Cytoplasmic streaming</u> and movements of food <u>vacuoles</u> are additional mechanisms of intracellular distribution of materials.

As organisms grew and evolved beyond unicells, transport by diffusion was supplemented by specialized structures that could deliver materials as rapidly as required by cells. Plants and animals responded differently to this challenge; despite the differences, however, there are many parallels in their solutions.

B. <u>Transport in Vascular Plants</u>.--The vascular plants include all plants <u>except</u> the bryophytes. Vascular plants are those with which most of us are most familiar; see Chapter 41 for examples. These plants possess <u>vascular tissues</u>, tissues specialized for transporting materials throughout the body (Fig. 24.1). Most of these plants are land-dwellers rather than aquatic. Interestingly, the <u>invasion of land</u> by plants was accompanied by the evolution of vascular tissues; in addition to transporting materials, some of these tissues (i.e., secondary xylem) are <u>rigid</u> and therefore <u>offer support</u>, much as does our own skeleton.

A conventional vascular plant (e.g., tree, dandelion) has roots, stems, and leaves. Water enters the body via the <u>root system</u>, specifically through <u>root hairs</u>. Water is needed in every cell of the plant. Water, the solvent of <u>sap</u>, is transported throughout the body through <u>xylem</u>, a tissue consisting of <u>tracheids</u> (only in gymnosperms) and <u>vessels</u>. The solutes of sap, primarily various minerals absorbed from the soil, likewise are transported via xylem. The efficiency of absorbtion of materials from the soil is tremendously enhanced by <u>root hairs</u>, hairlike extensions of the epidermis that provide an immense surface area; read in your textbook about the work of Dittmer.

Photosynthesis, the process by which light energy is trapped in organic form, does not occur everywhere in a vascular plant. Rather, this process generally is restricted to leaves and stems. Yet, every living cell requires glucose (some cells can directly use phosphoglyceraldehyde). Distribution of this "food" (usually in the form of sucrose) and of various other organic compounds involves a second vascular tissue, phloem. Phloem is composed of sieve cells and companion cells. Unlike xylem cells, these cells are alive when functioning.

Fig. 24.1: Gross view of conducting tissues in vascular plants

Be sure to return to Chapter 7 to review the characteristics of the cell types that are involved in transport of materials in plants. It will be difficult to understand function if you do not know cell structure.

C. Mechanisms of Transport in Vascular Tissues.--The processes that drive the movement of fluids in phloem and xylem are not fully understood. Several mechanisms have been proposed. You will be better able to understand these processes if you are familiar with structure of roots and stems; study text Figs. 24-5, 24-9, 24-11, 24-12, 24-14, and 24-18.

(1) Xylem. Movement of water from roots to the upper parts of a plant defies gravity. The current theory explaining how water rises within xylem is the cohesion-adhesion-tension theory: Transpiration, the loss of water vapor from the plant body, is an important player in this process. Evaporative loss of water occurs primarily from the inner surfaces of leaves; these surfaces must be kept moist so that carbon dioxide can diffuse into photosynthetic cells. This need results in a potentially detrimental loss of water. However, as this water loss creates a deficit of water in the leaf, additional water molecules are drawn into the leaf from adjacent vascular tissues. (Recall the attraction of water molecules to each other (cohesion) by hydrogen bonds; recall also that electrostatic bonds caues water molecules to adhere to hydrophilic surfaces, such as the walls of the xylem tubes.) This "pull" is exerted all the way down the water column into the roots. A deficit in the root system enables water from the soil

to enter the roots by osmosis. (Be sure to read about the two ways by which water and minerals enter the roots: apoplastic and symplastic pathways.)

Root pressure contributes a "push" that encourages water to move up the water column. As minerals are actively transported into root tissues, water osmotically follows (inwardly) the concentration gradient. The accumulation of water within a limited space causes hydrostatic pressure to build. Hence, water forces itself upward into the stems and leaves.

(2) Phloem. The pressure-flow hypothesis is the most widely accepted explanation for long-distance transport in phloem. The first step in the process is phloem loading; sucrose produced in leaves is actively transported into a sieve tube. Increasing the concentration of sucrose in the sieve tube causes a decrease in its water potential. Water then enters the sieve tube via osmosis from surrounding tissuess. The sucrose solution then is carried passively to some sink (e.g., root) where the sucrose is unloaded actively. The sucrose in the sink is either consumed by metabolic activities or is stored as a food reserve.

D. Transport in Animals.--As with plants, the fluid transport arrangements of animals vary according to body size and form. For many structurally simple animals, no specialized internal transport system exists. For example, the body of *Hydra* (a coelenterate) resembles little more than a sac with tentacles (see textbook Fig. 24-22). The walls of the sac are only two cells thick. The watery medium both surrounds the body and flows into the blind gut cavity. Needed materials and wastes simply diffuse into and out of the tissues because all cells are very close to the source of the materials. Flatworms (Platyhelminthes) are larger and somewhat more complex than *Hydra,* yet they, too, lack a specialized delivery system. It is their flat shape that achieves a high surface-to-volume ratio, and, hence enables diffusion to be the sole delivery mechanism.

Further increases in complexity and deviations from flat shape necessitated the evolution of specialized delivery systems. These systems involved internal spaces in which blood or some corresponding fluid moves. The simpler versions of such systems are referred to as open systems; in these, fluid simply sloshes about in internal sinuses. Such systems occur in insects, some worms, and many mollusks. Closed systems in which blood travels within a circuit of vessels characterize most worms and all vertebrates. Pumps to pressurize and direct blood flow accompany many closed systems. Let's examine the human circulatory system as an example of a highly specialized delivery system.

E. Human Circulatory System.--The human circulatory system is representative of this system in other mammals. The system consists of two connected circuits. The pulmonary circuit conducts blood from the right side of the heart, to the lungs via pulmonary arteries, through the lungs (where gas exchange occurs), and back to the heart via pulmonary veins. Freshly-oxygenated blood then flows to all tissues of the body via the systemic circuit. The aorta carries blood from the left side of the heart. Branches of the aorta distribute blood to major organs. These branches have smaller branches, called arterioles. In the tissues, these arterioles branch into very narrow diameter capillaries. Capillaries, which permeate every tissue of the body, are the only blood vessels where materials are exchanged with the tissues. On the opposite side of the capillary beds, the vessels coalesce into larger and larger diameter venules and veins. Ultimately, all veins below (inferior to) the heart converge into a single vein, the inferior vena cava. Similarly, all venous blood above (superior to) the heart flows into the superior vena cave. Both vena cavae carry blood into the right atrium. Blood entering the heart has then completed the systemic circuit and has entered the pulmonary circuit.

Vertebrate venous systems have at least one interesting "subcircuit" within the systemic segment. The hepatic portal system is a series of veins that carries blood from various abdominal organs that are "in contact" with the outside environment to the liver. The liver has many functions, including processing of materials that have just been absorbed by the intestines and other abdominal organs. Some of these materials might be toxic and dangerous; the liver can detoxify many harmful substances before they are distributed to the rest of the body. Also, the liver can synthesize many complex substances from the simple materials that have just been absorbed; many other organs are not able to synthesize such compounds. Hence, providing unprocessed blood to other tissues would not be beneficial to those tissues.

F. The Vertebrate Heart.--Use of such a heading implies that all vertebrates have similar hearts. Wrong! There is similarity in location and function, but design varies widely among vertebrates.

(1) Evolutionary trends. The general evolutionary trend in hearts of vertebrates has been an increase in the number of chambers from two (an atrium and a ventricle) to three (two atria and one ventricle) to four (two atria and two ventricles). The changes are largely associated with the invasion of land and a change in breathing strategies. Most fishes extract gases from water and exchange gases at gill surfaces; there is no pulmonary circuit and no separation of oxygenated blood from deoxygenated blood. Most land-dwelling vertebrates extract gases from air and exchange gases at surfaces of the lungs; pulmonary and systemic circuits are distinguishable. However, the degree of separation of oxygenated and deoxygenated blood varies with heart design: The three-chamber design of heart allows mixing of bloods in the ventricle, whereas complete separation of these bloodstreams occurs in the four-chamber design. Anatomically, how does one change a three-chambered heart into one with four chambers? What are the implications of mixing or not mixing of oxygenated with deoxygenated blood in terms of the level of activity that an animal can maintain?

(2) Heart function. We already have mentioned that a major role of the heart is to pressurize the bloodstream. The heart must apply enough pressure to the blood to ensure that blood flows completely through the circuit and returns to the heart. Drops in blood pressure occurs throughout the circuit, but especially at the capillary beds. In most non-fish vertebrates, there is enough peripheral resistance in the tissues that the blood must be pressurized twice for each complete circuit--once to push the blood through the pulmonary vessels and another time to get it through the systemic circuit. Blood pressure also must be maintained at an adequate level to encourage materials in the blood to move through capillary walls into the tissues; otherwise, tissues would starve for lack of needed raw materials. What other mechanisms explain how materials cross from the bloodstream into tissues . . . and in opposite directions?

Even when the heart is relaxed (the phase of diastole), some amount of blood pressure remains; this pressure measured in mm of mercury is the bottom number of the two-number blood-pressure reading taken with a sphygmomanometer. The larger of these figures is the pressure at systole, when the ventricles contract and force a surge of blood into the aorta. The rate at which the heart muscle contracts is dictated by specialize nodal tissues that build up and then discharge electrical charges; the atrioventricular and sinuatrial nodes, therefore, are known as pacemakers. How are these charges conducted throughout the heart muscle?

G. The Lymphatic System.--Blood vessels are not the only "pipes" in the bodies of vertebrates. A second set of tubes, the lymphatic system, courses throughout the tissues, collecting fluid from the tissues. What is the source of this interstitial fluid? Fluid from the bloodstream crosses into the tissues on the arterial side of the capillary beds; most of this fluid returns to the bloodstream on the venous side of these beds. However, about 10% of the fluid does not return directly to the bloodstream. This residual fluid finds its way into the lymphatic vessels that eventually return it to the venous system near the heart. The thoracic duct collects lymph from all parts of the body except the right arm, right shoulder, and the right side of the head and neck. Lymph from these areas drains into the right lymphatic duct. How does the condition of <u>elephantiasis</u> fit into this discussion? What vertebrates other than mammals have a lymphatic system?

TERMS TO UNDERSTAND: Define and identify the following terms:

diffusion

cytoplasmic streaming

food vacuoles

vascular plants

xylem

phloem

transpiration

guard cells

stomata

root hairs

epidermis

endodermis

casparian strip

pericycle

apoplastic pathway

symplastic pathway

root pressure

guttation

sap

trachieds

vessels

pith

cambium

cork cambium

suberin

dicotyledons

monocotyledons

cohesion-adhesion-tension theory

cavitation

sieve cells

sieve tube members

sieve plates

companion cells

pressure-flow hypothesis

phloem loading

heart

vessels

valves

sinuses

artery

vein

capillary

aorta

vena cava

ventricle

atrium

pulmonary circulation

portal circulation

diastole

systole

sinus venosus

gills

nodes of heart

electrocardiography

interstitial fluid

lymphatic system

thoracic duct

lymph hearts

edema

MATCHING

A. Match the process in the left column with the hypothesis or theory in the right column that has been proposed to explain how that process occurs.

	PROCESS		MECHANISM
____	1. movement of water into roots	A.	cohesion-adhesion-tension theory
____	2. translocation	B.	active transport
____	3. upward movement of sap in xylem	C.	osmosis
____	4. guttation	D.	pressure-flow hypothesis
____	5. movement of minerals into roots	E.	root pressure

B. Match the structure in the left column with its function in the right column.

	STRUCTURE		FUNCTION
____	1. guard cell	A.	prevents water loss
____	2. pericycle	B.	surrounds stomata
____	3. companion cell	C.	conduct sap
____	4. tracheid	D.	source of lateral roots
____	5. suberin	E.	provides metabolic support to sieve elements

PEOPLE TO KNOW: What is the significant biological contribution of each of these individuals as indicated in Chapter 24?

Stephen Hales

H. J. Dittmer

Marcello Malpighi

Ernst Munch

E. H. Starling

Sidney Ringer

DISCUSSION QUESTIONS

1. Summarize the major evolutionary changes of the heart among the vertebrates. Indicate the features of the heart in fishes, amphibians, reptiles, birds, and mammals. What advantages (if any) does a four-chambered heart have over other designs?

2. Compare and contrast the "plumbing" systems used by vascular plants and vertebrate animals.

3. What is the role of valves in the circulatory system of a vertebrate?

4. How is the surface-to-volume ratio of an organism related to the type of transport system found in that organism?

5. Label as many of the structures of the mammalian heart as you can in the drawing below. Feel free to consult references beyond your textbook to help you identify these structures.

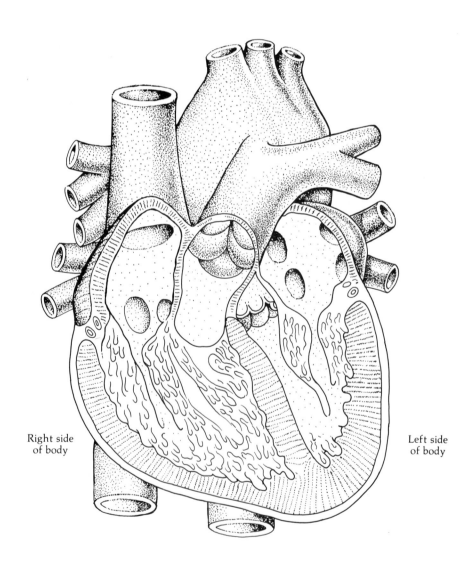

TESTING YOUR UNDERSTANDING

For each of the test items below, three of the lettered alternatives are true and the other is false. Determine which alternative is false and write its letter in the blank to the left of the question number. On the blank line below alternative D, write the corrected version of the false statement.

_____ 1. A. Cytoplasmic streaming occurs in one-celled organisms, but not in multicellular organisms.
B. Diffusion is a process that operates in all species.
C. Bryophytes are multicellular, nonvascular plants.
D. One of the most important functions of vascular tissue is support of the plant body.

Correction: _____

_____ 2. A. Three processes that are important in transport of water to the top of a tree are transpiration, active ion transport, and osmosis.
B. Plant bodies generally suffer greater water loss than do bodies of animals.
C. Loss of water vapor by a plant body is termed translocation.
D. Guard cells regulate the openings of stomata.

Correction: _____

_____ 3. A. Root hairs are fine extensions of the endodermis.
B. The endodermis of a plant root is only one cell thick.
C. Lateral roots grow from the pericycle.
D. Root hairs comprise a significant proportion of the absorptive surface area of the root system.

Correction: _____

_____ 4. A. Guttation results from positive root pressure.
B. The sap-conducting cells of gymnosperms consists only of tracheids.
C. The cortex of a plant stem is the region from which bark develops.
D. Suberin is a waxy substance that waterproofs cork cells.

Correction: _____

_____ 5. A. The small diameter of xylem conduits tends to prevent cavitation.
B. The strong cohesive attraction of water molecules for each other is a result of hydrogen bonds.
C. Water molecules are electrostatically attracted to the hydrophilic walls of xylem tissue cells.
D. The pressure-flow hypothesis explains how water is drawn up the xylem column.

Correction: _____

_____ 6. A. Unlike xylem cells, the cells of phloem tissue are alive when they are functional.
B. Because sieve cells have lost their nuclei and most mitochondria, they require metabolic assistance from companion cells.
C. Girdling will kill a tree because of the interruption of translocation.
D. Dittmer proposed the currently accepted hypothesis explaining how sucrose is transported throughout the plant body.

Correction: _____

_____ 7. A. Phloem loading decreases the water potential in the sieve tube and causes water to enter the sieve tube by osmosis.
B. Substances transported in phloem include sugars, nucleotides, and other organic compounds.
C. Sponges, corals, and fishes resemble nonvascular plants in that they lack vessels for internal movement of fluids.
D. Hearts are components of the circulatory systems and lymphatic systems of many vertebrate species.

Correction: _____

_____ 8. A. The following is a correct sequence of structures through which blood passes as it flows through the heart of an amphibian: right atrium, right ventricle, lungs, left atrium, left ventricle.
B. Valves in the heart and in veins of mammals help to keep blood flowing the proper direction.
C. Animals with open circulatory systems lack vessels to carry the blood.
D. Exchange of materials between the bloodstream and the tissues occurs only in capillary beds.

Correction: _____

_____ 9. A. The pulmonary artery carries deoxygenated blood from the heart to the lungs.
B. The liver receives blood from both the hepatic artery and the hepatic portal vein.
C. The aorta is part of the system circulation.
D. The two veins entering the left atrium of humans are the inferior vena cava and the posterior vena cava.

Correction: _____

_____ 10. A. Blood pressure can be measured by using a stethoscope.
B. Diastole is the phase of heart activity when the heart is relaxed.
C. Blood pressure is higher when the heart is contracted than when it is relaxed.
D. A murmur suggests that certain valves of the heart are defective.

Correction: _____

_____ 11. A. Most fishes have a two-chambered heart.
B. Most mammals have a four-chambered heart.
C. Most birds have a three-chambered heart.
D. Vertebrates that have a three-chambered heart have two atria and one ventricle.

Correction: _____

_____ 12. A. The presence of a high-pressure double circulation in mammals correlates with the high metabolic rates of mammals.
B. The rate at which a mammalian heart contracts is regulated by activity of the sinoatrial node.
C. If the sinoatrial node fails, surgery is required for implantation of an artificial pacemaker.
D. Coronary thrombosis is the condition in which flow of blood through the coronary artery is interrupted.

Correction: _____

_____ 13. A. Most of the resistance to blood flow occurs in small-diameter blood vessels such as the venules.
B. The fluid found between cells in the tissues is called interstitial fluid.
C. Wastes diffuse from interstitial fluid into plasma because waste concentrations in plasma are lower than in the interstitial fluid.
D. At the venous end of capillary beds the osmotic pressure exerted by plasma albumins causes water to flow from the tissues into the bloodstream.

Correction: _____

_____ 14. A. The right lymphatic duct returns lymph to the bloodstream via the large veins near the heart.
B. The lymphatic system is a closed system of vessels that circulates lymph throughout the body.
C. Edema can result if the lymphatic system does not drain enough fluid from the tissues.
D. Lymph flow is regulated by lymph hearts and by valves in the lymphatic vessels.

Correction: _____

CHAPTER 25--EXCRETION

CHAPTER OVERVIEW

Maintenance of internal constancy requires that waste products of metabolism be removed from the body. Homeostasis also requires that concentrations of salts and other materials be maintained within tolerable limits. Animals and protists have evolved a number of ways of achieving water balance and appropriate internal osmotic environments. The strategy used by a particular species corresponds with factors including body size and shape, the presence or absence of a closed or open circulatory system, and the osmotic conditions of the surrounding environment. Hence, the anatomy and physiology of osmoregulation is different for terrestrial vertebrates than for marine invertebrates than for freshwater fishes.

TOPIC SUMMARIES

A. <u>What Are Metabolic Wastes</u>?--Every chemical reaction yields one or more products. So it is with metabolic reactions of living organisms. Many metabolic products are needed and used in other metabolic reactions, whereas other products are not needed. Such unneeded materials are eliminated from the body by a variety of <u>excretory mechanisms</u>.

   (1) Nitrogenous wastes. One major category of metabolic wastes includes the <u>nitrogenous wastes</u>. These are formed from catabolism of compounds that contain nitrogen, for example, proteins, nucleic acids, and others. Accumulation of nitrogenous wastes within the body is dangerous because they are toxic. The chemical form in which nitrogenous waste is eliminated corresponds to the environment in which the organism lives and the availability of water.

   The amount of water available is an important factor because one way of keeping nitrogenous wastes from reaching toxic levels is to dilute water-soluble wastes with large amounts of water. This is the primary way of eliminating <u>ammonia</u>, a highly-toxic nitrogenous waste. <u>Urea</u> is another water-soluble form of nitrogenous waste, produced by animals under only moderate water stress. For species, and for stages of development, where no extra water is available for diluting nitrogenous waste, <u>uric acid</u> is excreted. Uric acid is a water-<u>insoluble</u> form of nitrogenous waste. We'll look at the environmental and taxonomic distribution of nitrogenous wastes in greater detail below.

   (2) Other metabolic wastes. Think back to a metabolic pathway that is present in some form in all living organisms--<u>respiration</u>. The essential products of this pathway are <u>water</u>, <u>carbon dioxide</u>, and <u>energy</u>. Energy is used in other reactions or is liberated as heat. Water is used in most physiological activities of the body; any excess is dispensed with in association with elimination of nitrogenous wastes or by other processes such as sweating (in mammals). Carbon dioxide is liberated to the external environment, generally at body surfaces that are components of the respiratory system . . . more on this in Chapter 27.

   Salts of various kinds, as well as certain other ions, must be maintained at certain levels. The osmoregulatory needs of an organism vary in accordance with the ion content of the environment they occupy.

B. <u>Correspondence of Environment with Osmoregulatory Strategy</u>.--As indicated above, the approach that species take to regulating water and ionic balance is closely tied to the enrinvonments that they occupy.

   (1) Aquatic animals. At first glance it would seem that any organism living in water must have easy access to all the water it could possibly need. This is true for some aquatic animals,

but not for others. In the following paragraphs, we'll examine the three possible osmotic relationships (what are they?) an organism can have with an aquatic environment.

Some organisms, particularly marine (occupying oceans and seas) invertebrates, do not maintain an internal concentration of salts and other ions different that their surroundings. Rather, their body fluid concentrations of these ions vary with solute concentrations of the surrounding medium. Such organisms are called osmoconformers (see Fig. 25.1). Most marine invertebrates (e.g., various coelenterates, echinoderms, etc.) are isosmotic with their environment. This condition is maintained by (a) diffusion and/or active transport of wastes and ions into, and out of, the tissues, and (b) subsequent osmotic flow of water into, or out of, the tissues along the concentration gradient established by ionic concentrations. Hence, no specialized excretory structures are needed.

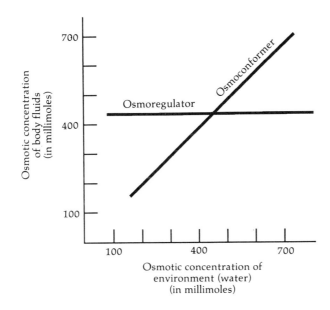

Fig. 25.1: Graph showing osmotic strategies found in aquatic organisms

Other aquatic animals, particularly vertebrates and primarily fishes, generally are not in osmotic balance with their surroundings. Most marine fishes, for example, have lower internal solute concentrations than does seawater. Hence, water concentration is higher inside their bodies than outside. Water, therefore, automatically and continuously leaves the body. Being hyposmotic to the hypertonic medium, these fishes suffer dehydration! Additionally, because solute molecules and ions are in greater concentration outside their bodies, solutes continuously enter the body. This problem of excess salt acquisition is compounded by continuous drinking, a behavior in response to dehydration. To avoid an overload of salts, for example, the fishes must continuously and actively secrete salts (against the concentration gradient) into the sea; much of the salt excretion (and water loss) occurs via the gills. How have the cartilaginous fishes resolved this problem?

Freshwater fishes face the opposite problem as that encountered by marine fishes. Freshwater fishes have body fluids that generally are hyperosmotic to the surrounding hypotonic water. Osmotic influx of water conceivably could drown such fish! And, continuous diffusion of salt and other ions to the exterior could result in a solute deficit. These fish excrete tremendous quantities of water as a very dilute urine. Because of the abundance of water, nitrogenous wastes remain in the potentially toxic form of ammonia; the abundance of water prevents ammonia from reaching toxic concentrations. Salts are absorbed actively at the gills and are ingested with food.

(2) Terrestrial animals. All land-dwelling animals (at least during those phases of their lives when they are not in water) experience water stress. Because the concentration of water within their bodies is immensely greater than that in air, water molecules continually move out of their bodies. Many external structural adaptations (e.g., skin and other organs of the integument) resist such water loss. Yet, dehydration still occurs, especially at the respiratory surfaces where gases are exchanged (e.g., lungs, gills). A behavioral way to combat dehydration is to drink at least as much water as is lost . . . but this approach can work only if enough water is available. What about organisms in deserts, where little water is available?

C. Excretory Mechanisms.--The rest of this chapter focusses on physiological adaptations that enable animals to conserve metabolic water and the water that they ingest. Kidneys of various designs are the primary organs involved in water conservation. Diffusion is a process associated with all designs of excretory structures. Inextricably associated with water conservation are processes that regulate the internal concentrations of solutes. We'll see that the relative efficiency of water conservation (expressed in terms of how dilute or concentrated the urine is) corresponds to the amount of water available in the organism's environment.

(1) Contractile vacuoles. These are probably the simplest of all structures that serve the role of osmoregulation. In some of the simplest of animals (e.g., freshwater sponges) and in some protists (e.g., *Paramecium)*, vacuoles pulsate as they fill with water (and, to a limited degree, with various ions and wastes) and contract to expel their contents to the exterior. This strategy works well as long as a high surface-to-volume ratio is maintained and as long as virtually all cells of the organism are located near the surface.

(2) Aquatic invertebrates. Marine sponges, coelenterates, and echinoderms generally lack structures specialized for osmoregulation. Those in marine habitats tend to be osmoconformers, hence water and many ions passively diffuse across membranes. Some ions that cannot move passively across membranes are actively transported in or out by membrane proteins.

Flatworms and certain other invertebrates possess simple tubular organs specialized for excretion and osmoregulation. Their protonephridial systems consist of a pair of bilaterally symmetric, blind tubule systems. Flame cells, so called because the beating of cilia suggests the motion of a flame, occur at the blind ends of tubule branches. Flame cells create negative pressure that draws extracellular fluids into the protonephridium. As the fluid passes along the tubules, it apparently is modified by secretion and excretion. The urine leaves the body via a series of excretory pores opening through the external body wall.

(3) Metanephridium. Earthworms (annelids) and a few other invertebrates possess a slightly more advanced nephridial system than seen in flatworms. This advance accompanies a closed circulatory system. Each body segment of an earthworm has a pair of nephridia. Fluid in the body cavity enters the nephridium via a nephrostome and travels along a thin tube that is wrapped by blood capillaries. During its passage along the tubule, the fluid is modified by excretion and absorption of materials out of and into the associated capillaries and surrounding tissues. The fluid collects in an enlarged region of the tubule (bladder) near its end at the body wall. Urine leaves the metanephridium via the nephridiopore.

(4) Malpighian tubules. The respiratory passageway system of insects contributes to the water stress problem they experience. A hard, waxy cuticle and the malpighian tubule excretory system function to conserve water. The nitrogenous waste product, uric acid, is not soluble in water, so very little water is wasted as a medium for elimination. A number of blind malpighian tubules extend from the hindgut into the body cavity. There, body fluids enter the

tubules. After entering the hindgut, most of the water and various solutes are reabsorbed into the body, leaving primarily uric acid (and feces) to be released from the gut tract via the anus.

D. Kidneys of Vertebrates.--The nephron is the functional unit of the kidneys of vertebrates. Kidney is the collective term for all the nephrons in a vertebrate animal. Early versions of nephrons are strikingly similar to the tubule-capillary arrangement of earthworms. Although the number of nephrons per body segment and the location of kidney vary among vertebrates, all versions of vertebrate kidneys have common ancestry and similar developmental sequences.

(1) Nephron design. The basic design of a nephron is shown in Fig. 25.2. The nephron is a tubule closed at one end, that is expanded into a cup called the glomerular (Bowman's) capsule. The cup tapers into a tube (proximal convoluted tubule) that at first follows a twisting path, then straightens into a hairpin loop of Henle, then again becomes twisted (distal convoluted tubule), and finally straightens into the collecting duct. The collecting tubules on opposite sides of the body coalesce into a pair of nephric ducts that eventually carry urine to the exterior via direct openings in the pelvic region or via anus or cloaca, depending on the species.

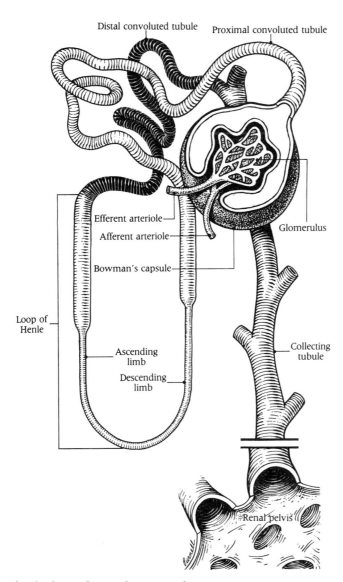

Fig. 25.2: Basic design of vertebrate nephron

A knot of capillaries (the glomerulus) occupies the glomerular capsule; this is the source of fluid that enters the nephron. Blood vessels extend from the glomerulus to follow the course of the nephron; these capillaries wrap around the all segments of nephron, particularly about the proximal and distal convoluted tubules. As fluid passes through the nephron, it is modified by excretion and reabsorption of various ions and by osmotic movement of water back into the tissues. It is the concentration gradient between the fluid in the nephron vs. that in the surrounding tissues that determines whether, and how much, water will be reclaimed from the urine. The ions that establish this concentration gradient are those that have been reabsorbed or have diffused from the blood filtrate within the nephron. Here, then, we see how water conservation is intimately associated with osmoregulation. We'll see below how structure and function of nephrons vary with the availability of water in the environment.

In fishes and amphibians, kidneys are long straplike organs running along either side of the vertebral column for much of the length of the abdominal and pelvic cavities. In reptiles, birds, and mammals, however, all nephrons are packaged into discrete bean-shaped organs.

(2) Freshwater fishes. More than any other vertebrates, freshwater fishes have need of a way to void large amounts of water. Their kidney design facilitates this need. Water tends to move very rapidly through their nephrons, due to the presence of "pumps" in the neck region (between glomerulus and proximal convoluted tubule) and in the intermediate segment (between proximal and distal convoluted tubules). These "pumps" are regions of ciliated cells that help propel the fluid rapidly. Additionally, these kidneys are very efficient at reabsorbing salts.

(3) Saltwater fishes. Continual osmotic loss of water to the surroundings and the need to excrete excess salts result in the production of small amounts of concentrated urine by saltwater fishes. Kidneys of saltwater fishes differ from those of freshwater fishes as follows: The ciliated pumps are absent, so fluid moves less rapidly and more water is reabsorbed. The distal convoluted tubule, where much salt reabsorption would occur, is lacking. Less fluid enters the capsule due to thicker capsular membranes and to lower blood pressure in the glomerulus. Additional salt is excreted at the gills.

(4) Mammals. Approximately one million nephrons are packed into each human kidney. The blood supply that is filtered to form urine enters the kidney via a renal artery, which ultimately branches into enough afferent arterioles to supply each glomerulus. The capillaries that entwine each nephron eventually coalesce into one renal vein leaving each kidney. The collecting ducts from all nephrons of a kidney contribute their urine to one duct (ureter) that carries urine to the urinary bladder where it is stored until a time appropriate for release (micturition).

The nephrons of mammals differ from those of freshwater fishes in the absence of the neck and intermediate segment and in the presence of hairpin loops of Henle. The concentration of ions in the tissues surrounding these loops of Henle are largely responsible for the ability of mammals to produce moderately to extremely concentrated urines, depending upon the ecologies of the species. For example, kangaroo rats occupy desert habitats where they might go month or years without ever seeing free drinking water; extremely long loops of Henle allow such efficient extraction of water from glomerular filtrate that these rats do not need drinking water! They can subsist on water contained in their diet and on metabolic water.

Kidney function is controlled by various factors including hormones. ADH (antidiuretic hormone, = vasopressin) decreases urine output by increasing the permeability of the walls of the collecting tubules to water; high ADH levels causes more water to be retained by the body. Salt has a similar effect; that is why someone who wishes to loose weight might opt for a low-salt diet. A variety of drugs and foods have the opposite effect; caffeine and alcohol are diuretics. Using coffee to help you stay up late to study increases urine production . . . if you didn't know that already, then you probably will soon!

TERMS TO UNDERSTAND: Define and identify the following terms:

nitrogenous wastes

urea

uric acid

ammonia

bile pigments

osmoconformers

osmoregulators

contractile vacuole

pseudocoelom

flame cell

excretory pore

nephridium

nephrostome

malpighian tubules

glomerular capsule

glomerulus

glomerular filtration

proximal vs. distal convoluted tubules

collecting tubule

nephron

aglomerular marine fish

cortex

renal pyramid

medulla

loop of Henle

renal pelvis

ureter

urinary bladder

urethra

renal artery

afferent arteriole

peritubular capillaries

renal vein

glomerular filtrate

diabetes mellitus

antidiuretic hormone (ADH)

countercurrent flow

two-solute model

vasopressin

osmoreceptor

aldosterone

adrenal cortex

tubular reabsorption

tubular secretion

DISCUSSION QUESTIONS

1. In this chapter, we have focused upon animals and protists. What about plants, fungi, and monerans . . . do they not produce metabolic wastes that must be removed from their bodies? If so, then what approaches do they use to excrete wastes?

2. Develop a chart indicating the primary nitrogenous waste product(s) for species in the major groups of animals. Discuss how a particular waste product represents a logical evolutionary solution for that species.

3. How are the processes of maintance of water balance and of maintenance of the proper internal osmotic conditions related to each other?

4. Salmon are an example of fish species whose life histories take them into aquatic situations with very different osmotic conditions. Salmon are born in freshwater. They live most of their adult lives in the ocean. They return to fresh water to spawn. How do they physiologically tolerate such changes in osmotic surroundings?

5. Label Figure 25.2 in this Study Guide to indicate the physiological processes that occur in each region of a nephron.

6. How are the nephridia of earthworms similar to, and different from, the nephron-capillary arrangement of vertebrates?

CHART EXERCISE

Organisms occupying different environments face different osmoregulatory problems. Complete the chart below for the five organisms listed across the top. Indicate whether the internal conditions are hypo-, hyper-, or isosmotic relative to the organism's surroundings. Indicate whether water will flow into or out of its body. What direction will various ions move? Summarize the osmoregulatory problems faced by that individual. Summarize the anatomical and physiological solutions that species has evolved.

	Amoeba	Mackerel	Rainbow Trout	Marine Sponge	Rattlesnake
Internal vs. external environment					
Direction of water flow					
Direction of ion movement					
Problems faced					
Anatomical and physiological responses					

TESTING YOUR UNDERSTANDING

For each of the test items below, three of the lettered alternatives are true and the other is false. Determine which alternative is false and write its letter in the blank to the left of the question number. On the blank line below alternative D, write the corrected version of the false statement.

_____ 1. A. Urea is a metabolic waste formed from the metabolism of nitrogen-containing compounds such as nucleic acids.
B. Uric acid is a water-soluble nitrogenous waste.
C. The primary nitrogenous waste produced by freshwater fishes is ammonia.
D. Most marine fishes produce small volumes of urine having solute concentrations slightly less than that of body fluids.

Correction: _____

_____ 2. A. Because of the relative osmotic conditions of their tissues and the surrounding waters, freshwater fishes face the problem of drowning whereas marine fishes face the problem of dehydration.
B. Two of the adaptations used by various land-dwelling tetrapods to combat dehydration are excretion of uric acid and presence of a waterproof integument.
C. Feces are one of the forms of wastes produced by metabolism of nitrogen containing compounds.
D. Bile pigments are the major product of the breakdown of hemoglobin from red blood cells.

Correction: _____

_____ 3. A. Organisms that are classified as osmoconformers generally possess specialized organs for regulating the concentrations of solutes in body fluids.
B. The protist *Paramecium* is an osmoconformer.
C. Some marine invertebrates, such as sponges, are isosmotic with the seawater surrounding them.
D. The gills are sites of salt absorption and/or salt excretion in various species of fishes.

Correction: _____

_____ 4. A. Nephrons are the functional units of the osmoregulatory structures in fishes, birds, and mammals.
B. The amniotic egg of reptiles solves the problem of dehydration that embryos of fishes and amphibian experience.
C. The allantois is a sac in which nitrogenous wastes are stored.
D. In protists, the contractile vacuole excretes wastes, whereas water balance is maintained by osmosis of water across the cell membrane.

Correction: _____

5. A. Flame cells are a component of the protonephridial system.
   B. Body fluids enter the protonephridial system of earthworms through the excretory pore.
   C. In flatworms, the force that drives fluids through the excretory system is provided by the movements of cilia.
   D. In insects, excretory wastes are released from the malpighian tubule system into the hindgut.

   Correction: _____

6. A. Each malpighian tubule has a blind end that lies in the coelom.
   B. The excretory product of insects is uric acid.
   C. Fluid from the body cavity enters the mammalian nephron through a nephrostome.
   D. Hydrostatic pressure in the arteries of birds drives fluid into the nephron by a process called glomerular filtration.

   Correction: _____

7. A. The fluid entering a nephron includes all of the solutes of blood plasma and lacks blood cells.
   B. Cilia in the neck and intermediate segments of the nephrons of freshwater fishes provide the force to drive the filtrate rapidly through the tubule system.
   C. The "degenerate" nature of nephrons in saltwater fishes represents an adaptation to the problems of dehydration and influx of excess salts.
   D. Much of the loop of Henle is located in the medulla of the kidney.

   Correction: _____

8. A. The following is a correct sequence of structures through which body fluids pass from the bloodstream to release from the body as urine: glomerular capsule, proximal convoluted tubule, ascending loop of Henle, renal pelvis, ureter, urinary bladder, urethra.
   B. Processes blood leaves the mammalian kidney via the renal artery.
   C. Intimate association of the mammalian excretory system with the circulatory system exists in the glomerulus and the peritubular capillaries.
   D. Human kidneys filter about 1,900 liters of blood per day.

   Correction: _____

9. A. The presence of urea in urine suggests the presence of diabetes mellitus.
   B. Once urine has entered the ureters, its chemical composition is not altered further.
   C. Movement of sodium ions from the filtrate into the tissues surrounding the nephron is via diffusion.
   D. Movement of chloride ions from the filtrate into the tissues surrounding the nephron is via active transport.

   Correction: _____

_____ 10. A. The walls of the descending loop of Henle are freely permeable to water.
B. The walls of the ascending loop of Henle are impermeable to water.
C. The countercurrent-flow arrangement of fluids through the descending and ascending loops of Henle help maintain differences in solute concentration along the entire loop of Henle.
D. Vasopressin is a hormone that increases the permeability of the walls of the collecting ducts to water.

Correction: _____

CHAPTER 26--NUTRITION AND METABOLISM

CHAPTER OVERVIEW

This chapter focusses on the raw materials that organisms need to carry out their metabolism and how these materials are obtained from the environment and introduced into their bodies. As we already have seen for many other aspects of life, the process of evolution solved nutritional needs of different groups of organisms in rather different ways. Indeed, organisms are broadly classified as either self feeders (autotrophs) or heterotrophs, those that must consume organic material ultimately produced by autotrophs. Because nutrients are a category of materials that are transported into and within the body, you should notice parallels in this topic with topics treated in the chapter (24) on transport.

Complexity of anatomical structures associated with digestion increases with increases in body size. This trend is in response to the inadequacy of diffusion alone to transport nutrients to cells distant from a body surface. Another trend you will observe is from no digestive cavity to a blind digestive cavity to a "flow-through" digestive system. We'll look at the human digestive system as an example the generalized tract of vertebrates, and then examine modifications of this system that reflect adaptations for specialized diets. We'll study some of the physiology associated with digestion and nutrient absorption, with particular emphasis on digestive enzymes.

A final topic in this chapter is thermoregulation. You already know that metabolism releases energy, largely in the form of heat. The physiology of most organisms is sensitive to heat; hence, most organisms have behavioral, anatomical, and physiological means of keeping their body temperatures within a suitable range.

TOPIC SUMMARIES

A. <u>Nutrition in Autotrophs</u>.--Unlike most members of the other four kingdoms, most plants are <u>photosynthetic autotrophs</u>. Compared to nutritive needs of heterotrophs, the materials needed by plants and other autotrophs to perform such syntheses are rather simple: carbon dioxide, water, and light (or other energy source) are required for photosynthesis; oxygen is needed for respiration. Plants synthesize virtually all of the organic compounds needed in their metabolism. The raw materials needed for organic synthesis are taken in by the roots by diffusion and/or active transport. The quantities in which these nutrients are needed are a criterion for their classification (see textbook Table 26-1). C, O, H, N, K, Ca, P, Mg, and S are <u>macronutrients</u>. Fe, Cl, Cu, Mn, Zn, Mo, and B are <u>micronutrients</u>. "Carnivorous" plants represent an interesting evolutionary offshoot in plant nutrition. Venus flytraps and sundew plants, for example, "consume" insects as a source of nitrogen; these and most other carnivorous plants reside in boggy, nitrogen-poor habitats.

As part of your study of plant nutrition, you should review material in several of the previous chapters: plant tissues in Chapter 7, photosynthesis in Chapter 10, and transport within plants in Chapter 24.

B. <u>Nutrition in Heterotrophs</u>.--Heterotrophs require organic molecules produced by other organisms--directly or indirectly by autotrophs. Absorptive heterotrophs (fungi and heterotrophic bacteria) release enzymes into the environment and then absorb the products of digestion. Humans are examples of ingestive heterotrophs (animals and nonphotosynthetic protists) that eat nutrients prior to digestion. In this chapter, we survey the diversity of approaches taken by heterotrophs in the ingestion, digestion, and assimilation of food.

C. <u>Ingestion of Food by Heterotrophs</u>.--Ingestion refers to the manner by which foods, nutrients, or other raw materials enter the body of an organism. As with so many other processes, different organisms use different ways to ingest food.

(1) By protists. Osmosis, diffusion, and active transport are the essential means of absorption of water and dissolved materials for many protists. In some other protists, such as *Paramecium,* an additional method is used. Solid food is taken into the cell and enclosed within a <u>food vacuole</u> in which chemical digestion occurs; such vacuoles are formed by <u>endocytosis</u>. Still other protists secrete digestive enzymes into the surrounding medium; there digestion of food occurs, with only the simpler, digested materials being absorbed.

(2) By animals. Animals generally have simple to highly-specialized cavities or passageways in which digestive activities occur. Sponges and coelenterates simply have an internal blind sac, the <u>gastrovascular cavity</u>, into which foods are placed. In *Hydra,* for example, protein-digesting enzymes are secreted into this cavity; extracellular digestion of proteins is followed by intracellular digestion of non-proteins.

With increasing complexity of the organism, digestive tracts tend to be more complex as well. A notable advance over a blind digestive system is a tubular system with an opening at each end; food enters at the <u>mouth</u>, and <u>feces</u> leave via the <u>anus</u>. Another trend in evolution of this system is that certain regions of the <u>alimentary canal</u> become specialized for certain activities (more on this below). The means by which food is taken into the alimentary canal depends largely on the nature of the food utilized.

Indeed, method of food procurement is one basis for classifying animals: <u>Bulk feeders</u> ingest relatively large food items which are mechanically broken into smaller pieces that will be further degraded by chemicals. Structures that assist in mechanical degradation include <u>teeth</u> (in most vertebrates and in some invertebrates) and specialized regions of the canal (e.g., <u>gizzard</u> and <u>proventriculus</u>) that muscularly churn and grind food. Most <u>filter feeders</u> are aquatic species (e.g., some mollusks, some whales, aquatic insect larvae, etc.) which screen food (e.g., algae, shrimp, etc.) from the surrounding medium (usually water). Various other animals obtain food in liquid form; <u>fluid feeders</u> ingest liquids such as blood (vampire bats . . . yes, three species occur in the New-World tropics), nectar (hummingbirds), and plant sap (aphids). Another category includes those animals that indiscriminantly ingest samples of their surroundings from which they extract and digest nutritive materials; earthworms are excellent examples of such <u>deposit feeders</u>. Many fishes are <u>suction feeders</u>; they rapidly open their mouths to develop a negative pressure that draws food into the mouth. Into which of these categories do humans fit? We'll examine the anatomy of the human digestive tract in part D below.

D. <u>Human Digestive System</u>.--The human digestive tract generally resembles that of most vertebrates. The two major components are the <u>alimentary canal</u>, a series of tubular structures through which food (or materials derived therefrom) pass, and a series of <u>accessory structures</u> that contribute digestive secretions or that help to mechanically breakdown food. Let's examine the specialized functions of these regions and structures by following the events that occur as a bolus of food passes through the tract.

Ingestion occurs at the <u>mouth</u>. Food is degraded mechanically by <u>lips</u>, <u>teeth</u> (the process of <u>mastication</u>), and the <u>tongue</u>. Chemical digestion begins in the mouth with the secretion of <u>saliva</u> from a series of <u>salivary glands</u>. Saliva includes an enzyme (<u>salivary amylase</u>) that digests starches such as in chips, french fries, bread, etc. Amylase, like virtually all digestive enzymes, is a hydrolase, an enzyme that breaks large molecules into smaller ones by cleaving and adding a molecule of water to the fragments of the large molecule. Mucus and water in saliva help form the food into a bolus suitable for swallowing (<u>deglutition</u>).

The esophagus is little more than a tube connecting pharynx (region at the rear of the mouth) with stomach. Activities and secretions of the stomach convert the bolus into chmye. The environment within the stomach is acidic due to the production of hydrochloric acid (HCl) by certain glands in the stomach lining. The action of HCl on pepsinogen produces pepsin, a strong proteolytic (protein-digesting) enzyme. A layer of mucus helps to protect the stomach lining from autodigestion by pepsin. Other components of gastric juice include intrinsic factor (a protein required for absorption of vitamin B_{12} in the small intestine) and rennin (an enzyme that curdles milk). Passage of chyme from the stomach into the small intestine is regulated by the pyloric valve.

Digestion continues within the first segment of the small intestine, the duodenum. Accessory structures, including the liver and pancreas, contribute digestive fluids to the chyme in the duodenum. Bile from the liver emulsifies lipids, thereby making them accessible to action by lipases, enzymes that hydrolyze lipids. What other functions does the liver have? Pancreatic juice is made of enzymes that digest proteins (proteases), carbohydrates (amylases), nucleic acids (nucleases), and lipids. Be sure to study the specific enzymes in each of these categories and to know the specific substrates (molecules and chemical groups) that they attack.

Very little absorption (assimilation) of nutrients occurs until food enters the small intestine. Absorption is enhanced by adaptations to the lining of the duodenum, jejunum, and ileum. Villi, fingerlike projections from the lining into the lumen, represent one of these adaptations that dramatically increase the surface area available for absorption. Waves of muscular contractions (peristalsis) encourage materials to move toward the posterior end of the tract.

By the time "food" reaches the large intestine, all of the nutritional absorption that will occur has occurred. Yet, the colon is more than merely a place to store feces. In the colon, considerable amounts of water are reclaimed from the feces. Additionally, this is a place where bacteria beneficial to nutrition are incubated. Once all processing of feces is completed, they are held in the rectum until a time appropriate for release through the anus.

It should be apparent that food and feces never were inside of the body. Rather, food has passed through a canal of outside space that the body surrounds.

E. Variations on a Theme: Adaptations in Other Animals.--The digestive system of humans reflects adaptations for a general (omnivorous or eurytrophic) diet including both plant and animal materials. Animals having different or stricter (stenotrophic) diets deviate from the general design in ways that facilitate efficient use of those diets.

In their alimentary tracts, many vertebrate herbivores (plant eaters) possess chambers in which microbes live. These microbes digest cellulose, the main carbohydrate of plants, for their hosts who cannot themselves digest cellulose. Hence, horses and deer can derive nutrition from plant material. The most specialized such arrangement is the ruminant stomach of ruminant artiodactyls (cud chewers). This stomach has four chambers; read in your textbook the role of each chamber in cellulose digestion. Many vertebrates (e.g., rabbits) have blind pouches (ceca) extending from their stomachs and/or intestines; these pouches are occupied by microbes that in various ways assist digestion. The teeth of herbivores also are adapted to this diet; numerous sharp cutting edges allow efficient shearing of the abrasive leaves of grasses and forbs.

Vertebrate carnivores (meat eaters) consume a diet that is much more easily digested than is plant material. Hence, carnivores generally lack ceca and other gut chambers. Their most notable digestive adaptations are the teeth, which are often blade-like to increase the efficiency of shearing meat.

Be sure to read in the textbook about digestive systems in invertebrate animals, including insects (Fig. 26-6), clams (Fig. 26-7), earthworms (Fig. 26-8), *Hydra* (Fig. 26-10), and flatworms (Fig. 26-14).

F. Dietary Requirements.--For a particular species, its dietary requirements are all of those materials it needs to generate energy and to construct its tissues. Clearly, autotrophs have fewer dietary requirements than do heterotrophs. The list of these requirements is different for each species. Humans, for example, can synthesize only a few of the basic materials; hence our dietary requirements list is longer than for many other species.

We must consume the proper balance of carbohydrates, fats, and proteins. Vitamins, too, must be provided in the diet because of their critical role in normal metabolism. Recall from Chapter 9 that vitamins are converted into coenzymes before they participate in cell metabolism. Vitamins are classified on the basis of their solubility in water or in lipids. Be sure to study textbook Table 26-4 for a list of the vitamins essential to humans, the food sources of these vitamins, and the symptoms of their deficiency in the diet. Another category of dietary materials essential to metabolism includes the mineral elements (textbook Table 26-5); notable examples are calcium, phosphorus, and iron.

Energy is meausred in calories. Most of the materials that organisms ingest contain not only elements and ions, but also energy. The energy contents for the three major categories of organic compounds in our diets are for carbohydrates, 4 Calories per gram, for proteins, 4 Cal/g, and for lipids, 9 Cal/g. The average male human requires about 3,000 Calories per day; females require slightly less, about 2,500 Cal/day. Sensible diets must take into account both caloric content and a proper balance of minerals and vitamins.

G. Body Temperature.--All matter, including bodies of organisms, has heat content, measurable as temperature. Life is restricted to a very narrow range of the temperatures that exist on Earth. As you recall from Chapter 8, enzyme function and rates of chemical reactions are temperature dependent. Also, the fluidity and elasticity of protoplasm are sensitive to temperature. It is apparent, then, that organisms must be cognizant of the heat content of the environments they occupy. Thermoregulation is a critical concern of organisms.

(1) Thermal stability of habitats. Those species that live in thermally stable environments, however, have not as great a need to actively regulate body temperature as do those in thermally variable situations. For example, the high specific heat of water generally prevents vast fluctuations in water temperature. Many aquatic organisms, then, are adapted to narrow temperature ranges; small changes in water temperature can have drastic effects on these species.

The temperature extremes in terrestrial environments are much greater than in water. Consequently, many land-dwelling species have adapted to operating over a broad range of temperatures. Anatomical adaptations include evolution of insulative materials, such as hair, feathers, blubber, etc. Physiological adaptations include evolution of a series of enzymes for each metabolic reaction, each enzyme in a series operating over a different temperature range. Also, blood-vessel diameter changes either to restrict blood flow to the interior of the body or to enhance blood flow to the periphery; this strategy of vasoconstriction and vasodilation works because blood carries body heat. Behavior is another means to avoid temperature extremes (or other unsuitable conditions). Animals move into or out of shade, or into or out of water. Some species take this approach to the extreme of migrating seasonally.

(2) Classification of organisms on the basis of body heat. The terms "warm-blooded" and "cold-blooded" mean little to scientists. Two other pairs of terms much better convey the information so poorly denoted in the previous pair of terms. One suitable way of categorizing

animals is to determine the source of their body heat. Those that obtain most of their body heat from the environment are called ectotherms, whereas those generating their own body heat via metabolism are endotherms. Alternatively, we can examine the degree of constancy of body temperature. Those with relatively stable body temperature are homeotherms; those that allow body temperature to vary with the environmental temperature are poikilotherms (=heterotherms). The graphs below should help you see how much more metabolically expensive it is to be a homeotherm or endotherm compared to being a poikilotherm or ectotherm.

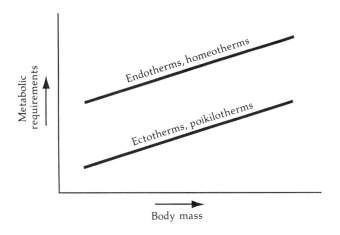

Fig. 26.1: Metabolic requirements compared to body mass for different metabolic strategies

These two pairs of terms are not synonyms, but the groups of vertebrates they indicate overlap greatly (but not completely). Both ectotherm and poikilotherm broadly denote fishes, amphibians, and reptiles, whereas both endotherm and homeotherm broadly indicate birds and mammals. Some exceptions exist; there are endothermic poikilotherms and ectothermic homeotherms.

TERMS TO UNDERSTAND: Define and identify the following terms:

autotrophs

heterotrophs

ingestion

digestion

assimilation

food vacuole

alimentary canal

bulk feeders

gizzard

proventriculus

filter feeders

fluid feeders

deposit feeders

macronutrients

micronutrients

accessory glands

mouth

pharynx

esophagus

stomach

duodenum

jejunum

ileum

cecum

colon

rectum

anus

salivary glands

pancreas

peristalsis

hydrolases

mastication

gastric juice

salivary amylase

chyme

pepsin

mucus

pepsinogen

rennin

omnivorous

herbivorous

cellulases

ruminant

coprophagy

proventriculus

pancreatic juice

bile

proteolytic enzymes

lipases

villi

feces

vitamins

calorie

basal metabolism

thermophilic bacteria

acclimation

homeothermic

poikilothermic

ectotherms

endotherms

countercurrent exchange

CHART EXERCISE

Complete the chart below to indicate (1) the organ or structure that produces each enzyme, (2) where the enzyme is released, (3) the enzyme's function, and (4) the identity of the precursor of the enzyme (if applicable).

ENZYME	PRODUCED BY	RELEASED IN	FUNCTION	PRECURSOR
Amylase				
Trypsin				
Carboxypeptidase				
Enterokinase				
Maltase				
Lipase				
Rennin				
Pepsin				

MATCHING

Match the diseases in the left column with the appropriate description in the right column.

	DISEASE	DESCRIPTION
_____	1. pellagra	A. painful joint disease resulting from vitamin C deficiency
_____	2. pernicious anemia	B. results from deficiency of nicotinic acid
_____	3. scurvy	C. abnormal appetite loss
_____	4. anorexia	D. due to absence of intrinsic factor that transports vitamin B_{12} across intestine wall

DISCUSSION QUESTIONS

1. Discuss the various anatomical adaptations that accompany mutualistic associations in which microorganisms assist host species in food digestion.

2. Discuss the structures and mechanisms by which nutrients and minerals enter the tissues of plants and how those materials are transported to sites of use.

3. Discuss the major evolutionary trends seen in the digestive anatomy from protists to simple animals to complex animals.

4. Discuss the chemical mechanism by which digestive enzymes work to convert large, complex food molecules into simpler molecules.

5. Summarize the sequence of events by which lipids are digested, absorbed, and transported into the body of vertebrate animals.

6. Summarize the roles of the liver and the pancreas in nutrition in vertebrates.

7. What factors should be considered when planning a diet for the purpose of reducing body mass?

8. What is the effect of body size and shape on the strategies by which nutrients are introduced into an organism's body?

9. Discuss the correlation between feeding mode used by a species and the type of environment occupied by that species.

10. Compare and contrast the nutritional needs of autotrophs and heterotrophs.

11. Endotherms and ectotherms have different metabolic strategies. Compare these strategies. Which is the better strategy?

TESTING YOUR UNDERSTANDING

For each of the test items below, three of the lettered alternatives are true and the other is false. Determine which alternative is false and write its letter in the blank to the left of the question number. On the blank line below alternative D, write the corrected version of the false statement.

_____ 1. A. Green plants are photosynthetic autotrophs.
B. Fungi are heterotrophs.
C. Humans have an omnivorous diet.
D. Plants must obtain as many nutrients from their environments as vertebrate animals must obtain from their surroundings.

Correction: _____

_____ 2. A. Ingestion refers to taking of food materials into an organism's body.
B. Assimilation refers to the mechanical and chemical processes of breaking large molecules into smaller molecules.
C. Materials absorbed through the walls of the jejunum of a mammal are transported throughout the body through blood vessels and lymphatic vessels.
D. Absorption of nutrients by tissues of an endotherm occurs in the stomach and the small intestine.

Correction: _____

_____ 3. A. In some protists, digestion of food occurs in the contractile vacuole.
B. Extracellular digestion characterizes some protists, some fungi, and some simple animals.
C. Humans can be correctly classified as bulk feeders.
D. Among the various structures that various animals use to mechanically degrade food are teeth, gizzards, and proventriculi.

Correction: _____

_____ 4. A. Organisms that obtain their food by screening the surrounding medium are classified as filter feeders.
B. Some mammals and some insects are fluid feeders.
C. Nutrients enter the roots of plants by diffusion and active transport.
D. Micronutrients required by plants include calcium and nitrogen.

Correction: _____

_____ 5. A. The soils occupied by carnivorous plants are deficient in nitrogen.
B. The alimentary canal includes the pharynx, pancreas, and ileum.
C. Amylases are released into the mouth and duodenum of humans.
D. In vertebrates, waves of coordinated contractions of muscle in the intestinal wall that move chyme towards the anus are called peristalsis.

Correction: _____

_____ 6. A. All of the following are structures associated with microbial digestion of cellulose in the guts of various vertebrates: rumen, omasum, and cecum.
   B. All of the following are produced by accessory structures of the vertebrate digestive system: saliva, pancreatic juice, and bile.
   C. Cooking facilitates digestive efficiency by denaturing proteins and by bursting granules of strarch.
   D. The primary role of the colon is to absorb nutrients from the feces.

   Correction: _____

_____ 7. A. The enzyme-mediated chemical reactions that occur in digestion are classified as condensation reactions.
   B. Gastric juice contains pepsin and hydrochloric acid.
   C. Pancreatic juice contains nucleases, proteases, and lipases.
   D. By emulsifying lipids, bile enables lipases to attack lipids.

   Correction: _____

_____ 8. A. The stomach and the rectum function as storage chambers.
   B. Intrinsic factor, a component of saliva, is a protein necessary for the absorption of vitamin B_{12} by the small intestine.
   C. The substrate of rennin is milk.
   D. Coprophagy is a behavior that increases digestive efficiency in some herbivores.

   Correction: _____

_____ 9. A. Both trypsin and enterokinase are enzymes that catalyze the conversion of trypsinogen into trypsin.
   B. All of the following are proteolytic enzymes: carboxypeptidase, chymotrypsin, and pepsin.
   C. All of the following are adaptations for increasing surface area available for absorption: villi, microvilli, and circular folds.
   D. Pancreatic juice is secreted into the ileum.

   Correction: _____

_____ 10. A. Before their use in metabolism, vitamin molecules are converted into coenzyme molecules.
   B. Sugar is stored in the liver as bile.
   C. One of the functions of the liver is to remove ammonia from the blood.
   D. The liver detoxifies many harmful substances.

   Correction: _____

_____ 11. A. Lipids have greater caloric content than do equal quantities of either protein or carbohydrates.
B. A calorie is the amount of heat needed to raise the temperature of one gram of water by one degree Celsius.
C. The daily caloric needs of an average male human performing average amount of work is about 4,000 calories.
D. The heat content of a substance can be measured by using a bomb calorimeter.

Correction: _____

_____ 12. A. Birds and mammals are classified both as ectotherms and homeotherms.
B. An endotherm of 10-kg body mass has a higher metabolic rate than does an ectotherm of the same body mass.
C. Aquatic environments have greater thermal stability than do terrestrial habitats.
D. All of the following are strategies for dealing with adverse temperature conditions: dropping leaves, depositing layers of adipose tissue, migrating, and vasodilation.

Correction: _____

CHAPTER 27--RESPIRATION

CHAPTER OVERVIEW

Respiration, a term you encountered earlier in the textbook, refers both to the biochemical pathways by which energy is released from organic compounds and to the topic of this chapter. Here, we examine various anatomical and physiological adaptations of organisms for exchanging respiratory gases (oxygen and carbon dioxide) with the environment and for transporting them between the respiratory surfaces and the tissues that need or produce those gases.

The need to exchange respiratory gases with the environment resulted in the evolution of many different types of respiratory surfaces, including skin, gills, tracheae, lungs, and others. Additionally, a number of different means of internal transport evolved as well. In many animals, internal transport of gases is facilitated by respiratory pigments, some of which are contained within blood cells. We'll examine the physiology of blood chemistry as it relates to transport of respiratory gases.

TOPIC SUMMARIES

A. <u>Two Definitions for Respiration</u>.--We already have studied in some detail (Chapter 9) the topic of <u>cellular respiration</u>--the metabolic pathways by which energy is liberated from organic compounds such as glucose. To refresh your memory, you probably should review the major points of cellular respiration in this Study Guide and in the textbook.

We turn now to a different, yet related, type of respiration. In this chapter, we study <u>gross-level respiration</u>--the anatomy and physiology of how respiratory gases are delivered to and from exchange surfaces and how these gases are delivered to and from the tissues within the body.

B. <u>Respiratory Surfaces</u>.--Many organisms require free oxygen from the environment to drive their metabolic processes. This oxygen must cross the membranes of one or more cells (the number depending on the size of the organism) to enter the organism and its internal transport system. As with other biological challenges, organisms have evolved a diverse array of structures and mechanisms for moving oxygen into (and the waste gas, carbon dioxide, out of) the body (Fig. 27.1). These sites of respiratory gas exchange are called the <u>respiratory surfaces</u>.

(1) In monerans, protists, fungi, and plants. Plants and some protists conduct both photosynthesis and respiration. Hence, they, like most other organisms, must acquire oxygen from their surroundings. No special respiratory organs are necessary in protists, monerans, and many aquatic plants; the surface-to-volume ratio is favorable enough and the distance from the surface to the exterior is small enough that simple diffusion is sufficient to transport gases.

Larger plants, particularly those living on land, have added a layer of morphological complexity to the process of diffusion. Recall that diffusion across biological membranes requires that the diffusing substances be dissolved in water. Terrestrial organisms face the problem of dehydration. Hence, anatomical arrangements have evolved (in plants and many animals as well) to minimize water loss from these surfaces while still allowing adequate exposure of the respiratory surfaces to enable adequate gas exchange. In larger plants, air enters the intercellular spaces within the leaves by passing through the <u>stomata</u>. These openings are regulated by <u>guard cells</u>. Diffusion into the leaves elsewhere is prevented by the impervious, waxy <u>cuticle</u>. In woody parts of plants, gas exchange also occurs via <u>lenticels</u>

(textbook Fig. 27-23), small regions of loosely packed cells that form in the cork layer of the periderm. The "knees" of cypress and mangrove trees represent adaptations to enhance gas exchange.

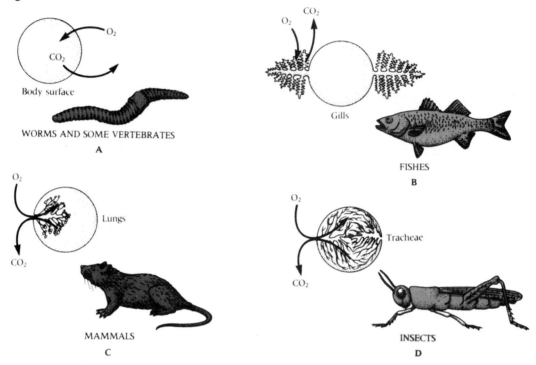

Fig. 27.1: Diversity of respiratory surfaces in several animals

(2) Simpler approaches used in animals. Many of the smaller, aquatic animals (e.g., many invertebrates, including *Hydra*) rely solely on diffusion across their outer layers of tissue for gas exchange. Other invertebrates have supplemented this trans-integument strategy with other means. Some worms, for example, experience diffusion of gases across the thin walls of the gut; other worms alternately take oxygen-carrying water into, and force water out of, the hindgut (textbook Fig. 27-3). In some echinoderms (e.g., sea cucumbers; textbook Fig. 27-24), water is pumped into the anus and circulated throughout a tubule system that can be thought of as a respiratory tree.

In many of the more complex animals (e.g., some vertebrates), diffusion of gases across the skin is an important contributor to the total gas supply. Although fishes and larval amphibians have gills, much gas moves across the skin (a process called cutaneous respiration). The adults of amphibian species that live in water use cutaneous respiration in addition to using lungs.

(3) More complex approaches used in animals. As animals evolutionarily became larger, cutaneous and other simple means of respiration became inadequate to meet metabolic oxygen demand. In response, several different respiratory structures evolved with the effect of tremendously increasing the surface area available for gas exchange.

Tracheae. Many terrestrial arthropods (insects, centipedes) have a novel breathing strategy. A system of branched tubes, the tracheae, extend throughout the body (textbook Fig. 27-9). The tracheae branch repeatedly, terminating in very narrow diameter tracheoles that deliver air to each body cell. No circulatory system is involved. Rather than having a single opening to the exterior by which air enters the system, there a numerous paired openings (spiracles) through the body wall.

Gills. Fishes, in general, and some arthropods, mollusks, and larval amphibians have gills of various designs. Internal gills of fishes are embedded or otherwise attached along gill slits through the body wall between the pharynx and the external surface just behind the eyes. Oxygen-laden water enters the mouth and flows rearward through the gill chambers to the outside. Alternatively, some fishes pump water in and out through the external gill openings. As water passes over the gill surfaces (the process of ventilation), gases are exchanged by diffusion. Dense blood capillary beds lie very close to the gill surfaces. Efficiency of gas exchange is enhanced by the countercurrent arrangement of blood flow and water flow (textbook Fig. 27-4).

External gills, such as in larval amphibians and some fishes, function on the same principle as internal gills. However, there are no gill chambers. Rather, these feathery outgrowths of the body wall simply extend into the surrounding water.

Lungs. According to many anatomists, swim bladders and lungs are essentially the same structures. These occur in most fishes, most adult amphibians, and in all reptiles, birds, and mammals. Both lungs and swim bladders develop as a branching series of air passageways extending from the pharynx. Air entering the nose and mouth passes through the pharynx before moving into the lungs. The surfaces of these passageways (trachea, bronchi, bronchioles, alveoli) are moist, and gases are exchanged across some of these surfaces. The degree of splitting of these passageways and the amount of exchange surface available varies according to major taxonomic group and metabolic demands. In all cases, however, the exchange surfaces are intimately associated with capillary beds.

Some fishes (especially the lungfishes) rely almost exclusively on gas exchange within the lungs. Other fishes use lungs merely to supplement gills. Still other fishes have swim bladders that do not open to the exterior; rather, the gas in these bladders is generated by special structures called gas glands (textbook Fig. 27-25). Clearly, such bladders have no respiratory function. Instead, they are used to passively maintain vertical position in the water column (neutral buoyancy), and to receive and/or transmit sound.

Ventilation of lungs involves muscular force. Most reptiles, birds, and mammals use negative-pressure breathing. In mammals, the diaphragm and rib cage muscles contract to expand the lungs, thereby creating a negative pressure into which air flows; this is the inspriation phase of breathing. Expiration involves some muscular force but also elastic recoil of the rib cage to force air out of the lungs. The design of the lungs of birds and the high efficiency of gas exchange in their lungs are related to the particularly high metabolic demands of flight. An alternate method of ventilating lungs, positive-pressure breathing, is used by many adult amphibians; air contained in the mouth cavity is forced into the lungs by contraction of muscles on the floor of the mouth (textbook Fig. 27-25).

C. Physiology of Gas Exchange and Transport.--For most animals, getting oxygen across the respiratory surface is only part of the total process of respiration. The rest of the story concerns how respiratory gases are moved between the respiratory exchange organs (skin, lungs, gills) and the tissues that use oxygen and produce carbon dioxide. Let's use humans as a model of the process.

Oxygen is transported in the bloodstream in two ways: some in solution in the plasma, but most bound chemically with hemoglobin (Hb) of erythrocytes (as oxyhemoglobin). Carbon dioxide is carried in three ways: a small amount dissolved in plasma, about 25% bound to hemoglobin as carbaminohemoglobin, and the majority as bicarbonate ions. Formation of bicarbonate ions is a two-step process: Carbon dioxide and water combine, in the presence of the enzyme carbonic anhydrase, to form carbonic acid (H_2CO_3). This acid then dissociates as hydrogen ions and bicarbonate ions (HCO_3^-). The affinity of Hb for oxygen is affected by changes in pH (acidity) of the blood. As more carbon dioxide enters this carbonic anhydrase system, more hydrogen ions are released, and blood acidity increases. An increase in plasma acidity causes Hb to release even more oxygen; see textbook Box 27-1 for discussion of the Bohr effect. This is beneficial because the events that increase carbon-dioxide pressure (e.g., exercise) consume oxygen and therefore leave working muscle tissues with an oxygen debt. Hence, production of carbon dioxide (and lactic acid from muscle metabolism) stimulates oxygen release from Hb.

In Chapter 22, you were introduced to the oxygen dissociation curve, which describes the affinity of oxygen for Hb and, thereby, indicates the tendency of Hb to unload oxygen in situations of different oxygen concentration. Study of textbook Fig. 27-14 shows the wide physiological range of oxygen concentrations occurring in the human body. Notice that the steep part of the curve is within the physiological range; this means that the process of unloading at the tissues is very sentitive to changes in oxygen pressure. You can read from the curve that Hb is fully saturated with oxygen at the oxygen pressures found in the lungs, and that Hb has released (at the capillary beds) the majority of its oxygen by the point the blood has entered the venous flow.

Fortunately, carbon dioxide differs from oxygen in its affinity for Hb. The carbon-dioxide dissociation curve (textbook Fig. 27-14) shows that very little carbon dioxide is released in the physiological range of carbon dioxide pressures. Virtually all carbon dioxide is unloaded at the alveoli of the lungs. The decrease in carbon dioxide concentration across the thin alveolar membrane is great--from about 30 torr in the blood to nearly 0 torr.

The human nervous system is sensitive to changes in blood chemistry. The respiratory center of the medulla oblongata is "wired" to a number of chemoreceptor structures in various arteries. These receptors detect falling oxygen levels, increases in carbon dioxide, and increases in acidity of the blood. Any of these stimuli cause increased respiratory rates.

These general principles hold for a wide range of vertebrate species. However, there are variations on the theme in accordance with environmental conditions. What changes in oxygen-dissociation behavior would you expect for species that live at high altitudes . . . or for air-breathing species that do not respire during prolonged underwater dives? What respiratory challenges are presented by space travel?

TERMS TO UNDERSTAND: Define and identify the following terms:

inspiration

expiration

oxidation

phosphagens

creatine phosphate

fermentation

gills

operculum

lungs

alveoli

bronchioles

ventilation

tracheae

spiracles

tracheoles

diaphragm

respiratory muscles

pneumothorax

respiratory center

chemoreceptor cells

acidosis

alkalosis

oxyhemoglobin

carbaminohemoglobin

bicarbonate ions

carbonic anhydrase

gas curves

Boyle's Law

nitrogen narcosis

lenticels

respiratory tree

positive-pressure breathing

negative-pressure breathing

gas bladder

gas gland

neutral buoyancy

DISCUSSION QUESTIONS

1. Describe the essential characteristics of a respiratory exchange surface.

2. Discuss the ways in which panting is important in maintaining homeostasis.

3. Discuss the physiological relationships of the following materials: creatine phosphate, lactic acid, glycogen, and pyruvic acid. Can you develop a single metabolic scenario in which these compounds are related?

4. The observation that really large insects never evolved has been blamed on their mechanism of respiration. Explain.

5. Recall from Chapter 26 that endothermy is a much more expensive lifestyle than is ectothermy. With this in mind, consider the following statement: Virtually all animal species that occupy aquatic environments (and that extract oxygen from the water) are ectotherms. Why are there so few (perhaps none) endothermic species that extract oxygen from water?

6. What gases are exchanged by plants, and how do these gases relate to physiological processes?

7. Compare and contrast the structure and function of the respiratory system in mammals and birds. Which system is more efficient? With what activity(ies) does greater respiratory efficiency correspond?

TESTING YOUR UNDERSTANDING

For each of the test items below, three of the lettered alternatives are true and the other is false. Determine which alternative is false and write its letter in the blank to the left of the question number. On the blank line below alternative D, write the corrected version of the false statement.

_____ 1. A. The oxygen dissociation curve of a diving mammal (such as a seal) is positioned to the left of the corresponding curve for a terrestrial mammal living at sealevel.
   B. The release of energy from carbohydrates is an oxidation reaction.
   C. Green plants conduct both photosynthesis and respiration.
   D. Lactic acid is a phosphagen, a compound that holds reserves of high-energy phosphate bonds.

   Correction: _____

_____ 2. A. Panting converts lactic acid into glycogen.
   B. Organisms with low surface-to-volume ratios rely on diffusion alone as their means of gas exchange.
   C. One of the shortcomings of the tracheae system of insects is that the rate of transport is slow.
   D. The operculum is a flap of tissue that covers the gills.

   Correction: _____

_____ 3. A. In the gills of a fish, blood in the capillaries flows in a direction opposite to the flow of water.
   B. Respiratory surfaces are places in which the blood stream is brought into close contact with a continuously renewed supply of oxygen across a thin membrane of extremely large area.
   C. In virtually all insects, diffusion of gases through the tracheae system is enhanced by rhythmic pumping of their abdomens and synchronized opening and closing of the spiracles.
   D. The diaphragm is a muscularized partition found only in mammals.

   Correction: _____

_____ 4. A. All of the following are structures through which air passes en route to the respiratory exchange surfaces of mammals: trachea, larynx, bronchioles, and pharynx.
   B. The respiratory center of mammals is located in the medulla oblongata.
   C. Chemoreceptor cells related to breathing respond to falling oxygen levels and increasing carbon-dioxide levels in the blood.
   D. An increase in the hydrogen-ion concentration in plasma causes the condition of alkalosis.

   Correction: _____

_____ 5. A. Hyperventilation causes excessive loss of carbon dioxide from the blood stream.
   B. Increasing the concentration of carbon dioxide in the blood causes an increase in the pH of the blood.
   C. Carbonic anhydrase catalyzes a reaction that converts carbon dioxide and water into carbonic acid.
   D. Carbonic acid dissociates in water into bicarbonate ions and hydrogen ions.

   Correction: _____

_____ 6. A. The majority of the carbon dioxide that is carried in blood is dissolved in the plasma.
   B. The majority of oxygen carried in blood is in the form of oxyhemoglobin.
   C. Increasing the acidity of plasma causes the oxygen dissociation curve to shift to the left.
   D. Fetal hemoglobin has a greater affinity for oxygen than does adult hemoglobin.

   Correction: _____

_____ 7. A. Increased production of erythropoietin by the kidney accompanies deprivation of oxygen.
   B. Increased production of red blood cells occurs when a person who normally lives at low elevations moves to higher elevations.
   C. About 21% of the composition of Earth's atmosphere is oxygen.
   D. The proportion of the atmosphere comprised by carbon dioxide is greater than that made up of nitrogen.

   Correction: _____

_____ 8. A. The "bends," a condition resulting from an excessively rapid ascent from a deep underwater dive, can lead to the formation of air embolisms.
   B. Guard cells, stomata, and lenticels are structures related to respiration in plants.
   C. Many plants have a highly branched conducting system, the respiratory tree, that carries the respiratory stream throughout the plant body.
   D. In some invertebrate animals, the hindgut is involved in both respiratory and digestive functions.

   Correction: _____

_____ 9. A. Vertebrates that use negative-pressure breathing rely on musculature of the floor of the mouth to force air into the lungs.
B. Bird lungs do not have alveoli.
C. The gas in swim bladders of some fishes is produced by a structure called the gas gland.
D. Functions of swim bladders of fishes relate to hearing and neutral buoyancy.

Correction: _____

_____ 10. A. Exchange of gases across the skin occurs in some amphibians.
B. The air sacs of the respiratory system of birds are sites of gas exchange.
C. All species that exchange gases with the environment rely on diffusion.
D. One consequence of lactic-acid accumulation is muscle fatigue.

Correction: _____

CHAPTER 28--CHEMICAL COORDINATION

CHAPTER OVERVIEW

Individual parts of an organism, regardless of the species, do not function independently of the rest of the organism. Rather, all parts compose a single organism whose parts function in a coordinated fashion to maintain homeostasis.

Such coordination suggests that some form(s) of communication among the parts must exist. In all organisms, some form(s) of chemical communication occurs. In the higher animals this chemical communication system, called the endocrine system, is complex. This chapter focusses on the endocrine systems of animals and the phytohormone system of plants. We'll examine the hormones produced, what organs produce them, their physiological effects, and the mechanisms by which they carry their messages. Because so much of the research and knowledge of hormones relates to humans, we'll concentrate on humans. In this Study Guide, we'll work at the broader level of principles and mechanisms, and leave much of the details to the textbook.

A second major type of communications network, a nervous system, exists in most animal species. The next two chapters (29 and 30) treat that topic.

TOPIC SUMMARIES

A. What Is a Hormone?--Hormones are chemical messengers. They are produced in one or several places in the body (by endocrine glands or tissues) and exert a particular physiological influence on a target structure somewhere else in the body. Endocrine secretions are not delivered directly to the target, but are transported generally throughout the body. In higher animals, the bloodstream is the transport mechanism.

Probably more than 50 different hormones are produced in the human body. Chemically, these hormones belong to four major groups: Many are steroids, a group of lipids. Some are amine or amino- acid derivatives. Others are polypeptides. The fourth group is a catch-all category of various other kinds of organic molecules. Return to Chapter 4 to refresh your memory of the characteristics of these chemical groups.

B. Hormone Receptors.--If hormones are present throughout the body, then how is it that some cells respond to the particular hormone molecules and other cells do not? The answer lies in the presence of chemical receptors within or on the surface of the target cells. Both the hormone molecules and the receptor molecules are highly specific in their ability to recognize and bind to other molecules.

Receptor molecules occur in three general areas of a cell: Receptors for lipid-insoluble hormones occur on or within the cell membrane. (Recall that the plasma membrane is made of phospholipids.) Some lipid-soluble hormone molecules pass through the cell membrane and directly into the nucleus where they bind directly with the chromatin, thereby stimulating or repressing production of certain mRNA molecules. Other lipid-soluble hormone molecules pass through the cell membrane and bind to a receptor molecule somewhere in the cell; the bound complex then moves into the nucleus to bind to the chromatin to affect mRNA production.

C. Mechanisms of Hormone Action.--Hormones affect metabolism in a variety of ways. They can alter rates of enzyme activity. They can stimulate enzyme synthesis and affect the synthesis of other proteins. They can alter cell-membrane permeability. They can stimulate the activity of target enzymes.

Just how hormones can do these things is explained by the "second messenger" mechanism. Several different second messengers have been identified, each regulating different cellular activities. Cyclic AMP was the first such messenger discovered: In this system, binding of a specific hormone and its membrane-bound receptor activates the enzyme adenylate cyclase. This enzyme catalyzes the conversion of ATP into cyclic AMP, which in turn stimulates certain protein kinases; these enzymes phosphorylate various other enzymes and thereby affect their rate behavior. As you should suspect (from the location of the receptor molecules), this mechanism works for hormones that do not enter the cell.

Calcium ions, in association with the protein calmodulin, is another widespread second messenger. The calcium-calmodulin complex activates certain enzymes that regulate a vast array of physiological activities, including muscle contraction, nerve impulse transmission, chromosome movement, and many others. Other second messengers are listed in textbook Table 28-2.

D. Growth Factors.--Like hormones, growth factors (GF's) are chemical messengers. Growth factors are polypeptides that bind to specific, high-affinity surface receptors. They exert effects at extremely low concentrations. All types of growth factors function as mitogens, agents that stimulate mitosis and cell division. The main differences between hormones and growth factors pertain to mode of synthesis and mode of delivery. Delivery is via diffusion rather than by the bloodstream or by ducts. Target cells can be the very cell which produces the GF (autocrine secretion) or nearby cells (paracrine secretion). GF's are stored in vesicles or granules of the cells that synthesize the factors. They function by protein phosphorylation. In many situations, several GF's cooperate to exert an effect.

An array of GF's is known and more are being discovered. Examples include nerve GF (stimulates growth of nerves, fibroblasts, leukocytes; stimulates healing of wounds), platelet-derived GF (stimulates growth of fibroblasts, smooth muscle cells, certain nervous system cells, etc.), epidermal GF, transforming GF-beta (the "master" GF because it works through intervention of other GF's), colony-stimulating GF (promote leukocyte and macrophage production), and tumor angiogenesis factor (stimulates vascularization of tumors).

E. Endocrine System of the Human Body.--As noted in the Chapter Overview, we'll not examine each hormone in detail, but will work at a broader level. Work with the textbook to learn the production site(s), specific secretions, and effects of each hormone.

As in most vertebrates, the human endocrine system consists of a number of organs located throughout the body. Fig. 28.1 below shows many of the endocrine structures in a silhouette of the human body; as part of your study, indicate the hormone(s) produced by labelled endocrine structure.

F. Adenohypophysis and Physiological Regulation of Endocrine Secretions.--Although not all endocrine structures are anatomically coordinated, many are coordinated physiologically. The nervous system interfaces physically and physiologically with the endocrine system at the hypothalamus and pituitary of the brain.

The anterior portion of the pituitary (adenohypophysis) often is referred to as the "master gland" because of the number of hormones it produces (at least 9) and because many of its hormones (the tropic hormones) influence growth and endocrine function of other endocrine organs. The four tropic hormones, and the structures they influence, are adrenocorticotropic hormone (adrenal cortex), thyroid-stimulating hormone (thyroid gland), follicle-stimulating hormone (gonads), and luteinizing hormone (gonads).

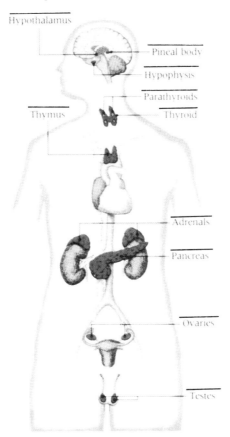

Fig. 28.1: Locations of major endocrine glands in human body

The adenohypophysis is regulated in two ways. One type of regulation, negative feedback, is characteristic of most endocrine organs. In the case of the tropic hormones of the adenohypophysis, the target organs of the tropic hormones directly influence the master gland. As the endocrine product of the target organ reaches appropriate levels, the adenohypophysis slows or ceases secretion of the hormone that causes the target organ to release its hormone.

The second means of regulating adenohypophysis activity emanates from the brain: The hypothalamus of the brain secretes hormones that pass through the hypothalamo-hypophyseal portal system. Some of these hormones, called releasing factors, stimulate the hypophysis to release certain hormones; others, called release inhibiting factors, inhibit the release of hormones from the hypophysis.

The other (non-tropic) endocrine secretions of the anterior pituitary directly influence non-endocrine tissues. Growth hormone is essential for normal growth. Prolactin stimulates breast development and lactation. The recently discovered beta-lipotropin probably is involved in formation of endorphins and enkephalins, a group of brain peptides that reduce pain and that alter mood and behavior.

G. Neurohypophysis.--This structure, also known as posterior pituitary, releases antidiuretic hormone (ADH, = vasopressin) and oxytocin. These secretions are produced in the hypothalamus (part of the brain) and transported to the neurohypophysis via neurons (nerve cells). Recall from the chapter on excretion that higher levels of ADH cause the kidneys to produce a concentrated urine; conservation of body water results from ADH's effect of increasing the permeability of collecting tubule walls to water. ADH also increases blood pressure.

Oxytocin has several effects related to female reproductive functions. During childbirth, oxytocin stimulates contractions of the muscles of the uterus. Oxytocin also causes ejection of milk from the breasts, and it hastens the onset of the estrus cycle.

H. Pineal Body.--Also known as the epiphysis, the pineal body is both a glandular structure and a photoreceptor. Although both regions function in many of the vertebrates, the light-receptive function has been lost in mammals. Melatonin is the hormone of the pineal. This hormone influences the rhythm of many physiological processes; as a major function, melatonin suppresses growth and function of gonads. In certain less-derived vertebrates (e.g., some cyclostomes, frogs, and reptiles), the pineal functions as a "third eye" that detects light intensity but that generally does not form images. The hormonal and photoreceptive functions of the pineal are related: the pineal is an example of a neuroendocrine transducer, an endocrine gland that secretes its hormone in respones to local neural influences. (What other structure already studied in this chapter is a neuroendocrine transducer?) Even in vertebrates in which the photoreceptive ability is lost, melatonin production is keyed to light intensity; the local neural influence is provided by neural connections to the "regular" lateral eyes.

I. Adrenal Glands.--Functionally, this pair of organs really is two pairs of glands. The adrenal medulla produces epinephrine and norepinephrine, whereas the adrenal cortex produces several adrenocortical hormones.

Function of the adrenal cortex is stimulated by one of the tropic hormones, adrenocorticotropic hormone. All endocrine secretions of this gland are steroids and can be grouped into three categories: Mineralocorticoids (e.g., aldosterone) act upon the content of various ions (sodium, potassium, chloride) in extracellular fluids. Aldosterone, for example, effects the amount of sodium excreted by the kidneys, and therefore influences urine concentration and volume, cardiac output, and blood pressure. Glucocorticoids, such as corticosterone, have an array of effects related to metabolism of carbohydrates, lipids, and proteins; one of these effects is elevation of blood sugar levels. Although the gonads produce most of the sex hormones, the adrenal cortex produces some androgens; these mimic the effects of testosterone (produced by testes), the development of secondary sexual characteristics in males.

The two hormones produced by the adrenal medulla, epinephrine and norepinephrine, are quite similar chemically and have similar effects. These hormones mimic the effects of stimulating the sympathetic nervous system--the "fight or flight" response (much more on this in Chapter 30). This response involves elevated physiological levels in response to situations of danger or emergency. Particular physiological effects include elevated blood pressure, increased heart rate, dilation of pupils of eyes, and dilation of bronchioles of the lungs.

J. Thyroid Gland.--This small gland produces two hormones, each of which contains iodine. Thyroxine (produced by follicular cells) has widespread effects, influencing the rates of virtually every aspect of intracellular metabolism in the body. Calcitonin (produced by parafollicular cells) regulates calcium metabolism.

K. Parathyroid Glands.--Coupled with calcitonin, parathyroid hormone (PTH) regulates blood levels of ionic calcium. Specifically, PTH increases blood calcium levels by degrading bone, by reducing calcium excretion in the kidneys, and by increasing calcium absorption in the intestines.

L. <u>Gonads</u>.--Many of the hormones that effect the activities of other sex organs are secreted by the gonads themselves. Both sexes produce and secrete both male and female sex hormones; "maleness" or "femaleness" corresponds to the relative amounts of the two sets of hormones produced. Let's examine similarities and differences of the sex hormones by sex:

(1) Females. The <u>ovaries</u> are the main sources of estrogens and progesterones in non-pregnant females. <u>Estrogens</u> have several effects including stimulation of secondary sexual characteristics, generally increased metabolic rate, changing the amounts of FSH and LH by the adenophyophysis, and others. Cyclic production of estrogens causes the proliferative phase of the uterine cycle. <u>Progesterones</u> have the general effects of preparing the uterus and breasts for pregnancy and lactation, respectively. High concentrations of progesterone inhibit ovulation. During pregnancy, these female hormones also are secreted by the placenta.

(2) Males. The primary male sex hormone is <u>testosterone</u>, produced by the <u>interstitial cells of Leydig</u> within the <u>testes</u>. The most obvious effects of testosterone are the development of primary and secondary sexual characteristics. Additionally, androgens affect bone growth, stimulate protein anabolism and gluconeogenesis, increase metabolic rate, and have other effects. Another androgen is <u>inhibin</u>, a hormone that decreases FSH secretion.

M. <u>Pancreas</u>.--In Chapter 26, we discussed one of the major functions of the pancreas, production of a digestive fluid called pancreatic juice. As a compound organ, the pancreas has endocrine functions as well. Two hormones, <u>insulin</u> and <u>glucagon</u>, are produced by different groups of cells in the <u>islets of Langerhans</u>. Interestingly, these hormones drive the same metabolic pathway--but in opposite directions. Insulin encourages the conversion of blood sugar (<u>glucose</u>) into animal starch (<u>glycogen</u>) that is stored in the liver. Glucagon causes glycogen to be converted into glucose.

N. <u>Heart</u>.--Only recently, it was discovered that the muscle cells of the atrium of the heart produce a hormone, <u>atrial natriuretic factor</u> (ANF). ANF is involved in water balance. Specifically, it promotes urine production by promoting sodium excretion and inhibiting aldosterone secretion. A consequence of secretion of ANF is a lowering of blood pressure.

O. <u>Hormones in Non-human Animals</u>.--In general, most other mammals have endocrine systems resembling that of humans. Yet, some differences exist. Hormones are often species-specific. For example, the growth hormone of humans is chemically (and therefore functionally) different than the growth hormone in other mammals. Research has revealed other interspecific inequivalencies of hormone effects and target organs.

In many aspects of biological design and function, invertebrates and vertebrates differ. Differences in endocrine systems are reasonably expected as well. <u>Metamorphosis</u> (development) of insects involves a series of <u>molts</u> in which the exoskeleton is shed. Three hormones are involved in this process: <u>Alpha- and beta-ecdysone</u> induce molting; <u>juvenile hormone</u> inhibits the metamorphosis of larvae and pupae. Another aspect of biology in many invertebrates (and in a few vertebrates) that is regulated by hormones is change in skin color. As a parting observation, it is interesting that the vast majority of invertebrates <u>lack</u> definitive endocrine structures; rather, secretions probably are produced by small clusters of cells dispersed within the body.

P. <u>Chemical Messengers in Plants</u>.--As in animals, physiological activities within plants are coordinated by chemical communication. We call these substances <u>phytohormones</u>, (members of a larger group of <u>plant growth regulators</u>), organic molecules (other than nutrients) whose presence in small quantities affects physiological processes. Phytohormones seem to affect a narrower scope of activities (generally those associated with development) than do animal hormones (associated with development and regulation). Another interesting contrast is that phytohormones are secreted by <u>un</u>specialized tissues rather than by specialized endocrine tissues as in animals.

Let's examine form and function of the four major categories of phytohormones:

(1) Auxins. <u>Auxins</u> are substances that are produced at stem meristems and that migrate to other regions of the plant. Their effect is to promote cell elongation. The most common auxin is <u>indoleacetic acid</u>. Auxins account for the phenomenon of <u>phototropism</u>, the bending of a plant towards light. Auxins cause such plant "movement" by redistributing auxin within the shoot: Auxin accumulates on the side away from light; those cells elongate while those cells on the light side do not elongate as greatly.

Other functions of auxins include inhibition of lateral bud growth, stimulation of fruit development, and regulation of leaf fall (<u>abscission</u>). Practical applications of auxins include production of seedless fruits and use as defoliants (<u>herbicides</u>).

(2) Gibberellins. More than 60 different <u>gibberellins</u> are now known. These activate certain genes to promote synthesis of certain enzymes. The presence of different enzymes, of course, alters the physiology of a plant. One response, disproportionately tall growth, reflects a change in the balance between elongating growth and leaf development. Other roles of gibberellins pertain to induction of specific enzymes in germinating seeds and in flowering and fruit development.

(3) Cytokinins. <u>Cytokinins</u> promote plant growth, but in a way different from auxins. Cytokinins provide "stop and go" signals for cell division. They are transported via xylem to the leaves from their sites of production, the shoots and roots.

(4) Inhibitors. Unlike the stimulatory phytohormones noted above, ethylene and abscisic acid regulate plant development by inhibiting growth. The action of ethylene is antagonistic to that of auxin. <u>Ethylene</u>, produced in lateral buds, is important in aging processes such as fruit ripening. In all likelihood, the "vine-ripened" tomatoes on your salad were picked green and were ripened via application of ethylene gas during shipment to your grocer.

<u>Abscisic acid</u> helps to prepare a plant for dormancy by slowing growth, by forming protective scales over buds, and by inhibiting cell division in the vascular cambium. Abscisic acid also aids a plant in coping with adverse conditions. When a plant is under water stress, abscisic acid accumulates within the leaves and causes the stomata to close.

TERMS TO UNDERSTAND: Define and identify the following terms:

hormone

growth factor

target tissues

hypothalamus

adenohypophysis (anterior pituitary)

receptors

second messenger

adenylate cyclase

cyclic AMP

protein kinases

calmodulin

phosphoinositides

prostaglandins

mitogens

endocrine secretion

nerve growth factor

colony-stimulating factors

neurohypophysis (posterior pituitary)

tropic hormones

adrenocorticotropic hormone

thyroid-stimulating hormone

follicle-stimulating hormone

luteinizing hormone

growth hormone

prolactin

releasing factors

release-inhibiting factors

somatostatin

endorphins

enkephalins

vasopressin

oxytocin

pineal body (epiphysis)

melatonin

adrenal glands

corticosteroids

mineralocorticoids

glucocorticoids

gluconeogenesis

epinephrine

norepinephrine

catecholamines

thyroxine

calcitonin

parathyroid glands

placental lactogen

chorionic gonadotropin

progesterone

testosterone

inhibin

islets of Langerhans

insulin

glucagon

atrial natriuretic factor

brain hormone

prothoracitropic hormone

prothoracic gland

ecdysone

juvenile hormone

phytohormones

auxins

gibberellins

cytokinins

ethylene

abscisic acid

indoleacetic acid

phototropism

kinetin

CHART EXERCISE

Several chemical messenger substances are listed in the left column. For each chemical listed, complete the chart to indicate (1) the groups of organisms in which this chemical occurs, (2) the site(s) of production and release, (3) its metabolic effects, and (4) the effects of excessive or under supply of this chemical. Not all chemical messengers presented in Chapter 28 are listed here; you will find it useful to continue this chart onto extra pages so that you can assemble this information for additional chemicals.

CHEMICAL	SPECIES	SITE	EFFECTS	SUPPLY
Auxins				
Oxytocin				
Epinephrine				

Growth hormone
_____

Estrogen
_____

Testosterone
_____

Gibberellins
_____

Glucagon
_____

Ethylene
_____

Atrial natriuretic
  factor
_____

DISCUSSION QUESTIONS

1. How do follicle-stimulating hormone and luteinizing hormone function to influence the gonads? Is this influence the same in both sexes?

2. Compare and contrast chemical messengers in animals and plants. Consider chemical composition, mode of transport, aspects of physiology regulated, etc. in your answer.

3. Select one of the tropic hormones. As part of your answer, complete the diagram below to show how production of that hormone by the adenohypophysis is regulated. Include in your answer the roles of releasing factors, release inhibiting factors, and the secretion(s) of the target organ.

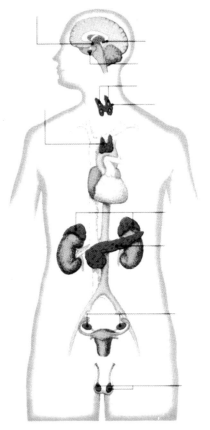

4. Discuss the mechanisms by which animal hormones exert their influence at a target cell.

5. Discuss the various hormones that are involved in gluconeogenesis. How do they interact to affect this process?

6. Discuss the various hormones that are involved in regulation of water balance of the body. How do they interact to affect this process?

7. Write a one-paragraph overview discussing the diversity of phytohormones, their mechanisms, and examples of practical applications.

TESTING YOUR UNDERSTANDING

For each of the test items below, three of the lettered alternatives are true and the other is false. Determine which alternative is false and write its letter in the blank to the left of the question number. On the blank line below alternative D, write the corrected version of the false statement.

_____ 1. A. Charles Darwin was one of the early researchers of phototropism.
B. Hormones, phytohormones, and plant growth regulators are examples of chemical messengers.
C. A hormone that is carried throughout an animal body via a closed circulatory system to its target cells is called an apocrine secretion.
D. Releasing factors and release-inhibiting factors are released by the hypothalamus.

Correction: _____

_____ 2. A. The pituitary is a primary site of interaction of the nervous and endocrine systems.
B. Hormones tend to exert their effects faster than does the nervous system.
C. Most hormones are classified as proteins or lipids.
D. Peptide hormones are released into the bloodstream as prohormones or preprohormones.

Correction: _____

_____ 3. A. Hormone receptor molecules specific for steroid hormones usually are located on the external surface of the cell.
  B. The intracellular receptors of the thyroid hormones are located within the chromatin of the target cells.
  C. One of the ways in which hormones operate is by altering the permeability of the cell membrane.
  D. Cyclic AMP is the "second messenger" for polypeptide hormones.

  Correction: _____

_____ 4. A. Regulation of the adenylate cyclase system involves two G proteins.
  B. The function of protein kinases is to phosphorylate various enzymes.
  C. Calcium functions as a "second messenger" when it binds to calmodulin.
  D. The conversion of glycogen to glucose is catalyzed by glycogen synthetase.

  Correction: _____

_____ 5. A. Growth factors are classified as mitogens because they stimulate mitosis and cell division.
  B. Distribution of molecules of growth factors from sites of synthesis to sites of action is via transport through the bloodstream.
  C. Releasing factors and release inhibiting factors are transported to the hypophysis through the hypothalamo-hypophyseal portal system.
  D. Tropic hormones are produced by the adenohypophysis.

  Correction: _____

_____ 6. A. Endocrine structures influenced by various tropic hormones include the ovaries, testes, thyroid, and adrenal medulla.
  B. Endorphins and enkephalins reduce pain and influence mood and behavior.
  C. Antidiuretic hormone is produced in the hypothalamus and is released by the neurohypophysis.
  D. All of the following hormones influence water balance in the body: vasopressin, aldosterone, and atrial natriuretic factor.

  Correction: _____

_____ 7. A. The condition of diabetes insipidus suggests that the hypothalamus or the neurohypophysis is not functioning normally.
  B. Oxytocin has numerous roles, including hastening the onset of the estrus cycle.
  C. In many vertebrates, the pineal body has both photoreceptive and endocrine functions.
  D. Calcitonin is the secretion of the pineal body.

  Correction: _____

_____ 8. A. The neurohypophysis and the epiphysis are examples of neuroendocrine transducers.
  B. Melatonin influences the rhythm of many physiological processes.
  C. The level of glucose in the bloodstream is influenced by several glucocorticoids.
  D. Male sex hormones are produced in males, but not in females.

  Correction: _____

_____ 9. A. The effects of androgens secreted by the adrenal cortex are very similar to the effects produced by testosterone.
B. Activity of the adrenal cortex decreases as stress levels increase.
C. Epinephrine causes elevation of certain aspects of physiology to "fight-or-flight" levels.
D. One of the hormones secreted by the thyroid is involved with regulation of calcium metabolism.

Correction: _____

_____ 10. A. Thyroid-stimulating hormone increases the amount of thyroxine produced and released.
B. Calcium is important to all of the following processes: nerve impulse transmission, muscular contractions, and blood clotting.
C. Development and maintenance of secondary sexual characteristics is promoted by androgens and estrogens.
D. In high concentrations, estrogens inhibit ovulation.

Correction: _____

_____ 11. A. The placenta and chorion are endocrine structures.
B. Leydig cells produce inhibin and Sertoli cells produce testosterone.
C. The pancreas has both endocrine and exocrine functions.
D. Insulin and glucagon catalyze the same reaction, but in opposite directions.

Correction: _____

_____ 12. A. The growth hormones of other mammals function in humans in the same way as does human growth hormone.
B. Molting, metamorphosis, and color change are some of the physiological processes that are mediated by hormones in invertebrates.
C. Ecdysone and juvenile hormone are hormones of certain invertebrates.
D. Most invertebrates lack definite endocrine structures.

Correction: _____

_____ 13. A. The role of phytohormones is developmental, not regulatory.
B. Auxins are produced by cells in root and shoot meristems.
C. Gibberellins direct the phenomenon of phototropism.
D. Indoleacetic acid is the chemical identity of one of the auxins.

Correction: _____

_____ 14. A. Removal of an apical bud allows lateral buds to grow because the supply of cytokinins has been removed.
B. Gibberellins influence the balance between elongating growth and leaf growth.
C. Cytokinins stimulate cell division.
D. Cytokinins are produced in the roots.

Correction: _____

_____ 15. A. Ethylene influences aging processes, such as ripening of fruit.
B. The actions of abscisic acid and ethylene are inhibitory rather than stimulatory.
C. Gibberellins help to prepare a plant for dormancy.
D. Closure of stomata as a guard against dehydration is influenced by abscisic acid.

Correction: _____

CHAPTER 29--NEURAL COORDINATION

CHAPTER OVERVIEW

This chapter presents the second of the two general means by which the internal activities of the more-complex animals are coordinated. Control via nervous systems generally is more rapid than is coordination via chemical means (Chapter 28). As noted in the previous chapter, the two systems do interact to achieve homeostasis.

In this chapter, we introduce the players (cell types of nervous systems) and the mechanisms by which they operate. Here, we examine the form and function of junctions between neurons and briefly study simple neural circuitry. In the following chapter (30), we will take a broader look at the nervous system.

TOPIC SUMMARIES

A. Responsiveness.--One of the characteristics of all living organisms is irritability, the ability to respond to a stimulus. Stimuli occur in many forms--light, heat, sound, chemical, etc. Each stimulus contains information about the environment; proper response to this information often is critical to the survival of the organism.

Three steps are involved in responsiveness. The organism must somehow perceive the stimulus; this involves some sort of receptor. The information received must be conducted to other parts of the organism. Finally, a response follows some degree of processing of the information. The complexity of the structures and pathways involved in responsiveness correspond with the complexity of the organism. A receptor-conductor-effector network of some sort is characteristic of animals. Much of the rest of this chapter addresses form and function of the nervous systems of higher vertebrates.

B. Kinds of Stimuli and Receptors.--A tremendous array of stimuli exists in the environment. These can be classified according to the physical nature of the stimulus. Indeed, receptors are sometimes classified according to the stimulus perceived. For example, organisms respond to various chemical stimuli; vertebrates have relatively complex and specialized chemoreceptors such as taste buds (the sense of gustation) and olfactory hair cells (sense of smell, olfaction) to sample the outside environment. The internal chemical environment also can be monitored; recall from Chapter 27 the specialized receptors (carotid bodies), found in the walls of certain arteries, capable of monitoring the oxygen and carbon-dioxide concentration in the bloodstream.

Other sorts of energy can be monitored as well. Thermoreceptors perceive warmth, cold, or changes in heat content. Electroreceptors perceive electrical fields. Baroreceptors perceive pressure. Photoreceptors perceive light (recall the pineal body from Chapter 28). In vertebrates, specialized sections of the inner ear (semicircular canal and duct system) are involved in perception of gravity (gravireception). The inner-ear apparatus also is an example of a mechanoreceptor, structures that perceive mechanical energy that is interpreted as motion, sound, or equilibrium. Although the mechanisms are not yet well understood, many species (some insects, some crustaceans, some birds, and perhaps some other vertebrates) apparently are capable of perceiving magnetic fields, even those so weak as that surrounding Earth. Magnetoreception is useful in homing and in migration.

We should note a couple of additional points here: Not all organisms can perceive all types of stimuli. Some organisms perceive different portions of the available stimulus; for example, not all those wavelengths of light that are visible to humans are visible to insects, or the some of the range of wavelengths audible to dogs is not audible to humans. The sensory abilities of each species represent advantageous adaptations for that species' lifestyle. Also, some stimuli exist (e.g., radio waves) for which there are no known biological receptors.

C. <u>Stimulus Intensity and Responses</u>.--For most stimuli, a certain minimum amount (the <u>threshold of intensity</u>) of that stimulus is required before a receptor can perceive the stimulus. Your textbook uses the example that some minimum amount of salt must be dissolved in water before our taste buds detect the presence of the salt. Once the threshold is reached, the factors determining the particular response that will be elicited depends on both the duration of the stimulus and its rate of change.

Response to a stimulus is not instantaneous. Rather, there is a (usually) brief <u>latent period</u> from the onset of stimulus to initiation of response. For many tissues and structures (e.g., nerves and muscles), there is an interval (<u>refractory period</u>) immediately following a response when no stimulus is strong enough to cause another response.

D. <u>Cells of the Nervous System</u>.--We'll examine vertebrates as we begin our anatomical and physiological studies of the nervous system. One feature shared by all receptors (sensory structures) is that they convert (transduce) stimuli into electrical energy that is carried to processing centers via nerves. The design of any particular sensory structure reflects the nature of the stimulus it perceives. Every sensory structure consists of at least one (sometimes millions) specialized nerve cell (neurons).

<u>Neurons</u> are the only type of nervous-system cell that can carry electrical energy (i.e., transmit a nerve impulse). In general, a neuron has three major parts. The nucleus and most other organelles reside in the <u>cell body</u>. One or many cytoplasmic processes extend from the cell body. Most of these processes (<u>dendrites</u>) are relatively short and highly branched. One (or perhaps a very few) process, the <u>axon</u>, is much longer than the dendrites. At its end, the axon branches into a spray of <u>terminal fibers</u>. The dendrites provide a large surface area for receiving nerve impulses from other neurons; <u>transmisison</u> of the impulse within a neuron proceeds from dendrites and/or cell body, away from the cell body along the axon, and finally to another cell at the terminal fibers. Transmission of the impulse from one cell to another involves specific chemical events in the gap (<u>synaptic gap or cleft</u>) lying between the two cells.

The nervous system consists of more cell types than merely neurons. All other nervous cells are collectively called <u>neuroglial cells</u>. These do not conduct impulses, but perform an array of <u>support roles</u>. Be sure to learn the functions of each of these types of neuroglia: <u>astrocytes</u>, <u>oligodendrites</u>, <u>Schwann cells</u>, <u>microglia</u>, and <u>ependyma</u>.

E. <u>Nerve Impulses</u>.--A nerve impulse is an <u>electrical event</u>, yet this event is chemically-mediated in that the electrical charges are carried by ions of sodium, potassium, and chlorine. These ions are distributed across the plasma membrane of a neuron in such a way that there is a difference in the electrical charges on the inside and outside surfaces of the membrane. It is changes in the distributions of ions, and consequently in the differences in charges, across the membrane that constitute the events of a nerve impulse. Let's look at several phases of propagation of a nerve impulse (Fig. 29.1).

An unstimulated neuron is said to be at rest. Even at rest, however, cations (sodium and potassium) and anions (chloride) have a particular distribution across the membrane. This distribution produces a difference in electrical charges across the membrane; this difference is referred to as the <u>resting potential difference</u>, measured in millivolts. In this polarized state, the potential difference across the membrane is -60 mV; the interior surface is negatively charged relative to the external surface.

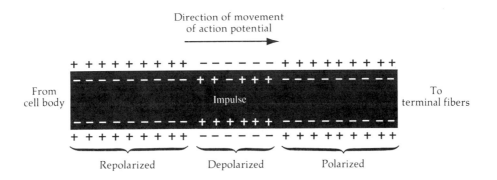

Fig. 29.1: Propagation of a nerve impulse along a segment of a neuron

The potential difference results from two characteristics of the membrane: The membrane is selectively permeable; potassium is free to move through the membrane, but when at rest the membrane does not allow passive diffusion of sodium. Hence, even though sodium concentration outside the membrane is higher than inside, sodium does not diffuse inwardly. The ability of the membrane to maintain an ionic concentration gradient also involves a pump, the sodium-potassium pump. ATP is spent to pump sodium to the exterior. Because the efflux of sodium is only partially balanced by influx of potassium, the resting potential remains negative.

Stimulation of a neuron causes the action potential--a shift in distribution of ions (and, therefore, of charges) as the impulse sweeps along the neuron. This shift of charges is called depolarization; at the peak of depolarization, the potential difference across the membrane is about +40 mV. This change of charge results primarily from a rapid influx of sodium ions; during the action potential, the membrane briefly becomes permeable to sodium, which moves to the interior through sodium channels and gates.

The zone of depolarization continues down the neuron. Then, as rapidly as the sodium channels opened, they close. The proper resting distribution of cations is restored through operation of the sodium-potassium pump; this process is repolarization. The time interval (only about 4 milliseconds) during which repolarization is occurring also is referred to as the refractory period. During the first part of this period (the absolute refractory period), no stimulus, no matter how strong, can cause depolarization of the membrane. During the later phases (the relative refractory period), depolarization is possible only with a stronger-than-normal stimulus.

It should be apparent, then, that a nerve impulse is a narrow zone of changing differences in electrical potential that sweeps along a neuron. At any given time, only a small region of a neuron is involved in transmitting an impulse.

F. Neural Circuits.--No complete neural circuits (connecting receptor and effector) consist of only one neuron; some consist of only two (a simple reflex arc, Fig. 29.2), but most involve many neurons. Therefore, to understand the physiology of nerve impulse transmission, we must be concerned with action potentials within (in part E above) and between neurons.

Synapses are the functional (but not physical) unions of adjacent neurons of a neural circuit. In most species, transmission of an impulse from one neuron (the presynaptic neuron) to another neuron (postsynaptic neuron) is a chemical event. (A few invertebrates and fishes have electrical junctions.) The process of propagating a new nerve impulse in the postsynaptic neuron is not directly related to distributions of sodium and potassium, but to sending of chemical messengers (<u>neurotransmitter molecules</u>) from the terminal fibers of a presynaptic neuron to the dendrites and/or cell body of the postsynaptic neuron.

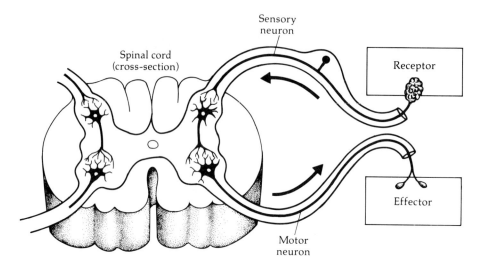

Fig. 29.2: Diagram of simple reflex arc

Synaptic vesicles in the terminal fibers of the presynaptic neuron contain molecules of neurotransmitter substance. (The particular neurotransmitter secreted [e.g., <u>acetylcholine</u>, <u>norepinephrine</u>, <u>serotonin</u>, <u>dopamine</u>, <u>gamma-aminobutyric acid</u>] depends on the particular neural circuit to which that neuron belongs . . . more on this in Chapter 30.) The action potential arriving at the end of the axon causes an influx of calcium ions that, in turn, causes <u>calmodulin</u> to interact with the synaptic-vesicle membranes, ultimately to cause release of neurotransmitter molecules into the <u>synaptic cleft</u>. These molecules bind with specific receptors on the postsynaptic membrane and then activate the cyclic-AMP "second messenger" system.

What happens to the neurotransmitter molecules once they have done their job? Do they remain in the cleft and continue stimulating the postsynaptic neuron? No. One means of clearing the synapse of acetylcholine, for example, is through the hydrolytic action of the enzyme <u>acetylcholinesterase</u>.

Our discussion to this point has been on mechanisms that stimulate propagation of the impulse in the postsynaptic neuron (<u>excitatory synapses</u>). We must remember, however, that <u>inhibitory synapses</u> also exist.

G. <u>Other Properties of Synapses</u>. Every synapse represents a point at which a "decision" can be made; the decision is whether or not to stimulate (or inhibit, depending upon the nature of the synapse) the postsynaptic neuron. Passage of an action potential to the end of an axon and the release of neurotransmitter substances into the cleft does <u>not</u> ensure that the stimulatory (or inhibitory) effect will be passed on to the postsynaptic neuron. Enough neurotransmitter molecules must bind with enough receptors in a short enough period of time for the effect to occur; this is the concept of <u>summation</u>. The phenomena of <u>fatigue</u> and <u>facilitation</u> also effect synaptic behavior.

TERMS TO UNDERSTAND: Define and identify the following terms:

stimulus

receptor

effector

threshold

latent period

refractory period

nerve impulse

neuron

nerve

glial cell

cell body

dendrite

axon

terminal fiber

axonal transport

synapse

polarity

axolemma

myelin

neurilemma sheath

astrocytes

oligodendrites

Schwann cells

microglia

ependymal cells

electrical potential

polarized

resting potential

sodium-potassium pump

action potential

gating currents

all-or-none reaction

absolute vs. relative refractory periods

nodes of Ranvier

saltatory conduction

sensory neuron

motor neuron

reflex arc

acetylcholine

synaptic vesicles

synapsin

fodrin

spectrin

calpain

acetylcholinesterase

excitatory vs. inhibitory postsynaptic potential

norepinephrine

cholinergic

adrenergic

serotonin

GABA

dopamine

spatial summation

temporal summation

facilitation

association neuron

motor endplate

CHART EXERCISE

Complete the chart below to compare characteristics of chemical and nervous means of coordination in various animals. Indicate (1) the nature of the messengers, (2) where and how the messengers are produced, (3) the types of activities mediated by these messsengers, (4) the relative speed of response to messages, and (5) the groups of organisms in which these messengers occur.

CHARACTERISTIC	CHEMICAL SYSTEM	NERVOUS SYSTEM
Nature of messenger		
Sites and methods of production		
Activities mediated		
Relative speed of response		
Taxonomic distribution		

PEOPLE TO KNOW: What is the significant biological contribution of each of these individuals as indicated in Chapter 29?

Camillo Golgi

Santiago Ramon y Cajal

Theodor Schwann

Otto Loewi

DISCUSSION QUESTIONS

1. Discuss the interaction of the following materials: calcium ions, synapsin, calmodulin, acetylcholine, fodrin, acetylcholinesterase, and calpain.

2. Compare and contrast the form and function of chemical and electrical synapses. Which of the two types of synapse represents the better evolutionary answer to the need to transmit nerve impulses between neurons?

3. Discuss the role of sodium ions in transmission of nerve impulses.

4. In the space below, outline a classification of sensory structures that is based on the physical nature of the stimulus perceived. For each category, provide examples of the sensory structures and indicate the groups of animals in which those particular structures occur.

5. Why is it sensible that no neural circuit can consist of only one neuron?

341

TESTING YOUR UNDERSTANDING

For each of the test items below, three of the lettered alternatives are true and the other is false. Determine which alternative is false and write its letter in the blank to the left of the question number. On the blank line below alternative D, write the corrected version of the false statement.

_____ 1. A. The sensitivity of plants to light is attributed to the presence of phytochrome.
B. No organisms are known to be able to perceive and respond to gravity.
C. The senses of taste and smell represent types of chemoreception.
D. Hearing is a type of mechanoreception.

Correction: _____

_____ 2. A. A stimulus whose intensity is below the threshold value will not be perceived by a sensory structure.
B. The refractory period of a neuron immediately precedes the phase of repolarization.
C. The following is a correct sequence of structures through which information flows from receptor to effector: sensory neuron, synapse, motor neuron.
D. All sensory structures contain at least one sensory neuron.

Correction: _____

_____ 3. A. All of the following are examples of neuroglial cells: ependyma, astrocytes, neurons, and Schwann cells.
B. Saltation of nerve impulses can occur because of the presence of myelin sheaths around the axon.
C. Phagocytosis is the function of one of the types of neuroglial cells.
D. Neuroglial cells occur in both the central and peripheral regions of the nervous system.

Correction: _____

_____ 4. A. During its polarized status, the cell membrane of a neuron is impermeable to sodium ions.
B. At the peak of depolarization, the outer surface of a neuron's plasma membrane is more negatively charged than is the inner surface of the membrane.
C. Release of neurotransmitter molecules from synaptic vesicles requires that the cell membrane become permeable to calcium ions.
D. When a neuron is transmitting a nerve impulse, the entire cell is depolarized.

Correction: _____

_____ 5. A. White matter appears white due to the presence of myelin.
B. Returning sodium ions to the exterior of a neuron during repolarization requires expenditure of energy.
C. A stimulus of greater-than-normal intensity is required to establish a nerve impulse during the absolute refractory period.
D. Nodes of Ranvier are regions along an axon that lack an insulative covering.

Correction: _____

_____ 6. A. Neurons located between the receptor and the sensory neuron are called association neurons.
B. A simple reflex arc lacks association neurons.
C. Transmission through the gap junction of an electrical synapse is more rapid than is transmission across the cleft of a chemical synapse.
D. Transmission across an electrical synapse can be in either direction, whereas impulses can travel in only one direction across a chemical synapse.

Correction: _____

_____ 7. A. All of the following are neurotransmitter substances: norepinephrine, acetylcholine, synapsin, and dopamine.
B. At least one neurotransmitter substance is also produced and secreted by an endocrine structure.
C. Current research suggests that calpain and fodrin interact to restructure the cytoskeleton in ways that facilitate learning and memory.
D. One of the ways of removing neurotransmitter molecules from the synaptic cleft is by enzymatic hydrolysis.

Correction: _____

_____ 8. A. Two types of synapses, excitatory and inhibitory, have been identified within the nervous system.
B. The cell body of a single motor neuron may receive incoming signals from thousands of modulating neurons.
C. In some cases, failure of an impulse to cross a synapse can be due to fatigue.
D. Continual and repetitive stimulation of a presynaptic terminal causes the phenomenon of facilitation.

Correction: _____

_____ 9. A. One of the advantages of electrical synapses is that their efficiency is not significantly effected by low environmental temperatures.
B. The junction between a motor neuron and a muscle is called a motor endplate.
C. Transmission of impulses from a neuron to an effector structure is usually by an electrical means rather than by means using neurotransmitters.
D. The postsynaptic membrane of a muscle cells has palisades.

Correction: _____

_____ 10. A. Both endocrine glands and skeletal muscles can be considered as effectors.
B. The transmitter substance gamma-aminobutyric acid is used exclusively in inhibitory synapses.
C. Dopamine mimics the actions of norepinephrine.
D. The intensity of the impulse within a neuron depends on the intensity of the stimulus.

Correction: _____

CHAPTER 30--NERVOUS SYSTEMS

CHAPTER OVERVIEW

The previous chapter (29) introduced to us the anatomy and physiology of the cells and tissues from which nervous systems are constructed. In this chapter (30), we assemble those cells and tissues into intact nervous systems. We will survey the form and function of receptors, structures that are specialized for perceiving certain types of stimuli in the environment and within the body. We will examine the brain and spinal cord, the structures in which information is processed. We will review major trends that characterize the evolution of complex nervous systems, such as our own.

TOPIC SUMMARIES

A. Receptors.--As we will see below, receptors may be quite simple or extremely complex in their construction. As noted in Chapter 29, the receptor cells that actually perceive a stimulus are neurons, modified to a lesser or greater degree in order to be able to respond to a particular stimulus. Regardless of degree of specialization, all receptors share the ability to convert (transduce) some particular form of energy into electrical energy. Also as noted in Chapter 29, receptors may be classified on the basis of the nature of the stimuli they perceive. In the sections below, we will survey receptors found in organisms of Kingdom Animalia.

B. Photoreceptors.--The lateral eyes that we humans (and most of our vertebrate counterparts) possess do not represent the only structural solution to the need to see our environment. Other designs of "eyes," however, do not have all the capabilities that vertebrate eyes have.

(1) Non-vertebrates. All light receptors incorporate some sort of light-sensitive pigment molecules. Many invertebrate animals possess various arrangements of lenses and light-sensitive cells (see textbook Fig. 30-2). The compound eyes of insects are made of many semi-independent ommatidia. Each of these tubular structures receives light from a different portion of the field of view. Hence, the insect has a composite image of its surroundings, and it can detect movement of objects in its environment. Some other invertebrates, such as snails and other gastropods, have eyes with designs much more like our own than found elsewhere among animals. Such eyes have a lens with many light-sensitive cells arranged into a retina.

(2) Human eye as an example of vertebrates. The vertebrate eye is a complex and elegant structure whose abilities will never be matched by any camera despite textbooks' use of the "eye-as-camera" analogy. This Study Guide provides only an overview of structure of the eye (and other sensory structures); be sure to work closely with the textbook to master the details.

Anatomically, the eye consists of three more-or-less spherical layers. The outer layer, the sclera, is an opaque, tough, protective layer, except for the front surface where it is transparent (cornea) to allow passage of light. The next layer in, the choroid, aids to reflect light back to the light-sensitive layer and, towards the front of the eye, forms the iris. The iris surrounds the pupil; the musculature of the iris enables it to regulate the diameter of the pupil and, thereby, to control the amount of light entering the eye. The innermost layer, the retina, contains the millions of light-sensitive cells (rods and cones) and the neural circuitry (involving bipolar cells, ganglion cells, horizontal cells, amacrine cells) that carries information to the brain's visual cortex via the pair of optic nerves, optic chiasma, optic tracts, and lateral geniculate nuclei. The lens focuses entering light onto the retina.

Physiologically, light is absorbed by pigments (several different opsins contained within sensory cells) that are sensitive to light of different wavelengths and intensities. Light energy causes reversible changes in the chemical configuration of the pigment molecules (Fig. 30.1). Pigment that has been stimulated by light is said to be bleached. Transduction from bleached pigment molecule to nerve impulse involves cyclic GMP, transducin, GTP, and GDP in the scheme outlined in textbook Fig. 30-11. Return of bleached pigment to the configuration which can again be stimulated by light is catalyzed by retinal isomerase.

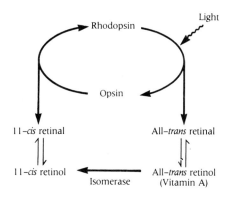

Fig. 30.1: The visual cycle of rhodopsin in the vertebrate eye

C. Mechanoreceptors.--Mechanoreceptors are sensory structures that respond to mechanical forces. Depending on the design of the structure, these forces can be perceived by the brain in myriad ways: sound, equilibrium, touch, pain, etc. All mechanoreceptors operate on the principle of mechanical deformation (pressing, bending, etc.) of sensory neurons.

(1) Sensory corpuscles. The diversity and the complexity of mechanoreceptors present in an organism have increased through evolutionary time. Many vertebrates have one or more types of pressure-sensitive sensory corpuscles, such as Pacinian corpuscles. Application of pressure to these corpuscles sends signals interpreted variously as pain, light touch, etc. In humans, Pacinian corpuscles occur in the dermis of the skin over most of the body and deeper in association with joints and visceral organs.

(2) Lateral line organs. Most fishes possess lateral line organs, consisting of series of mechanoreceptors located in shallow canals on the head and along the body. Moving water in these canals deflects the "hairs" (really dendrites) of sensory cells in the canals. The direction and strength of deflection informs the fish of its velocity and of movement of other objects in the immediate vicinity. A lateral-line system as such is not found in terrestrial vertebrates (except in aquatic larvae of amphibians); however, the senses of hearing and equilibrium probably derived from modification of lateral-line structures that sank into the skull.

(3) Hearing. Three regions are recognized in the ears of higher vertebrates. Transduction of mechanical vibrations in air or water occurs in the inner ear. The outer ear is little more than a cup-like structure (pinna) that funnels vibrations into a canal (external auditory canal). These vibrations impact the eardrum (tympanum) and cause it to vibrate. The middle ear contains three small bones (the auditory ossicles: malleus, incus, stapes) linked together with the malleus attached to the tympanum and the stapes contacting the inner ear at the oval window. Depending on the tightness of linkage of these ossicles (as determined by middle-ear muscles), vibrations can be either enhanced or damped as appropriate.

The inner ear consists of sensory receptors for both equilibrium (more below) and hearing. The sensory cells themselves are contained within the cochlea and are bathed in fluid. Vibration of the fluid (imparted by stapes at oval window) in turn deflects the sensory hair cells that together form the <u>organ of Corti</u>. Nerve impulses from the organ of Corti are carried to the auditory cortex of the brain via the cochlear branch of the <u>auditory nerves</u>.

(4) Equilibrium. This sense permits orientation of the body within a field of gravity. The arrangement in invertebrates is rather simple: The <u>statocyst</u> contains a small stone of calcium carbonate (<u>otolith</u>) and an array of hair cells. Change in the position of the otolith stimulates sensory hairs, and, thereby informs the animal of its orientation relative to gravity.

Verbetrates have adopted a very similar arrangement. The same principle of stones and hair cells in a fluid medium operates. The portion of the inner ear that is not cochlea is the <u>vestibular apparatus</u>. It consists of chambers housing organs for both static (the maculas of the <u>utricle</u> and <u>saccule</u>) and dynamic (the cristae of the <u>semicircular canals</u>) equilibrium. Information on body orientation, velocity, and acceleration is sent to the <u>cerebral cortex</u> and <u>cerebellum</u> via the vestibular branches of the <u>auditory nerves</u>.

D. <u>Chemoreceptors</u>.--The ability to perceive the chemical quality of the environment is found in all organisms. Even unicells possess membrane receptors that enable <u>chemotaxis</u>. Chemoreceptors represent a more complex approach to chemoreception. Chemoreceptor cells are modified dendrites of sensory neurons with the ability to recognize one or more particular chemicals. The primary chemical senses are the "common chemical sense," smell (olfaction), and taste (gustation).

(1) Gustation. The tastes we associate with food or drink really are a combination of sensory responses by both our senses of smell and of taste. Taste involves chemoreception by <u>gustatory hair cells</u> located in <u>taste buds</u>, which in turn are situated in <u>papillae</u>. These hair cells are specialized for reception of only four flavors: sweet, sour, salty, and bitter. In humans and closely related species, taste buds are located primarily on the tongue, but also on the palate and tonsillar regions. In other animals, taste buds sometimes have more widespread locations, even on the external surface of the head and on "whiskers" of catfish and other species that scavenge their food from detritus.

(2) Olfaction. The sense of smell is important to humans, but not nearly as critical as for many other species. For many blind (or nearly so) species, olfaction is the primary sense. For many fishes, olfaction is the means by which spawning grounds are identified.

In humans, olfaction is restricted to the <u>olfactory epithelium</u> of the nasal cavity. Much as for gustatory cells, <u>olfactory cells</u> bear sensory hairs which recognize specific chemicals inhaled in the airstream and dissolved in the <u>mucus</u> that bathes the sensory dendrites. The "pair" of <u>olfactory nerves</u> really is many axonal fibers ascending upward through a perforated region of bone into the cranial cavity. These coalesce into the pair of <u>olfactory tracts</u> that carry impulses to various parts of the brain.

E. <u>Other Receptors</u>.--Not all organisms possess all types of senses or sensors. Some specialized senses not found in humans include the heat-sensitive pit organs of certain snakes and the humidity receptors of many insects.

F. <u>Trends in Evolution of Nervous Systems</u>.--The brain represents the culmination of a long sequence of evolutionary changes. Let's briefly survey the various degrees of development of nervous systems (and brains, where present) en route to a survey of form and function of the vertebrate brain.

Brains and nerves are multicellular structures. Therefore, neither can be present in any unicellular organism (e.g., <u>monerans</u>, <u>protists</u>). Even so, all cells still possess the quality of irritibility (refer back to Chapter 29). All animal species (except <u>sponges</u>) possess nerves that conduct information from one point to another within the body. Coelenterates possess a nerve net of neurons ramifying throughout the body; the main response to stimulation in a nerve net is contraction and, hence, movement of parts of or the entire body. Flatworms have a nervous system that represents a next step in the evolutionary progression: They have a nerve net serving the outer regions of the body as well as a rudimentary central nervous system consisting of a series of nerve cords and an enlargement towards the anterior end of these cords that minimally qualifies as a brain.

The situation in flatworms and all more-advanced animals illustrates the phenomenon of cephalization, the concentration of organs of special sense and of information processing towards the anterior of the body. Cephalization is functionally correlated with the more active lifestyles exhibited by these animals; clearly, placing sensory and processing structures at the end of the body that first enters previously unsampled areas of environment is adaptively advantageous.

In addition to cephalization, four other trends characterize the evolution of the nervous system: (a) The system becomes centralized by concentrating cell bodies of neurons into a brain, spinal cord, and ganglia; hence, a peripheral nervous system consisting mostly of axons is recognizable. (b) The peripheral system evolved to carry messages in two directions; afferent fibers carry impulses from receptor to central nervous system, and efferent fibers carry instructions to effectors. (c) The ability to associate information increased via the inclusion of association neurons between the afferent and efferent segments of neural circuits. (d) Sensory structures became much more complex, and, unlike in the earliest nervous systems, the sensors became connected with the central nervous system.

G. The Vertebrate Brain.--Embryologically and evolutionarily, the first divisions apparent in the vertebrate brain are the forebrain, midbrain, and hindbrain. The forebrain is associated with olfaction, the midbrian with vision and eye musculature, and the hindbrain with receptors more scattered through the body and with various head muscles. These regions differentiate to various degrees during their development depending on the species.

The brains of fishes and amphibians often are called "smell brains" because the olfactory portions of the forebrain are proportionately better developed than is the rest of the brain. From this condition, development of the forebrain and hindbrain proportionately greatly exceeded midbrain development. Consequently, olfactory capabilities were enhanced and, first appearing in reptiles, was the neopallium, a region of the forebrain that pertains to receptors and activities not related to olfaction. The gray matter of the neopallium is the neocortex.

Birds and mammals both arose from reptiles, but from different groups of reptiles. This is apparent in many features, including the evolutionary pathways by which their brains arose. Birds lack the regions of the brain that mammals use to in associative activities, whereas mammals lack brain regions used by birds for complex, unlearned behaviors such as courtship and nest-building. Also in (most) mammals, the olfactory portions of the cerebrum diminished and neopallium grew more expansively and became convoluted.

H. Function of the Mammalian Brain.--Mammalian brains are primarily associative and coordinating centers. It is organized into centers, nuclei, motor areas, sensory areas, and association areas (see textbook for definitions of these terms). Activities of the evolutionarily older portions of the mammalian brain (e.g., brainstem, midbrain) pertain to basic maintenance functions such as breathing and digestion. Most of the more complex functions (e.g., coordination of sensory input, initiation of appropriate behavior) occur in the neocortex. In your studies, be sure to review the specific areas of the brain (e.g., Broca's area, prefrontal lobe, reticular formation, etc.) and the functions associated with them.

I. **Coordination of the Internal Environment.**--In Chapter 28 we studied chemical messengers as one of the communication systems for regulating and coordinating the internal environment. Here, we focus on that part of the nervous system that coordinates internal events--the autonomic nervous system. This segment of the nervous system bears this name because it indeed seems to be "self-ruling;" at least it operates without our conscious awareness.

Two sets of efferent visceral nerves are present in the autonomic system: Each visceral effector is dually innervated by the autonomic system. The motor outflow to each effector consists of two neurons which synapse in a ganglion outside of the spinal cord. These divisions are called the sympathetic and parasympathetic divisions of the autonomic system. The action of the effector depends upon which division stimulates the effector. The sympathetic division secretes epinephrine (or a related neurotransmitter) and causes the changes associated with "flight-or-flight" physiology. Stimulation by the parasympathetic division (secretes acetylcholine) reverses sympathetic effects to restore normal physiology. Be sure to learn the specific physiological changes associated with sympathetic and parasympathetic stimulation.

The hypothalamus can be thought of as the coordinator of visceral activities. This region of the brain contains many nulcei associated with various visceral activities (e.g., water balance). The hypothalamus plays a key role in emotional behavior, and, as we know from Chapter 28, is an interface between the endocrine and nervous systems.

TERMS TO UNDERSTAND: Define and identify the following terms:

transducer

photoreceptors

compound eyes

lens

ommatidia

sclera

retina

rod

cone

fovea

blind spot

stereopsis

photochemical effect

rhodopsin

retinal

carotenoids

retinal isomerase

transducin

bipolar cells

ganglion cells

horizontal cells

amacrine cells

visual cortex

mechanoreceptors

phonoreceptors

nociceptors

Pacinian corpuscle

lateral-line organs

tympanic membrane

middle-ear ossicles

oval window

perilymph

cochlea

round window

organ of Corti

auditory cortex

statocyst

otolith

utricle

saccule

semicircular canals

otoconia

maculas

statoreceptors

chemoreceptors

gustatory sense

taste buds

olfactory membrane

microvilli (olfactory "hairs")

heat-sensitive sensory pits

pheromones

nerve net

nerve cord

cephalization

peripheral nervous system

ganglion

olfactory bulbs

cranial nerves

neopallium

archepallium

neocortex

inducers

reticular formation

autonomic nervous system

sympathetic division

parasympathetic division

CHART EXERCISE

The vertebrate brain is divisible into many regions and structures, each associated with specific functions. Not all structures are present in all vertebrates. Complete the chart below to indicate (1) the region of the brain (forebrain, midbrain, hindbrain) in which each structure is found, (2) the function(s) associated with each structure, and (3) the vertebrate species in which each structure occurs.

STRUCTURE	REGION	FUNCTIONS	TAXON. DISTRIB.
Cerebrum			
Hypothalamus			
Thalamus			
Medulla oblongata			
Optic lobes			
Olfactory bulbs			
Neopallium			
Prefrontal lobe			

PEOPLE TO KNOW: What is the significant biological contribution of each of these individuals as indicated in Chapter 30?

George Wald

H. K. Hartline

Paul Broca

Sigmund Freud

John Papenheimer and Manfred Karnovsky

DISCUSSION QUESTIONS

1. What characteristics are shared by all sensory structures?

2. Compare and contrast the photosensitive pigments that are involved in the processes of photosynthesis and vision.

3. What is the value of having some organs of special sense (e.g., eyes, ears) present in pairs rather than as individual organs?

4. In what ways is olfaction used by various species of vertebrates?

5. What major changes occurred in the brain in the course of evolution from reptiles to mammals?

TESTING YOUR UNDERSTANDING

For each of the test items below, three of the lettered alternatives are true and the other is false. Determine which alternative is false and write its letter in the blank to the left of the question number. On the blank line below alternative D, write the corrected version of the false statement.

_____ 1. A. Despite their differences, all receptors share the ability to transduce a stimulus into electrical energy.
B. All of the following are photoreceptors: pineal body of some vertebrates, lateral eyes, and ommatidia.
C. Compound eyes allow detection of light intensity and movement of objects in the field of view.
D. Compound eyes occur in a variety of invertebrate and vertebrate species.

Correction: _____

_____ 2. A. A compound eye consists of many ommatidia.
B. In the vertebrate eye, the first cell layer of the retina that light strikes is made of photoreceptive cells, the rods and cones.
C. The visible spectrum of some insects includes ultraviolet light.
D. Among mammals, only a few species have color vision.

Correction: _____

_____ 3. A. The choroid is a layer of tough tissue that encloses the retina and the sclera.
B. Cones are concentrated within the fovea.
C. Because of the positioning of the retina, there is a blind spot in eyes of vertebrates, but not in the eyes of invertebrates that possess a retina.
D. Bleached rhodopsin is renewed to its active cis-isomer form by retinal isomerase.

Correction: _____

_____ 4. A. Because animals cannot synthesize retinal, it must be obtained in the diet.
B. Binding of transducin with GTP produces cyclic GMP which opens sodium channels in the outer membrane of a rod.
C. Lateral transfer of information within the retina is facilitated by bipolar cells and ganglion cells.
D. Nerve impulses generated in the retina pass, in sequence, through the optic nerves, optic chiasma, and the optic tracts en route to the visual cortex of the brain.

Correction: _____

_____ 5. A. All of the following are examples of mechanoreceptors: Pacinian corpuscles, organ of Corti, macula, and lateral-line organs.
B. Phonoreceptors operate on the principle of deformation of sensory "hairs."
C. In the vertebrate ear, mechanical energy is transduced into nerve impulses in the middle ear.
D. Perilymph is a fluid located in the inner ear.

Correction: _____

_____ 6. A. Both binaural hearing and stereopsis allow location of sounds and images in space.
B. All of the following are involved in the sense of equilibrium: otolith, otoconia, statocyst, utricle, and saccule.
C. Sensory structures for the sense of hearing are located in the semicircular canals.
D. Lateral-line organs of fishes probably were the evolutionary forerunners of the inner-ear apparatus.

Correction: _____

_____ 7. A. Olfactory hair cells are located in clusters, called buds, in the nasal cavity.
B. The value of gustation is the role it plays in the selection of food.
C. Gustatory hair cells are located on the tongue, palate, and, for some species, on the outer body surface.
D. Without the assistance of olfaction, gustation can recognize only four sensations: sweet, sour, salty, and bitter.

Correction: _____

_____ 8. A. Although humans do not have sensory structures that can perceive infrared radiation and humidity, such receptors exist in certain other species.
B. Evolution of a skull accompanied the phenomenon of cephalization in vertebrates.
C. The nervous system of sponges consists of only a primitive nerve net.
D. The simplest version of a brain is found in planarians.

Correction: _____

_____ 9. A. The olfactory regions of the brain are located in the midbrain.
B. The medulla oblongata is the portion of the brain that is closest to the spinal cord.
C. The forebrain became elaborately developed during the evolution of vertebrates.
D. The neopallium is better developed in mammals than in any other vertebrates.

Correction: _____

_____ 10. A. Mammalian brains have greater numbers of associative neurons than do brains of other vertebrates.
B. The motor areas of the cerebral cortex send messages to muscles all over the body.
C. Mechanisms proposed to explain memory traces pertain to neural RNA molecules, synaptic facilitation, and modification of the cytoskeletons of neurons.
D. The nuclei of the hypothalamus are primarily concerned with the regulation of functions of skeletal muscles.

Correction: _____

_____ 11. A. Physiological events associated with the "fight-or-flight" condition include elevated heart rate, increased blood pressure, and diversion of blood flow away from the stomach and intestines.
B. Norepinephrine is a neurotransmitter of the parasympathetic division of the autononuc nervous system.
C. The adrenal medulla produces hormones that are sympathomimetic.
D. The hypothalamus plays a key role in emotional behavior.

Correction: _____

CHAPTER 31--EFFECTORS

CHAPTER OVERVIEW

Effectors are those structures that act. Effectors, such as muscles and glands, are found at the end of efferent neural pathways. Therefore, study of effectors is a logical next topic in our continuing study of integration and control. In this chapter, we will examine the form and function of muscles, at both the cellular and gross levels. Because muscles can cause motion only if attached to some framework, we will study also the various types of skeletons. We'll conclude this chapter with a survey of the more curious effectors found in some animals and in some plants.

TOPIC SUMMARIES

A. Types of Muscle.--Although all types of muscle share the ability to contract and all contain contractile microfilaments, several different designs of muscle occur in animal bodies. We will classify muscle according to cytological characteristics and the degree of voluntary control the animal has over the function of muscle.

(1) Skeletal muscles. The bulk of the muscle mass in most animals' bodies is composed of striated muscle, muscle that has a streaked appearance when viewed microscopically. Most striated muscle is associated with the skeleton, as skeletal muscle, and causes body movements. For motion to occur, a skeletal muscle usually is attached to the skeleton (or other structures) so that the attachment points span a joint, a movable connection between two bones or other skeletal elements. Attachment is usually by strands of connective tissue called tendons. The part of the muscle located between the tendons, the body, contains the contractile apparatus of the muscle (more on this in part B below).

Muscles act as groups to cause movements; rarely does one muscle act alone. Muscles cause movement only by contraction: Return of a body part to an original position does not involve expansion of an already contracted muscle; instead, a different set of muscles (called antagonists) contracts to "undo" the action of the other set of muscles (the agonists).

(2) Cardiac muscle. Because of its similarity to skeletal muscle, both types often are grouped as striated muscle. Cardiac muscle is striated, but it has a branching pattern not seen in skeletal muscle. Discrete sarcomeres are not defined in cardiac muscle; indeed, cell boundaries are difficult to define. Hence, cardiac muscle "cells" sometimes are said to be multinucleate.

Cardiac muscle occurs only in the heart, and to some degree, in the walls of the major blood vessels in the immediate vicinity of the heart. Innervation is not necessary to stimulate contraction. Electrical charges that stimulate contraction develop inherently in nodal tissues; cardiac muscle is, therefore, myogenic. Close electrical contact between adjacent fibers is maintained by intercalated disks.

(3) Smooth muscle. Because of its association with visceral organs, this muscle type also is called visceral muscle. It commonly occurs in the walls of structures such as arteries, alimentary tract organs, reproductive conduits, and others. It is largely involuntary and is stimulated by the autonomic nervous system (refer to Chapter 30).

Histologically, smooth muscle consists of discrete, spindle-shaped cells containing longitudinally oriented myofilaments. The single nucleus is located centrally in the broadest part of the cell. Contraction is thought to operate by a variation of the sliding-filament mechanism presented below for striated muscle.

B. How Striated Muscle Contracts.--Our understanding of how muscles contract has developed over only the last four decades. This mechanism is called the sliding filament model. It involves interaction between several different kinds of protein molecules, two of which (actin and myosin) are long filaments. These filaments are located within each sacromere, the contractile units of the linear myofibrils (see textbook Fig. 31-3 for the organization of a skeletal muscle). The sacromeres of a myofibril are stacked end to end, much like logs cut from a tree trunk.

(1) Sarcomeres and Sliding Filaments. Fig. 31.1 below shows a sacromere in relaxed (above) and contracted (lower) conditions. Notice how the degree of contaction corresponds to the degree of overlap of the thick filaments (myosin) and thin filaments (actin). Notice, too, that the sarcomere shortens as the degree of filament overlap increases. Interestingly, these filaments do not decrease in length. Shortening occurs because the actin molecules are attached, at one end, to the end of the sacromere. Hence, as the actin filaments are drawn to the center of the sarcomere, the end walls of the sacromere are drawn closer together. Adjacent sarcomeres are attached to each other. The more sacromeres of a myofibril that are involved in shortening, the shorter the entire muscle becomes, and the greater is the amount of tension generated by the entire muscle.

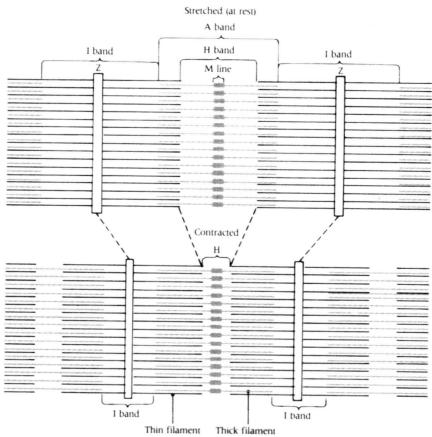

Fig. 31.1: Sarcomeres in relaxed (above) and contracted (below) conditions

Also labeled on Fig. 31.1 above are several bands. These represent various regions of sarcomeres that are visible in microscopic examination . . . hence, the label "striated" muscle. Specifically, these regions are the I band (actin not overlapped with myosin), A band (the zone spanned by myosin filaments), H band (the central part of the A band in which myosin, but no actin, is present), and Z lines (the narrow band defining the ends of a sarcomere). Which of these bands and lines change, and which remain constant, as a sarcomere contracts?

(2) The role of ATP. How can we explain the changes in overlap of actin and myosin filaments? From your earlier study, you know that ATP is involved, but how? Research revealed that heads extending from the myosin molecules interact to form cross-bridges with adjacent actin filaments. Interestingly, the role of ATP is not to form these connections, but to break them so that the actin molecules can advance another "notch" to form new bridges. The role of other associated proteins (tropomyosin, troponin) is to regulate calcium ions, and thereby influence cross-bridge formation.

(3) Excitation. What instructs a sacromere to contract? Whether a muscle is under neural control or not, electrical impulses initiate contraction. Assuming neural control, the impulse crosses from the axon at a neuromuscular junction to the muscle. Sufficient stimulation establishes an electrical impulse that depolarizes the sacrolemma (the plasma membrane of the muscle). This wave of depolarization sweeps across the surface of the muscle fiber and is transmitted to each sarcomere within the fiber via the sacroplasmic reticulum and the transverse tubular system (T system). Stimulation of the T system releases calcium ions that stimulate ATPase in the cross-bridges to initiate the sliding filament mechanism.

C. Functional Patterns of Muscular Activity.--If the ends of a muscle are attached to bones or other structures, tension develops as the muscle contracts. Such a contraction in which shortening occurs, and in which tension reaches the required level and then becomes constant, is called isotonic. This is the more normal situation (e.g., lifting objects) in which we use our skeletal muscles. An alternate application of muscular activity is one in which muscle length does not change, but muscle tension does change; this is an isometric contraction.

(1) Isometric twitch. Stimulation of a muscle by a single, maximum-strength impulse produces a sudden twitch (Fig. 31.2). Subsequent stimulation might or might not produce twitching, depending on the amount of time since the first stimulation and upon the amount of time between subsequent repeated stimulation. A stimulus immediately after the first does not cause another twitch. By slightly increasing the interval after the first twitch and by decreasing the time between impulses, the twitch pattern changes. Notice in Fig. 31.2 that the force generated is greater with greater frequency of impulses. Note also that the force generated is additive up to a maximum amount for that stimulation frequency. This additive phenomenon is called summation. Tetanus occurs where individual twitch responses cannot be distinguished.

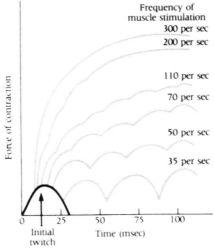

Fig. 31.2: Summation and tetany in vertebrate muscle

(2) Fatigue. We all know that muscles tire after heavy or repeated activity. Such <u>fatigued muscles</u> can no longer perform work until the accumulated waste products are removed and the oxygen deficit is satisfied. Do you recall from Chapter 9 the relationship between accumulation of <u>lactic acid</u> and an <u>oxygen deficit</u>? If not, go back and review it now!

D. <u>Skeletons: Materials and Designs</u>.--All organisms are subject to the force of Earth's gravitational field. To keep from collapsing, organisms must have some structural framework adequate to resist gravity. Skeletons fill such a role. Evolution has produced several versions of skeletons.

(1) Hydrostatic skeleton. In a functional context, you might think of some organisms (e.g., many marine invertebrates) as little more than a sac of fluid. Even at great ocean depths, such sacs of internal fluid withstand high pressures without specialized skeletons. Such survival occurs because liquids, such as body fluids, are noncompressible. Application of force against this fluid, then, can produce motion. Such <u>hydrostatic skeletons</u> afford body support not only in many aquatic species (squids, snails, sea anemones, etc.), but also in a few terrestrial forms (earthworms, caterpillars, etc.). Even vertebrate species rely upon hydrostatic skeletons for some bodily functions: external genitalia stiffen due to engorgement by blood.

(2) Exoskeleton. An external skeleton of <u>chitin</u> was the next evolutionary response by animals to the need to provide greater body support in a terrestrial setting. Not only are gravitational forces greater out of the water, but on land the large amount of water needed to maintain a hydrostatic skeleton was not available. Hence, a rigid skeleton not based on water was needed. An additional feature of the <u>exoskeleton</u> is that it is <u>jointed</u>. Spanning such joints with muscles allowed greater mobility than did a hydrostatic skeleton. The exoskeleton surrounds the soft tissues; muscles attach to inner surfaces of the exoskeleton. Because it encases a growing body, the exoskeleton must be periodically shed. Insects, among others, possess exoskeletons. The bony armor of some vertebrates (e.g., extinct armored fishes, armored dinosaurs, armadillos) is properly considered an exoskeleton.

(3) Endoskeleton. The <u>endoskeleton</u>, such as we humans possess, evolved in the subphylum Vertebrata. Although <u>bone</u> (or <u>cartilage</u>) of this skeleton generally is embedded within soft tissues, the reverse is true in some regions of the body; for example, the skull surrounds and protects the brain. The bone and cartilage of a living vertebrate are living tissues that remodel themselves according to stresses to which they are subjected. Growth of organisms having endoskeletons does not require shedding or replacement of bones. With proper positioning of musculature (i.e., spanning joints), the jointed nature of an endoskeleton facilitates movements.

One segment of the field of <u>functional morphology</u> involves study of how skeletons reflect adaptations to certain lifestyles. Bone-muscle systems can be thought of as <u>machines</u> in which the bones are <u>levers</u>. Analyses of relative lengths of portions of the lever to either side of a pivot point (the joint) can indicate the type of locomotion (e.g., running vs. digging) to which the animal is adapted. This is one of those situations in which your study of basic principles of physics is critical to understanding of biology.

E. <u>Other Effectors</u>.--To this point, we have focused our attention on types of effectors that occur in organisms more familiar to us. Now, let's look at a few other interesting effectors.

(1) Missle-firing systems. Evolution has independently produced an array of projectile-firing systems in diverse groups of organisms. The coelenterates (corals, hydras, jellyfish) are allied by their possession of <u>cnidoblasts</u>, specialized cells containing tiny harpoons called <u>nematocysts</u>. The nematocysts are tipped with poison that can repel attackers (e.g., humans) and kill or stun prey. As a defensive adaptation, bombardier beetles produce and eject a hot

cloud of corrosive fluid from the rear of the body. Some ciliated protists and some fungi own projectile mechanisms, too.

(2) Electric organs. In many groups of fishes, certain regions of muscle tissue have developed into organs specialized for production of electricity rather than of tension. These <u>electric organs</u> are composed of disk-shaped cells called <u>electroplaques</u>. Uses of these organs range from general orientation, communication, and species recognition to prey capture and predator repulsion.

(3) Bioluminescence. Most major categories of organisms include at least a few species that produce light. A visit to your local pet store will provide examples of <u>bioluminescent</u> fishes. Probably you have seen fireflies on a summer evening. Among the applications of biological light are species recognition, courtship, luring of prey, and diversion of predators.

(4) Effectors in plants. Operation of conventional effectors generally involves movement. Most parts of plants, of course, are encased and immobilized by cellulose. The mobility that we normally associate with plants (e.g., bending towards light, wilting) involves slow movements and can be explained by plant growth substances or by changes in water content (turgidity). No nervous tissue is found in plants.

TERMS TO UNDERSTAND: Define and identify the following terms:

microfilaments

microtubules

smooth (=visceral) muscle

striated muscle

skeletal muscle

cardiac muscle

tendon

flexor

extensor

adductor

abductor

rotators

myofiber

myofibril

sarcomere

thin filaments (actin)

thick filaments (myosin)

tropomyosin

troponin

cross-bridges

A band

I band

H band

Z lines

creatine phosphokinase

myosin-ATPase cycle

rigor mortis

sarcoplasmic reticulum

transverse tubular (T) system

fast muscles

slow muscles

catch mechanism

intercalated disks

myosin light chain kinase

isotonic

isometric

summation effect

subtetanus

tetanus

compression elements

tensile elements

hydrostatic skeleton

exoskeleton

endoskeleton

mesoglea

haemocoel

lever

joint

in-force

out-force

lever arm

cnidoblasts

nematocysts

trichocysts

electroplaques

bioluminescence

luciferine

PEOPLE TO KNOW: What is the significant biological contribution of each of these individuals as indicated in Chapter 31?

H. E. Huxley and Jean Hansen

A. F. Huxley and R. Niedergerke

A. V. Hill

R. Buckminster Fuller

CHART EXERCISE

Histologically and functionally, the three major groups of muscle tissue found in vertebrates differ in many ways. Complete the chart below to indicate the following characteristics of skeletal, cardiac, and smooth muscle: (1) multinucleate vs. uninucleate, (2) location of nucleus, (3) presence of branching pattern, (4) presence of striations, (5) locations in body, (6) source of nerve supply, and (7) whether control is voluntary and/or involuntary.

FEATURE	SKELETAL	CARDIAC	SMOOTH
Number of nuclei			
Location of nuclei			
Branching pattern			
Presence of striations			
Location in body			
Nerve supply			
Voluntary vs. involuntary			

DISCUSSION QUESTIONS

1. Compare and contrast the contractile apparatus and mechanisms found in the three major types of vertebrate muscle tissues.

2. What major function does the skeleton (any kind) of an animal share with the supportive vascular tissues (e.g., secondary xylem) of a plant? Discuss how each accomplishes this function.

3. The contractile filaments of muscle are found much more widely among organisms than merely in muscle. What is the identity of these myofilaments? What is their taxonomic distribution? Discuss the functions of these filaments in the various species and cell types in which they occur.

4. What is the role of calcium in the contraction of muscles?

5. Using the terminology of a biophysicist, compare and contrast the skeletal design of the limbs of different mammals adapted for running and for digging.

TESTING YOUR UNDERSTANDING

For each of the test items below, three of the lettered alternatives are true and the other is false. Determine which alternative is false and write its letter in the blank to the left of the question number. On the blank line below alternative D, write the corrected version of the false statement.

_____ 1. A. Operation of "effectors" in plants is more likely based on changes in water pressure in cells than on stimulation by nerve impulses.
B. The limited diversity of "effectors" in plants reflects the limitations imposed by cell walls made of chitin.
C. The contractile filaments of animals are made of proteins including actin and myosin, but not of tubulin.
D. Contractile filaments occur in muscle cells and in other cell types as well.

Correction: _____

_____ 2. A. Both smooth and skeletal muscle can be classified as striated muscle.
B. Z lines mark the ends of adjacent sacromeres.
C. Actin molecules are attached to the ends of a sarcomere, whereas myosin filaments are located in the middle of a sarcomere.
D. The amount of tension generated within a muscle is related to the degree of overlap of actin and myosin filaments and to the number of sacromeres contracting.

Correction: _____

_____ 3. A. Strenuous muscular activity causes accumulation of lactic acid.
B. The cross-bridges between actin and myosin molecules are formed by molecules of troponin and tropomyosin.
C. The role of ATP in muscle contraction is to break cross-bridges.
D. During contraction of sacromeres, the I bands and the H bands become narrower.

Correction: _____

_____ 4. A. For the sliding filament mechanism to operate, calcium ions must be present.
B. Rigor mortis occurs due to depletion of the supply of ATP in muscle tissue.
C. The transverse tubular system of skeletal muscle fibers represents an extension of the sarcolemma into the fiber.
D. "Slow" muscles have a better-developed sarcoplasmic reticulum and T systems than do "fast" muscles.

Correction: _____

_____ 5. A. Cardiac muscle is myogenic, whereas skeletal muscle is not myogenic.
B. Intercalated disks are extensions of the sarcolemma that facilitate electrical contact between adjacent smooth muscle cells.
C. Impulses from the vagus nerve slow the rate of contraction of heart muscle.
D. Cardiac muscle is branched, whereas skeletal muscle is not branched.

Correction: _____

_____ 6. A. The type of muscle found in the walls of arteries is smooth muscle.
B. Troponin is present in skeletal muscle, but absent from smooth muscle.
C. The role of both calmodulin and troponin in muscle contraction is to bind calcium ions.
D. Inactivation of protein kinase causes smooth muscle to contract.

Correction: _____

_____ 7. A. After tension adequate for lifting a weight is reached, the tension in an isotonically contracting muscle remains constant while the muscle shortens.
B. The summation effect of repeated muscle stimulation can be due to an increase in the number of motor units that are contracting simultaneously.
C. Stimulation of a muscle immediately following an initial maximum stimulus results in tetanus.
D. Heat is liberated both by the contraction of a muscle and by the resynthesis of ATP.

Correction: _____

_____ 8. A. Fatigue of a muscle can be recognized by increased concentrations of lactic acid and by an inadequate supply of oxygen.
B. The need for a rigid skeleton is greater for a land-dwelling vertebrate than for a deep-sea invertebrate.
C. Mobility is greater for an animal with a jointed skeleton than for one with a hydrostatic skeleton.
D. Hydrostatic skeletons characterize many aquatic species, but no terrestrial species.

Correction: _____

_____ 9. A. Mesoglea is the main supportive material for the hydrostatic skeleton of a sea anemone.
B. Endoskeletons characterize vertebrate species as well as many terrestrial invertebrates.
C. An animal whose limb bones have relatively long in-lever arms probably is better adapted to digging than to running.
D. The firing of a nematocyst is a highly specialized version of exocytosis.

Correction: _____

_____ 10. A. A nematocyst is a multicellular structure housed within cnidoblasts.
B. Electric organs develop from muscle tissue.
C. One of the functions of bioluminescence is communication between individuals of opposite sex.
D. Production of bioluminescent light involves oxidation of the molecule luciferase.

Correction: _____

CHAPTER 32--BEHAVIOR

CHAPTER OVERVIEW

We now come to the point in this six-chapter section on integration and coordination where the entire organism responds in a coordinated way to environmental stimuli. Behavior is the result of integrated operation of sensory structures, information processing centers, and effectors. Of course, communication within the body by neural and/or chemical messengers is a critical part of behavior. Behavior appropriate to certain stimuli is an integral component for survival.

In this chapter, we will survey various types of responses that organisms have towards stimuli. We will distinguish between behaviors that are learned and those that are not learned (innate). We will examine behaviors that cycle over periods of days and seasons. We will study orientation and navigation and the cues that guide such movements.

TOPIC SUMMARIES

A. <u>What Is Behavior</u>?--We'll define <u>behavior</u> simply and broadly as being "any directed activity" of any organism. Behaviors are activities that, on average, benefit the species to which the organism belongs; that is, behaviors are <u>adaptive</u>. The activity might be complex (as solving calculus problems) or simple (as a kneejerk reflex). Very simple behaviors often result from activity of a <u>simple reflex arc</u>; in such situations, no associative involvement of thought or memory need be involved. More complex behaviors, however, generally involve sensors, <u>associative centers</u>, and effectors. Scientific study of behavior is a young field. Efforts to explain behaviors involve manipulation of stimuli, observation of behaviors, and examination of activity in associative centers; the technique of <u>electroencephalography</u> permits objective evaluation of associative activity.

B. <u>Behavior in Plants</u>.--The behaviors of plants are classified as <u>tropisms</u>. Movements associated with tropisms are growth responses whose direction is determined by differences in the intensity of the stimulus on opposite sides of the structure (organs such as roots, leaves, etc.) that moves. The turning response is innate and inflexible, and it cannot be modified or controlled by the organism. The name of a tropism corresponds to the stimulus towards or away from which the organ turns. In the context of chemical coordination (Chapter 28), we were introduced to <u>phototropism</u>. Other stimuli to which plants respond include gravity (<u>geotropism</u>), contact (<u>thigmotropism</u>), chemicals (<u>chemotropism</u>), and water (<u>hydrotropism</u>).

Similar directed movements in animals and other non-plants are called <u>taxes</u>. These represent unlearned behaviors. Examples include movements of leukocytes to sites of injury and attraction of mates by pheromones.

C. <u>Innate and Learned Behavior</u>.--Behavioral biologists generally agree that there are two <u>modes of behavior</u>. <u>Innate behavior</u> is instinctive, inherited, and cannot be modified, whereas <u>learned behavior</u> can be modified by experience. Probably, only innate behavior occurs in plants, protists, and coelenterates. The ability to learn exists in flatworms nad earthworms and perhaps in echinoderms. The ability to learn corresponds with the increasing complexity of the nervous system.

Historically, these modes of behavior have been studied via two approaches. <u>Ethology</u> is the scientific study of behavior; ethologists study behavior via direct observation of animals in their natural surroundings. The alternate approach is <u>experimental psychology</u>; such psychologists study animal behavior in laboratory settings.

D. <u>Basic Components of Behavior</u>.--Several basic components contribute to behavior; these are sign stimuli and releasers, fixed action patterns, and motivations and drives. These responses are <u>adaptive</u> in that an organism can quickly and automatically respond to environmental stimuli, usually in appropriate ways.

(1) Sign stimuli and releasers. Certain specific stimuli can trigger particular behavior patterns. The textbook offers as an example a fish that exhibits aggressive territorial behavior when it sees objects that have red undersides. The object need <u>not</u> be associated with another fish (which is the naturally-occurring object with a red underside). Similarly, male robins defend their territories against other male robins; recognition of a male robin is simply by the presence of a tuft of red feathers, whether the tuft is associated with a bird or not. Because they release specific behaviors, these <u>sign stimuli</u> sometimes are called <u>releasers</u>.

Such response is attributed to an exaggerated neural sensitivity to the sign stimulus. The adaptive value of recognition of sign stimuli lies with the rapid, automatic response. This is particularly important to survival when, for example, the releaser is a predator.

(2) Fixed action patterns. Sometimes the behavior elicited by a stimulus proceeds in an all-or-none fashion. That is, the behavior continues to completion whether or not the full response is necessary. For example, a toad will go through all of the motions of swallowing, shutting its eyes, and wiping its mouth even if it fails to capture a worm with the tongue strike. These behaviors are <u>fixed action patterns</u>.

(3) Motivations and drives. Sometimes, the same stimuli elicit different responses. The response can be modified by internal forces, called <u>drives</u>. The effect of a drive is to <u>motivate</u> different responses at different times. The means by which a drive operates appear to involve alteration of stimulus thresholds. Hence, our drives for food and drink lessen as we consume a meal. Several different drives may compete simultaneously; usually one drive will prevail, and the behavior associated with that drive will occur. The drive to escape from danger usually has priority over other drives.

E. <u>Learning</u>.--Many situations are too complex for innate responses to be appropriate. The need for flexibility of behavior in such complex (or unanticipated) situations is met by <u>learning</u>. The degree of learning that an animal is capable of generally corresponds with anatomical features (e.g., the complexity of its nervous system) and with life-history characteristics (e.g., life span). Several levels of learning are recognized:

(1) Habituation. This is a process by which an animal decreases the frequency of response when a particular stimulus is applied. For example, young animals often respond to stimuli that are not significant to survival. If, over time, these animals learn that nothing detrimental follows the stimulus, then they cease to respond to that stimulus. This "response-waning" phenomenon frees the animal's nervous system to deal with more meaningful stimuli.

(2) Associative learning. This process involves association of multiple stimuli with a particular reward or punishment. <u>Conditioning</u> of the individual has occurred when any one of the multiple stimuli elicits the behavior. The renowned work of Pavlov demonstrated that dogs could be conditioned to respond nonsensically (to salivate) to the sound of a bell. The <u>classical conditioning</u> principles of <u>associative learning</u> are used widely in both research and treatment programs.

(3) Trial-and-error learning. <u>Operant conditioning</u>, a term coined by the late B. F. Skinner, occurs as a result of <u>trial-and-error learning</u>. Such <u>operant behavior</u> is molded by the <u>consequences</u> of that behavior; this differs from reflexive behavior in which events before the behavior elicit the behavior. Operant conditioning involves three phases: (a) stimulus, (b)

unorganized trial-and-error responses, at least one of which proves successful, and (c) subsequent repetition of successful responses. Repeated achievement of the reward <u>reinforces</u> operant behavior.

(4) Imprinting. Unlike some of the behaviors we have surveyed above, <u>imprinting</u> does <u>not</u> involve reward or punishment. It is an innate behavior that is learned during a critical time interval, usually early in life. There are intervals before and after this critical period during which imprinting cannot occur. Interestingly, imprinting is irreversible. Common examples of imprinting are of newborn or newly hatched vertebrates that come to recognize certain other individuals as parents. Recognition of parents is critical to survival of the young.

(5) Insight learning. This process, apparently occurring only in primates, involves reasoning and generalization. Insight reasoning involves (a) use of memory from earlier experiences unrelated to current stimuli and (b) subsequent application of what was learned previously to solve novel problems in new situations. You use this ability to solve problems and to answer questions on exams.

F. <u>Biological Clocks and Circadian Rhythms</u>.--The physical world is characterized by many cycles--tides, day and night, hot and cold, etc. Because organisms live in such a cyclic world, they, too, must (and do) cycle in their activities. The rhythms within organisms somehow track these environmental cycles. The cues to which organisms respond are varied: Day length affects the flowering of plants. Spawning in certain fishes correlates with the tides. Migration in some birds is thought to be cued by changing length of night.

Experimentation has shown, however, that many organisms maintain their normal (or nearly normal) rhythm in the absence of environmental stimuli. Usually these activity cycles have a period of about 24 hours, hence the name <u>circadian rhythms</u>. Apparently, some sort of internal <u>biological clock</u> operates. In vertebrates, the <u>pineal body</u> (introduced in Chapter 28) seemingly serves as the seat of biological clocks that cycle in response to light-dark cycles. Recall that there is a connection of the eyes with the pineal body via the hypothalamus. The pineal body produces the hormone <u>melatonin</u> in response to stimulus by light. Melatonin, then, stimulates the circadian cycle.

Fig. 32.1: Migratory routes of the American golden plover

G. <u>Orientation and Navigation</u>.--In their daily and seasonal movements, animals demonstrate remarkable ability to find their way about within their surroundings. Animals placed anywhere in their familiar surroundings generally have no trouble finding their way home. In many species, long-distance returns to home occur, even if the animal is displaced (by the researcher or a natural event such as a storm) in an unfamiliar, distant place. Hence, animals somehow maintain <u>orientation</u> within space and can <u>navigate</u> even in unfamiliar situations. Some seasonal <u>migrations</u> cover immense distances (Fig. 32.1); see your textbook for examples.

What mechanisms explain such migratory and homing behavior? A relatively simple approach is the <u>light-compass reaction</u> used by many invertebrates: ants and bees use the sun, keeping the angle between the direction of movement and the direction of light constant. <u>Olfactory imprinting</u> on molecules in a particular stream cues the return of salmon to that stream to spawn. Migration in some birds probably involves both <u>internal timing</u> and <u>landmarks</u> (visual and otherwise?) in the flight path. <u>Celestial cues</u> (sun and other stars) guide certain other migrating birds. Homing involves more than mere navigation. An animal displaced into unfamiliar surroundings must somehow determine where it is in relation to home; how this <u>map sense</u> operates is unknown. There is some evidence that ability to detect position in Earth's magnetic field might play an role in navigation and map sense.

TERMS TO UNDERSTAND: Define and identify the following terms:

behavior

anthropomorphism

Morgan's canon

behaviorism

tropisms

taxes

adaptive

innate behavior

learned behavior

ethology

ethogram

instinct

releaser

supernormal stimuli

heterogeneous summation

fixed action pattern

chain reactions

motor programs

drives

learning

habituation

associative learning

conditioned reflex

trial-and-error learning

operant conditioning

parental imprinting

insight learning

entrainment

photoperiodism

biological clock

circadian rhythms

pineal gland

melatonin

navigation

orientation

homing

nerve-net behavior

PEOPLE TO KNOW: What is the significant biological contribution of each of these individuals as indicated in Chapter 32?

Lloyd Morgan

John B. Watson

Konrad Lorenz

Niko Tinbergen

Ivan P. Pavlov

B. F. Skinner

Fred A. Urquhart

DISCUSSION QUESTIONS

1. Compare and contrast tropisms, taxes, and simple modes of navigation.

2. Select a human behavior, and analyze it in the context of the types of learning presented in this chapter.

3. Define "behavior." Provide an example of behavior for some species in each of the five kingdoms of organisms.

4. Discuss the adaptive nature of behavior.

5. Distinguish learning from mechanisms such as releasers, fixed action patterns, and drives.

6. Behaviors of many organisms differ according to time of day or to season of the year. Provide several examples. How is the timing of such behaviors regulated? How (if at all) is such variable behavior of value to the organism practicing that behavior?

7. Discuss the various environmental stimuli that can serve as cues that direct navigation and orientation in various organisms.

TESTING YOUR UNDERSTANDING

For each of the test items below, three of the lettered alternatives are true and the other is false. Determine which alternative is false and write its letter in the blank to the left of the question number. On the blank line below alternative D, write the corrected version of the false statement.

_____ 1. A. Behavior can be defined as any directed activity of an organism.
B. In animals with simple nervous systems, behaviors tend to be triggered by immediate external stimuli, whereas behaviors in animals with complex nervous systems frequently are stimulated by associative processes.
C. The law of parsimony indicates that the behavior of an animal should be interpreted in terms of the simplest possible mental processes.
D. Anthropomorphism is an approach to the study of behavior in which all behaviors are thought to correspond directly to specific stimuli.

Correction: _____

_____ 2. A. Electroencephalography is a technique that allows an investigator to determine if brain activity occurs in response to a stimulus.
B. In plants, a stimulus that invariably produces the same automatic response is called a taxis.
C. In tropisms, the turning response is an innate and inflexible part of the organism's genome.
D. The turning response in thigmotropism occurs as a result of physical contact.

Correction: _____

_____ 3. A. If a particular behavior is only occassionally detrimental to an individual, the behavior will be maintained in the behavioral repertoire of the species.
   B. Appropriate responses to stimuli are critical to the survival of a species.
   C. Innate behavior is that which is performed without prior learning.
   D. Several species of protists are known to possess the ability to learn.

   Correction: _____

_____ 4. A. The ability to learn corresponds to the structural complexity of the nervous system of the organism.
   B. Ethologists produce ethograms by conducting behavioral experiments in the laboratory.
   C. Sign stimuli are features of a stimulus that trigger a particular behavior.
   D. The adaptive value of releasers is that they automatically and rapidly evoke a specific, critical behavioral response.

   Correction: _____

_____ 5. A. Fixed action patterns are elaborate sequences of meuromuscular events.
   B. When the stimulus of a fixed action pattern is removed, the behavior stops at that point in the fixed-action-pattern sequence.
   C. Drives are internal forces that cause an individual to respond differently to the same stimulus on different occassions.
   D. In the event that a drive to quench a thrist is occurring at the same time as the drive to escape from danger, only the behavior associated with the escape drive will be performed.

   Correction: _____

_____ 6. A. Innate behavioral mechanisms are flexibile and can be modified by learning.
   B. Learning is defined as modification of behavior by experience.
   C. Imprinting occurs during a brief interval during early development.
   D. Imprinting can involve visual or olfactory cues.

   Correction: _____

_____ 7. A. In habituation, the absence of a negative consequence causes the animal to cease a behavior.
   B. In a conditioned reflex, the response to one stimulus becomes associated with another.
   C. Pavlov's study of dog salivation behavior is an example of operant conditioning.
   D. Operant behavior is behavior that is guided by consequences of the behavior.

   Correction: _____

_____ 8. A. Reinforcements are an important part of associative learning.
   B. Insight learning involves application and adaptation of existing knowledge to solving new problems in a different context.
   C. Among vertebrates, the pineal body is the site of biological clocks.
   D. Function of the pineal body involves routing of nerve impulses from the eyes to the hypothalamus.

   Correction: _____

_____ 9. A. Circadian rhythms in most vertebrates cycle with approximately a 24-hour period.
B. All of the following are environmental cues by which various animals navigate: sun, stars, landmarks, chemical composition.
C. Use of the light-compass reaction to orient involves keeping constant the angle between the direction of movement and the direction of the light stimulus.
D. Map sense refers to the ability of an animal to return home by way of a familiar route.

Correction: _____

_____ 10. A. The approach by which the study of behavior was reduced simply to correlations between stimuli and responses is known as behaviorism.
B. Releasers are sign stiumli that trigger specific behaviors.
C. Some of the elaborate fixed action patterns are called chain reactions because each step is the stimulus for the next step.
D. In innate behavior patterns such as drives, an animal will respond to the same stimulus in the same way.

Correction: _____

CHAPTER 33--SOCIAL ASPECTS OF BEHAVIOR

CHAPTER OVERVIEW

In this chapter, we move from the level of the individual organism to study various types of social behavior in which many organisms interact for a common cause. We begin by demonstrating that behaviors can be inherited and that evolution of behaviors is directed by natural selection.

Social behaviors (and some nonsocial behaviors) require communication between individuals. Channels of communication include chemical, visual, auditory, electrical, and tactile stimuli. Social behaviors evolved because such behaviors enhanced survival and reproductive success of those practicing the behaviors. Social behavior evolved several times among the insects and among primate mammals as well. However, the operations of societies of insects and of primates are significantly different in terms of reproduction, development, division of labor, and other characteristics.

TOPIC SUMMARIES

A. <u>Genetic Aspects of Behavior</u>.--It is logical to expect that behavior, like other phenotypic traits, is heritable. Study of white-crowned sparrows demonstrated that the ability of a young bird to sing the song of its species is inherited, although proper performance of the song requires learning. Similarly, selection of foods by garter snakes is genetically-based.

If behavioral traits can be inherited, then they must also be subject to the forces of evolution. Those behaviors that aid in the survival of an individual likely will be passed on to other generations. Study in this area is the field of <u>behavioral ecology</u>. We'll talk much more of evolution and natural selection in Chapters 34 and 35.

B. <u>Communication between Individuals</u>.--Animal communication involves at least two parties--the signaller and the signal receiver. Evolution of a signalling system requires that the signal receiver benefits while the signaller either benefits or, at least, is not harmed. Various forms of communication can be used by an individual that desires to alter the behavior of another individual:

(1) Chemical channels. The most ancient form of communication probably involved production and release of chemicals. Chemical signals carry many types of messages, including indication of receptiveness for reproductive activity (e.g., <u>pheromones</u>), identification of individuals of the same species, and warning of the presence of danger.

(2) Visual channels. Certain information about the physical and biotic surroundings can be perceived by photoreceptors, such as eyes. Color, size, motion, and other sorts of information can be detected quickly through visual channels.

(3) Auditory channels. Sound and other sorts of vibrations can convey important information, especially in situations that are dark or in which vision is obscured. Languages and vocalizations often are very elaborate means of communication.

(4) Electrical channels. Another alternative means of sending messages or of sensing the environment involves the production of, and sensing of, electrical fields. This strategy is used by many species of bony and cartilaginous fishes, especially those that occupy dark and turbid waters.

(5) Tactile channels. In many species, the sense of touch is particularly well developed. Examples include many fishes that are adapted to the lightless conditions of caves and underground waterways; in many such species, the eyes evolutionarily have been lost or are vestigial.

C. <u>Social Behavior</u>.--Organisms act not only as individuals, but in many species they act as members of a group or a <u>society</u>. Social interaction is favored evolutionarily when such behavior leads to survival and greater reproductive production--hence, the adage "safety in numbers."

(1) Evolution of social behavior. Two hypotheses have been offered to explain how social behavior evolved. The <u>familial pathway</u> holds that the animals of the species originally bred solitarily and did not care for their young. Then <u>parental care</u> emerged. Young of an early litter remained with parents, eventually for long enough periods that older offspring were still present when later litters came along. Older offspring took on roles of assisting in the rearing of younger offspring. Hence, social groups consisted of only close relatives.

The <u>parasocial pathway</u> is an alternative explanation. It, too, begins with a solitary species. Unrelated individuals of the same generation formed groups around resources (food, water, etc.). Cooperative interaction between individuals provided a greater-that-normal share of resources to those cooperating. Eventually, different individuals within the group took on specialized roles (<u>division of labor</u>), with the result that per capita benefits increased further.

(2) Patterns underlying social behavior. Animal societies are organized (and classified) around six principal patterns of behavior. <u>Cooperation</u> can facilitate more efficient prey capture and can be effective in reducing the likelihood of becoming prey. Individuals (or groups) might establish and patrol <u>territories</u>, areas which contain adequate amounts of necessary resources. Operation of <u>dominance hierarchies</u> (pecking orders) serves to reduce competition and tension within the group. <u>Leadership</u> refers to long-term direction of group activities, and, in some societies, is related to dominance. Organization of many societies is based on factors such as <u>parental care</u> of young and <u>reciprocal care</u>; societies based on these two patterns usually are associated with long-lasting, complex social bonds.

D. <u>Insect Societies</u>.--Although most insect species are not organized into societies, many are highly social. The social insects include many <u>hymenopterans</u> (ants, bees, wasps, termites). That such insect societies have evolved independently several times attests to the species survival value of social organization. In such arrangements, the good of the individual is sacrificed for the good of the group--the concept of <u>altruism</u>. Insects societies feature high degrees of <u>division of labor</u>, such that each individual belongs to a <u>caste</u>. Individuals in different castes have very different roles (e.g., obtaining food, defending the group, laying eggs, fertilizing the queen, etc.). Indeed, roles are so specialized that no one caste alone could ensure continuation of the species. A complex interplay of genetics and of developmental events determines the caste of an individual.

Considering that insects have the most complex nervous systems among invertebrates, it is reasonable that social behavior evolved so extensively in insects. These insects inherit many complex social behaviors; social insects exhibit a greater ability to learn than do solitary insects.

E. <u>Primate Societies</u>.--An array of types of social organizations exists among primates. The type of society found in a particular species corresponds to factors such as group size, size of home range, dominance behavior, and degree of territoriality.

The pinnacle of primate social evolution prevails in humans. Like other primate societies, individuals in human societies have specialized roles and individuals integrate their roles with those of others to accomplish tasks and to achieve goals. Human societies have other features that set them apart from those of non-human primates: Permanancy of male-female bonds is greater. The nuclear family is a central element in most human socities. Both sexes cooperate in rearing of young. Generally, there is sexual exclusivity rather than promiscuity. Human societies are associative rather than strictly familial. Individual humans have much greater control over their fates, being able to select their own roles in society rather than having roles determined by genetic or developmental factors. Learning is a key in preparing for the selected role. Individual humans are sufficiently self-sufficient to survive outside of the society.

TERMS TO UNDERSTAND: Define and identify the following terms:

behavioral ecology

chemical channels

visual channels

auditory channels

larynx

syrinx

electrical channels

tactile channels

displacement behavior

social behavior

familial pathway

parasocial pathway

cooperation

territoriality

dominance

leadership

parental care

mutual stimulation

social insects

subsocial

altruism

trophallaxis

coefficient of relatedness

inclusive fitness

tribal behavior

polygamy

monogamy

polygyny

associative societies

mimicry

PEOPLE TO KNOW: What is the significant biological contribution of each of these individuals as indicated in Chapter 33?

    Peter Marler

Stevan Arnold

Theodosius Dobzhansky

Karl von Frisch

W. D. Hamilton

T. H. Clutton-Brock and P. H. Harvey

H. Wu and G. Sackett

DISCUSSION QUESTIONS

1. a. Compare and contrast insect societies and human societies. Do both arise in the same genetic and developmental manners?

   b. Are there differences in the degrees of occupational flexibility allowed members of these societies?

   c. Are there differences in opportunities of individuals to reproduce?

2. a. In what way(s) is behavior important to the survival of an individual?

b. How does survival of the individual promote survival of that species?

c. How should the answers to these questions differ (if at all) if you consider solitary versus social species?

3. a. List the various types of mating systems that occur in societies of nonhuman primates. Give examples of a species for each type of mating system.

b. What are the advantages of each type of mating system?

c. Propose scenarios by which one type of mating system might have evolved from another.

TESTING YOUR UNDERSTANDING

For each of the test items below, three of the lettered alternatives are true and the other is false. Determine which alternative is false and write its letter in the blank to the left of the question number. On the blank line below alternative D, write the corrected version of the false statement.

_____ 1. A. Like other traits, behavior is heritable.
B. In birds, the ability to sing a species-specific song is inherited.
C. Proper performance of a species-specific song requires learning.
D. Signalling systems evolve only if the signaler benefits.

Correction: _____

_____ 2. A. Signals can be in a variety of forms, including chemical, visual, and auditory.
B. Redirection of actions toward nearby objects is called displacement behavior.
C. Societies evolved because individuals operating independently fared better than those participating in group activities.
D. In parasocial groupings, the society involves active cooperation of unrelated individuals.

Correction: _____

_____ 3. A. A territory is an area that includes an adequate supply of needed resources.
B. Although considerable stress might be involved in establishing a dominance hierarchy, over time such hierarchies reduce competition and tension within the group.
C. A society in which parental care of young occurs usually also is characterized by long-lasting social bonds.
D. Most species of insects are social.

Correction: _____

_____ 4. A. Insects have the most-complex nervous systems among the invertebrates.
B. The caste of a social insect is determined by its learning experiences.
C. Insects societies are altruistic.
D. Sociality probably evolved independently several times among the insects.

Correction: _____

_____ 5. A. The role of an individual in human society is determined during an early developmental stage.
B. A variety of ecological factors (such as food availability and predator pressure) are important in molding the social structures of primates.
C. Although a human in a society generally occupies a particular role, the individual is able to change that role via learning experiences.
D. The size of an insect society is limited, largely because of the limited number of progeny that one, or only a few, females can produce.

Correction: _____

_____ 6. A. Monogamy probably evolved in a situation in which males could secure only one mate.
B. Monogamy probably evolved in a situation in which males were able to leave more offspring by helping the female to rear the young.
C. Harems are formed in some polygamous mating systems.
D. Monogamy is more common among primate species than is polygyny.

Correction: _____

_____ 7. A. Altruism and trophallaxis are features of societies of many species of nonhuman primates.
B. Cooperation, in the form of cooperative hunting, was an important element the evolution of human societies.
C. In the familial hypothesis of evolution of social behavior, all group members are closely related.
D. Early steps in the evolution of complex communicative behaviors probably involved exaggeration of various preparatory behaviors.

Correction: _____

_____ 8. A. Anatomical structures involved in communication by auditory channels include ears, a larynx, and a syrinx.
B. The abilities to produce and to perceive electrical fields is beneficial to animals residing in aquatic environments with low levels of turbidity.
C. Sociobiology is the study of the social biology of animals.
D. Chemical signalling is probably the ancestral means of communication between individuals.

Correction: _____

_____ 9. A. Deceit and mimicry are strategies that are used widely in communications between species.
B. Because behaviors are learned rather than inherited, behaviors are not subject to the effects of natural selection.
C. A primary value of social behavior to a prey species is its effectiveness in thwarting predation.
D. A primary value of social behavior to a predatory species is its effectiveness in enhancing predatory effeciency.

Correction: _____

_____ 10. A. Subsocial behavior in insects includes activities such as parental care.
B. Queens and workers develop from fertilized eggs, and genetically are females.
C. Sociality in insects results from extended relationships of parents and offspring.
D. Solitary behavior probably evolved from social behavior.

Correction: _____

CHAPTER 34--ELEMENTARY PROCESSES OF EVOLUTION

CHAPTER OVERVIEW

To this point, our study of biology has focussed largely on components and their functions and on various processes occurring within individual organisms. We now enter the final part of the textbook where our study is at the level of the whole organism and groups of organisms. This chapter introduces fundamental processes related to the concept of evolution. We will build upon these in the remaining chapters of the textbook.

Evolution is a property of populations, not of individuals. Evolution refers to genetic changes from one generation to the next. Natural selection operates to narrow the genetic diversity of a population. Natural selection selects those phenotypes that are best adapted to the environment. Favored phenotypes contribute proportionately more to the next generation, and, thereby, are said to be more fit.

TOPIC SUMMARIES

A. <u>Units of Life</u>.--For unicellular species (e.g., those in kingdoms Monera and Protista and some species of plants and fungi), study at the <u>cellular level</u> is the same as studying the entire organism, the <u>individual</u>. Study of the individual in other species, however, pertains to the entire package of all cell types and physiological processes. We are now beginning study of processes that characterize only entire individuals and groups of individuals.

Individuals can be grouped in various ways: A <u>species</u> is <u>all</u> individuals, no matter where they are, of the same kind, i.e., "a group of organisms so similar in structure and heredity that their demes tend to intergrade, fuse, and replace each other without changing the nature and role of the group as a whole." A species (e.g., all cougars in the Western Hemisphere) is divided into <u>populations</u>, all members of a particular species occupying a particular area at the same time. All cougars present in, say, North America form a population. A <u>deme</u> is a localized portion of a population, wherein interbreeding occurs. Hence, those cougars in mainland California and those occupying adjacent offshore islands can be thought of as belonging to separate demes (assuming little or no movement between islands and mainland).

Because demes are defined as reproductive units, it is logical to expect individuals from one deme to more resemble others from its own deme than individuals from other demes. Breeding among individuals from different demes generally has predictable phenotypic results. For example, cougars from deme A usually have lightly colored fur, whereas those from deme B have darker fur. Mating of individuals of the two demes produces progeny, at least some of which have intermediate pelage colors. Such individuals are <u>intergrades</u>. When studying the geographic distribution of such traits, the intergrades usually are found in the areas where the two demes overlap or come close to each other.

B. <u>Sources of Variation</u>.--Evolution is a process that operates at the level of the population, <u>not</u> at the level of the individual. A key element to the theory of evolution is the presence of <u>variation</u> among individuals in the way virtually every character is expressed in virtually every population of every kind of organism. A quick glance around the spectators at the football stadium verifies the application of this statement for humans. The textbook presents statistical means of quantifying variation within and among demes.

In previous chapters, we have examined various sources for genetic variation within individuals. These include <u>mutation</u> of one allelic form to another, <u>chromosomal mutations</u> (inversions, translocations, etc.) <u>crossing-over</u> during meiosis, new combinations of genes due to random assortment, recombination during zygote formation, relocation of genes (transposons), cell hybridization, chromosome-mediated gene transfer, and genetic engineering. Processes contributing to variation at the level of the population are discussed next.

C. Population Genetics.--Study of variation at the level of populations falls within the field of population genetics. For population geneticists, an operational definition of evolution is "a change in a population's genetic makeup." Such change can be tracked through changes in allelic frequencies from one generation to another. The following pertains to diploid species.

(1) Allelic Frequencies. Consider, for example, a gene represented by two alleles. The gene pertains to color of feathers. The dominant allele (F) codes for red feathers, whereas the recessive allele (f) encodes blue feathers. An individual may be heterozygous (Ff) or homozygous (FF or ff). Assume that we know that the population is in Hardy-Weinberg equilibrium. For any generation, the sum of the proportions of loci occupied by F (p) and by f (q) must always equal one ($p + q = 1$). You can determine q by calculating the square root of the proportion of the population that is homozygous recessive. The proportion of F always equals one minus the proportion for f ($p = 1 - q$). If homozygous dominant individuals are distinguishable from heterozygotes, you can compute p by summing the proportion of the population that is homozygous dominant and the half of the proportion that is heterozygous (p = [number having FF + 0.5 (number having Ff)] / total number in population). Of necessity, if p changes, so must q.

If allelic frequencies (values for p and q) for a generation differ from p and q of a previous (or later) generation, then evolution has occurred. In the absence of various genetic disturbances, however, p and q do not change; the processes of sexual reproduction tend to maintain the genetic status quo; that is, the population is in combinatorial equilibrium. This statement embodies the Hardy-Weinberg law, a tenet of population genetics. This law is modelled mathematically by expanding the binomial equation presented in the paragraph above:

expands to
$$p + q = 1$$
$$p^2 + 2pq + q^2 = 1,$$

where p^2 is the proportion of the population that is homozygous dominant, $2pq$ is the proportion heterozygous, and q^2 is the proportion homozygous recessive. What conditions are essential for Hardy-Weinberg equilibrium to prevail? Departure from Hardy-Weinberg equilibrium involves operation of one or more of the following agencies of genetic change at the population level.

(2) Processes that alter allelic frequencies. Mutations of various sorts occur spontaneously and due to the influence of various mutagens (e.g., various chemicals, radiation). Mutations can occur in both forward and backward directions: an allele that has mutated from dominant to recessive can revert to dominant. Because mutation is a chemical reaction, the process is reversible, and an equilibrium (theoretically) can be reached. Usually mutational equilibrium does not occur, largely because other processes intervene before the back mutation occurs. Nevertheless, mutation pressure can be one of a suite of factors driving evolution.

Genetic drift, too, can influence allelic frequencies from one generation to the next. Genetic drift refers to changes in allelic frequencies that occur entirely by chance, not because one allele causes individuals to be more fit than does the other allele. This process can be thought of in terms of the effects of sampling error: In large populations, deviations from random sampling generally have little effect. However, deviations from random can cause significant changes in small populations. Drift is well illustrated by the founder principle:

Suppose you are studying birds with the feather color gene introduced above in part (1) above. Suppose that both p and q are 0.5. in a huge (500,000 birds) mainland population, in which 75% of the birds have red feathers and 25% are blue-feathered. Suppose that a storm blows 50 birds out to sea. Purely by chance, 40 of these birds are blue-feathered. All 50 birds land on an island, far from the mainland, and establish a reproducing population. The allelic frequency of this population (before breeding) is markedly different from the mainland population, with q = 0.89 (from square root of 0.8 . . . from 40/50) and p = 0.11. After one generation of panmictic breeding and reproduction, what will be the values for p and q? This is an example of the marked effects the drift can have.

Earlier in this chapter, we noted that demes of a species are more-or-less distinct from each other. For example, most members of one deme might have a particular color of hair or might be of a particular height, whereas most members of other demes might have hair of a different color or might be taller or shorter. This degree of variation is not enough for the demes to be recognized as different species, but they might comprise different subspecies. A correlate to such distinctiveness of demes is the degree of isolation (by distance or by barriers to movement of the individuals) between demes. Isolation can be measured in terms of gene flow, the movement of alleles between demes. These genes, of course, are carried by individual organisms. So, we are interested in the number of individuals moving between demes. The degree of difference between demes is related to the amount of immigration into and emigration from those demes. A newly evolved allele can spread between demes by gene flow.

The fourth major factor that can influence allelic frequencies in populations is natural selection, "the perpetuation in the next generation of certain genotypes as a result of a shift in frequencies of certain phenotypes (and their underlying genotypes)." Horizontal gene transfer (textbook Box 34-1) is now considered an established means of genetic change.

D. Natural Selection. We have listed in this and previous chapters many processes which produce variation in genotype and phenotype. In each generation, Nature "selects" those phenotypes (and, consequently, genotypes) that are best adapted to the immediate environmental conditions at the time that selections acts. It is these "fit" individuals that contribute a disproportionately large amount to the next generation. Those individuals that are less fit pass a disproportionately small amount of their genome to the following generation. Over time, then, allelic frequencies within a population shift towards the more-fit genotype. Hence, we see that this evolutionary factor, natural selection, operates by differential reproduction (related to nonrandom fecundity).

It should be apparent, then, that physical fitness is not necessarily the same as evolutionary fitness. To be sure, evolutionary fitness demands some degree of physical fitness, but also requires an individual to do other things (e.g., food gathering, nest building, behavioral interactions, courtship rituals, etc.) in an appropriately fit manner. We must note here that longevity is a factor; to contribute to the next generation, an individual must live long enough to become reproductively mature. The longer an individual lives, the greater can be that individual's reproductive contribution to the next generation.

Nonrandom mating is a component of natural selection. Recall that the Hardy-Weinberg law requires, among other things, panmixia. Completely random mating probably never occurs in natural populations. One reason is purely logistical: The distance between prospective mates might be too great for a mating encounter to occur . . . or for it to occur often enough that fertilization results. For example, the likelihood of two humans from a small country town meeting and mating with each other certainly is greater than for two humans, one of whom is from Maine and the other from New Zealand. Suppose, however, that we restrict our discussion to a local deme. Even when all individuals are equally accessible to all others as prospective mates, panmixia still does not often occur. This is because individuals use phenotypic features (e.g., attractiveness, height, hair color, etc.) to select the "better" prospective mates from the pool.

In the preceding paragraphs, we addressed production of offspring from the viewpoint of an individual member of a deme; this phenomenon is call <u>individual selection</u>. An individual can promote inclusion of some of its own alleles by an additional (or alternative) strategy as well: <u>kin selection</u>. Any activity of an individual that promotes the fecundity of its own relatives (siblings, etc.) will indirectly enhance the representation of that individual's alleles in the next generation. Can you think of some specific examples? Together, individual selection and kin selection determine an individual's <u>inclusive fitness</u>.

E. <u>Ways in Which Variation Is Expressed</u>.--Populations often include different morphs or forms; if such forms persist over generations, then the population is <u>polymorphic</u>. Examples of traits in which various species are polymorphic include human blood types, directions of coiling of snail shells, and plumage color of snow geese; these polymorphisms are based on <u>discontinuous phenotypes</u>. Other polymorphisms pertain to <u>continuous characters</u>, such as height or color, that grade one phenotype into another. The genetic basis for polymorphisms involves several mechanisms, including (1) the actions of single genes where different alleles produce different phenotypes and (2) <u>heterosis</u>. Heterosis is a situation in which the heterozygote has a phenotype different from either homozygote and that phenotype confer a particular adaptive advantage. Neither allele can be eliminated from the population when the heterozygote is the most fit form. Maintenance of polymorphisms in a steady-state is called <u>balanced polymorphism</u>. <u>Transitional polymorphism</u> occurs while an advantageous allele is replacing other alleles.

F. <u>Forms of Natural Selection</u>.--Three means are recognized to describe the ways in which natural selection can act on the variability within a population.

(1) Stabilizing selection. This is the form of selection that acts on most traits in most populations of organisms. Simply, <u>stabilizing selection</u> favors those individuals that occupy the middle of the distribution and sacrifices those at the extremes (Fig. 34.1A). Yet, individuals of extreme phenotypes are produced in each generation due to the polygenic mode of inheritance. As an example, let's suppose a population of mouse includes pelage colors ranging from brown to white, with various shades of tan being intermediate and most abundant in the population. The soil on which these mice go about their activities is tannish in color. The tannish mice blend with the soil so that visual predators (e.g., hawks) do not see them easily. However, the white and brown mice stand out plainly against a background of tan soil. Hence, the hawks selectively feed on brown and white mice, but not often on tan mice.

(2) Directional selection. In this process, selection favors phenotypes at one end of the distribution and selects against those at the opposite end (Fig. 34.1B). The result of <u>directional selection</u> is to shift the phenotypic mean of the population towards a particular end of the distribution. Generally, this will be a transient process because eventually the allele(s) producing the nonfavored phenotype will lost from the genome. An example might pertain to the same mice as in the previous paragraph: Suppose that a volcano erupted some safe distance from these mice. Yet, dark brown ash from the volcano fell and blanketed the habitat of the mice. Assuming that the same predators survived and still used visual means for sighting the mice, then tan mice would no longer be favored. Rather, the dark brown mice would survive because of camouflaging. The curve would shift to the left. In time, allelic frequencies would stabilize in the dark brown range, and the period of directional selection would be concluded.

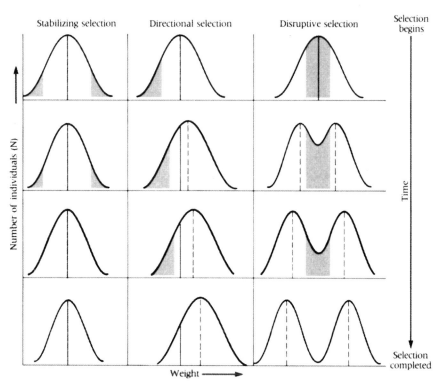

Fig. 34.1: Three forms of selection

(3) Disruptive selection. In this situation, the intermediate phenotype is not favored whereas phenotypes at opposite ends of the distribution are both favored (Fig. 34.1C). This was probably the mechanism that promoted the adaptive radiation of finches on the Galapagos Islands (see textbook). This mechanism might be involved in the production of polymorphisms. Apparently, disruptive selection is less common than either stabilizing or directional selection.

TERMS TO UNDERSTAND: Define and identify the following terms:

individual

population

deme

glade

cluster

region

distributional range

species

population genetics

frequency distribution curve

range

mode vs. mean

gene pool

microevolution vs. macroevolution

mutation pressure

random (genetic) drift

gene flow

natural selection

horizontal gene transfer

genetic equilibrium

mutational equilibrium

combinatorial equilibrium

gene (allelic) frequency

Hardy-Weinberg equilibrium

founder principle

immigration

introgression (introgressive hybridization)

fecundity

panmixia

group selection

species selection

individual selection

kin selection

fitness

polymorphism

morph

discontinuous phenotypes

heterosis

balanced vs. transitional polymorphism

frequency-dependent selection

assortative vs. disassortative mating

truncation selection

stabilizing selection

directional selection

disruptive selection

PEOPLE TO KNOW: What is the significant biological contribution of each of these individuals as indicated in Chapter 34?

G. H. Hardy

W. Weinberg

Charles Darwin

R. A. Fisher

DISCUSSION QUESTIONS

1. Some biologists have defined "evolution" simply as "change through time." Does this definition agree with the definition that would be used by a population geneticist? Why, or why not?

2. Distinguish between "individual selection" and "kin selection." How does each contribute to inclusive fitness?

3. List and briefly define the basic causes of genetic change in a population. Which of these causes pertain to changes that relate to the fitness conferred by certain alleles? Which of these factors are simply random events not related to the fitness conferred by those alleles?

4. Let's suppose that you are working in the laboratory of a population geneticist who studies evolution in a species of minnow. You have been asked to help analyze data collected during recent field and laboratory experiments. You are trying to determine the frequencies of the alleles that determine the presence or absence of an extra dorsal fin just anterior to the tail. You are told that presence of the fin is the homozygous recessive condition and that inheritance is by complete dominance.

   a. The samples collected in the field totalled 712 fish without the extra fin and 427 with the fin. What are the frequencies for the dominant and recessive alleles? Show how you calculated the values.

   b. A random sample of 10 males and 10 females taken from the field collection [in (a) above] was placed in an aquarium to breed. The progeny of these 20 fish were 364 without the fin and 157 with the fin. Calculate the allelic frequencies for this filial generation, and compare the frequencies with those of the field sample. If these values are different, can you say that evolution has occurred? Why, or why not?

TESTING YOUR UNDERSTANDING

For each of the test items below, three of the lettered alternatives are true and the other is false. Determine which alternative is false and write its letter in the blank to the left of the question number. On the blank line below alternative D, write the corrected version of the false statement.

_____ 1. A. A deme can include several species, but a species cannot contain several demes.
   B. "All of the large-mouthed bass in a farmer's stock pond" is a proper example of a population.
   C. A panmictic population is one in which each individual is equally likely to mate with any individual of the opposite sex.
   D. Evolution is a characteristic of a population, not a feature of an individual.

   Correction: _____

_____ 2. A. All of the following are ways by which the genetic makeup of a population can occur: crossing-over, recombination, and gene mobility.
   B. Populations with low genetic diversity have a greater risk of extinction than do populations with high genetic diversity.
   C. Natural selection is an important process in both microevolution and in macroevolution.
   D. The founder prinicple is an example of the process of genetic drift.

   Correction: _____

_____ 3. A. Although back mutations can occur, mutational equilibrium is rarely achieved in natural populations.
   B. Panmixia is a condition required to acheive Hardy-Weinberg equilibrium.
   C. In a situation of complete dominance, the allelic frequency for the recessive allele can be computed as the square root of the proportion of homozygous dominant individuals in the population.
   D. The proportion of heterozygotes in a population at Hardy-Weinberg equilibrium can be computed as two times the product of the allelic frequencies.

   Correction: _____

_____ 4. A. If a sample taken from a population in Hardy-Weinberg equilibrium is allowed to breed randomly, then the progeny of the sample should have the same gene frequencies as the original population.
   B. Balanced gene flow into and out of a population at Hardy-Weinberg equilibrium will not disturb the equilibrium.
   C. Evolution can be thought of as a deviation from genetic equilibrium.
   D. Because random mating normally occurs in natural populations, genetic equilibrium is maintained from one generation to the next.

   Correction: _____

_____ 5. A. The position effect explains how chromosomal rearrangements can have major phenotypic effects.
B. Mutations are required for evolution to occur.
C. The effects of genetic drift are more likely to be significant in smaller populations than in larger populations.
D. Genetic diversity is important in allowing a population to change as the environment changes.

Correction: _____

_____ 6. A. Nonrandom mating implies that individuals chose mates by phenotype.
B. Sexual selection refers to the sexual acceptance or refusal of one individual by another of its own species.
C. Fecundity refers to the degree to which hybrid individuals are able to breed back with the parent populations.
D. Fitness is measured in terms of the genetic contribution that an individual makes to the next generation.

Correction: _____

_____ 7. A. Kin selection is a component contributing to one's inclusive fitness.
B. Most natural populations are polymorphic.
C. Directional selection produces polymorphism in populations.
D. Sharply dissimilar phenotypes can result from the actions of single genes.

Correction: _____

_____ 8. A. Heterosis is the situation in which the heterozygote has an advantageous trait not present in either homozygote.
B. Steady-state preservation of variation in a population is called balanced polymorphism.
C. Selection in which individuals possessing phenotypes in the middle of phenotypic distribution are favored is called stabilizing selection.
D. Heterosis is common in monomorphic populations.

Correction: _____

_____ 9. A. Natural selection can introduce novel features into populations.
B. Natural selection operates on phenotypes, not directly on genotypes.
C. Individuals of different demes of the same species usually can interbreed, whereas individuals of different species generally cannot interbreed.
D. Macroevolution refers to events in which major groups of organisms have evolved and differentiated.

Correction: _____

_____ 10. A. If the frequency of the dominant allele is 0.3, then the proportion of the population expressing the dominant phenotype is 51%.
B. If the frequency of the recessive allele is 0.6, then the proportion of the population expressing the recessive phenotype is 36%.
C. That evolution occurs shows that not all populations are in genetic equilibrium.
D. High fecundity is favored in species where each individual's survival is likely.

Correction: _____

CHAPTER 35--ADAPTATION

CHAPTER OVERVIEW

In the previous chapter, we saw that evolution proceeds by a variety of mechanisms that operate through differential reproduction of different genotypes. In this chapter, we will examine the principle of adaptation, the vast array of means by which organisms enhance their chances of leaving proportionately more offspring than do other members of their deme. We'll find that adaptive features can be behavioral as well as anatomical or physiological. We'll examine various hypotheses that have been presented to explain how adaptive characters can be passed from one generation to the next. We also will look at how organisms presented with new ecological opportunities can diversify and radiate to occupy the available opportunities.

TOPIC SUMMARIES

A. <u>Nature of Adaptations</u>.--Any feature of an organism has the potential to be an <u>adaptive character</u>, a character that somehow or other enables the organism to produce relatively more progeny than do organisms with other genotypes. The relationship of that character to enhancing reproduction might, or might not, be directly apparent: Production of greater numbers of eggs seems a logical feature to enhance reproduction, whereas presence of a certain design of mouth parts seems distant from reproductive value (Fig. 35.1). Yet, design of mouth parts might increase the efficiency of extraction of nutrients from food, and, thereby, enhance reproduction. Hence, all adaptive characters directly or indirectly promote the highest evolutionary purpose--<u>to survive to reproduce</u>. Therefore, adaptations promote not only the welfare of the individual, but also the welfare of the species.

B. <u>Inheritance of Acquired Characteristics</u>.--Only those adaptations that have a genetic basis (i.e., are inheritable) are evolutionary adaptations. Changes in an organism that the organism cannot pass on to the next generation are not a significant part of this evolutionary picture. The ability to speak a foreign language, or the presence of only a stub tail in boxer dogs, are both <u>acquired characters</u> that cannot be inherited.

However, early in the history of the study of evolution, <u>inheritance of acquired characters</u> was an accepted hypothesis for explaining adaptation. Lamarck is perhaps the best-known champion of this doctrine. Testing of this hypothesis led to its rejection on five counts:

(1) For this mode of inheritance to occur, genetic information must be passed from modified body (somatic) cells to the germ cells; there is no known mechanism for this to occur.

(2) Experimental efforts to pass acquired characters to the next generation have failed.

(3) It is not understood how an inheritable adaptation could be caused by the efforts of the organism, or how the environment could produce such an adaptation.

(4) Some acquired characters cannot be passed to the next generation because the individuals having those characters do not reproduce. Examples include the non-reproducing worker and soldier castes of social insects.

(5) If acquired characters can be inherited, why are the detrimental characters (e.g., injured or malformed body parts) of one generation not passed on to the next?

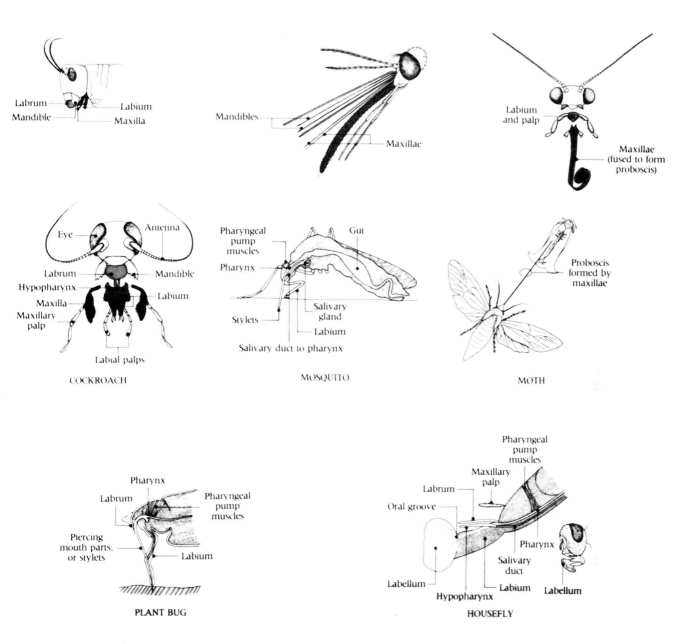

Fig. 35.1: Morphological adaptations of mouth parts of insects

C. <u>Seemingly Nonadaptive Characters</u>.--The principle of adaptation holds that, when a particular feature evolved and became incorporated into the genome, it was then an adaptively valuable feature. Yet, as time passed and environmental conditions changed, the adaptive character was kept--not because it was useful, but because it was not harmful. Our own tonsils, appendix, and wisdom teeth are such now <u>nonadaptive characters</u>. If, in time, these features become significant liabilities (i.e., <u>maladaptive characters</u>), perhaps they will be removed from the genome.

D. <u>Historical Opportunism</u>.--Evolution is replete with tales in which biochemical or anatomical or other opportunities were exploited repeatedly to produce series of new characters. For example, the three small bones of the <u>mammalian inner ear</u> represent a specialized adaptation of bones that in many other vertebrates are not involved in hearing, but that are simply part of the lower jaw. Evolutionary experimentation probably had arranged these bones in many different fashions before this opportunistic arrangement was inadvertantly encountered. (We must be careful here not to become anthropomorphic in our reasoning!) The panda's "thumb" is another example of opportunism; this thumb represents independent evolution of a structure analogous to the thumb of primates. Natural selection enables refinement of such accidentally encountered structures.

E. Adaptive Coloration.--In various ways, the external color pattern of an animal can enhance its chances of surviving. Cryptic coloration helps an individual to blend in with the background, thereby concealing it from predation. Industrial melanism in moths (textbook Box 35-2) is a classic example of directional selection in alleles affecting body color in an environment where the background color changed during the Industrial Revolution of England.

The opposite approach, having strikingly visible coloration, also can thwart predation. Such warning (or aposematic) coloration advertises undesirable traits (e.g., noxious chemicals in monarch butterflies) to a prospective predator. The effectiveness of this strategy requires that a predator survive at least one unpleasant encounter . . . and that it learns from this experience to avoid such animals in the future. There are several situations in which palatable prey species have "free-loaded" from such arrangements. In such a case, the palatable species mimics the unpalatable model species. What are the differences between Batesian and Mullerian mimicry?

F. Behavioral Adaptations.--Behaviors of many kinds are another category of adaptations that can significantly affect fitness. Obvious examples are behaviors related to reproduction: mate selection, courting, nest construction, mating, and parental care of young. Deceitful behavior that mimics the behavior of another species or of another sex can provide the mimic access to resources (mates, food, parental care in the nest of another species) that it would not otherwise obtain. All behaviors have costs; natural selection determines whether benefits of the behavior outweigh the costs.

G. Diversification and Adaptive Radiation.--Once a population has increased to the limit that the environment can support, then new pressures act on the population. As resources are used, only those genotypes that are most efficient in resource use will maintain the characteristics of that population. Perhaps the less efficient members of the population will die . . . or, perhaps, some of them happen to possess some feature that allows them to do something a little bit differently than the other individuals of the population. That different feature might allow the organism to satisfy, say, energetic requirements in a different way, using some resource that is abundant in the environment. Hence, there is pressure to diversify; natural selection can facilitate that diversification provided that sufficient variation exists within the gene pool.

For organisms to successfully enter new such environments, several conditions must be met. Physical access to the new situation must be available (physical opportunity). The design of the organism must meet the requirements of the new surroundings (design opportunity). Also, the level of competition facing the entering phenotype must be sufficiently light for the invader to establish its presence (ecological opportunity).

On many occasions during the history of life, evolution has produced novel characters at times and in places when the levels of opportunity were high enough for diversification to be highly successful. Such rapid bursts of diversification of many species from one ancestral species are called adaptive radiations. The finches of the Galapago Islands are a classic example of how a few birds from the South American mainland blundered into an isolated area with many open niches. These finches diversified from one form of bill and feeding mode to numerous forms of beaks and behaviors adapted to many feeding styles.

Examples of major adaptive radiations are those that produced land-dwelling vertebrates and flying vertebrates (birds and bats from separate radiations). Most such radiations are fueled by key innovative featues (e.g., legs for walking, lungs for extracting oxygen from air rather than water, wings for flying). We'll study these and other radiations in the later chapters on diversity of life.

H. New Approaches to the Study of Evolution.--With new, molecular-level technologies at hand, evolutionists have renewed their pursuit of many basic questions about the evolutionary history of life. Molecular evolution involves study of not merely the phenotypic expressions of the genome, but the genome itself. These new techniques go directly to the source: They determine the sequences of amino acids in proteins or even the sequence of nucleotides in the genes themselves. Similarities in sequences between different species can suggest either common ancestry (i.e., related lineages) or convergent evolution in unrelated lineages.

TERMS TO UNDERSTAND: Define and identify the following terms:

inheritance of acquired characters

Lamarckism

nonadaptive characters

maladaptive characters

historical opportunism

cryptic coloration

industrial melanism

warning coloration

aposematic coloration

mimicry

Batesian mimicry

mimic

model

Mullerian mimicry

behavioral mimicry

deceit

cost:benefit ratio

opportunity

adaptive radiation

niche

transitional form

amino-acid sequencing

molecular evolution

homology

common ancestry

convergent evolution

phylogenetic relationship

mutation distance

DNA-DNA hybridization

pseudogenes

processed genes

PEOPLE TO KNOW: What is the significant biological contribution of each of these individuals as indicated in Chapter 35?

Charles Darwin

Jean Baptiste Lamarck

G. D. H. Carpenter

H. W. Bates

Fritz Muller

DISCUSSION QUESTIONS

1. Two primary hypotheses (natural selection and inheritance of acquired characters) have been offered to account for how adaptations arise and how they are passed on to the next generation. Compare and contrast these two hypotheses. For which hypothesis is there substantial supportive evidence? What are the main problems with the hypothesis that is not accepted currently?

2. In Batesian mimicry, the model species is noxious whereas the mimic species is harmless. Learning, on the part of the predator, is an important component of this mimicry system. What can you infer about the relative numbers of individuals of the model and mimic species that must be maintained for the system to protect both prey species?

3. Discuss the saying, "There is more than one way to skin a cat," in terms of the concept of historical opportunism. Use examples.

4. Why are the techniques of sequencing of amino acids and of deoxyribonucleic acids appropriate for determining the evolutionary history of life?

5. Suppose that several mice of the same species rafted to a large island where no other mouse-like or rat-like vertebrates lived. In their former habitat, these mice fed on seeds. These mice find plenty of food, shelter, and other resources in their new island home, and therefore they are able to establish a thriving population. Suppose that we, as evolutionary field biologists, return to this island 10,000 years later. We then find not one, but nearly a dozen species of mice. We determine that all of these species evolved from the mouse population above. What happened to produce this radiation? Describe the specific niches to which each species might be adapted. [The answer to this question should be quite lengthy. Write your answer onto other sheets of paper.]

TESTING YOUR UNDERSTANDING

For each of the test items below, three of the lettered alternatives are true and the other is false. Determine which alternative is false and write its letter in the blank to the left of the question number. On the blank line below alternative D, write the corrected version of the false statement.

_____ 1. A. The human thumb and the panda's thumb represent an example of convergent evolution.
B. The primary difficulty with the hypothesis of inheritance of acquired characters is identifying a mechanism by which an adaptation in somatic (body) cells can become incorporated into the sex cells of that individual.
C. Lamarck was a proponent of the hypothesis of natural selection.
D. If the doctrine of inheritance of acquired characters actually operates, then both beneficial and maladaptive harmful traits would accumulate in the genome as generations pass.

Correction: _____

_____ 2. A. Both warning coloration and cryptic coloration can be effective strategies against predation.
B. Industrial melanism in moths is an example of aposematic coloration.
C. Warning coloration is an adaptation used in Batesian mimicry.
D. In Batesian mimicry, the model species must be at least as abundant as the mimic species.

Correction: _____

_____ 3. A. Mimicry strategies benefit both the model and mimic species because the predators usually die after an encounter with the model.
B. In behavioral mimicry, both the model and the mimic can be of the same species.
C. To be an evolutionarily viable strategy, the benefits of a behavioral adaptation must exceed the costs of that behavior.
D. Risk of injury is a factor that must be included in determining the cost of a behavior.

Correction: _____

_____ 4. A. Pressure for a species to diversify often occurs when population levels are high.
B. Ecological opportunity means that competition in the new habitat must be slight enough to permit survival of the new invader.
C. Niche refers to the way in which an organism makes a living.
D. Historical opportunism refers to the rapid development from one ancestral species of many species that "spread out" to occupy many different niches.

Correction: _____

_____ 5. A. The many species of finches occupying the Galapagos Islands probably evolved via adaptive radiation from one species that originated on the South American mainland.
B. A second invasion of land by fishes probably would not be successful because of the modern-day presence of land-dwelling vertebrate competitors and predators.
C. The great radiations of mammals occurred before the extinction of dinosaurs.
D. The highest evolutionary purpose is survival to reproduce.

Correction: _____

_____ 6. A. Adaptations that serve the welfare of the individual generally promote the welfare of the species as well.
B. Adaptations can be structural, physiological, or behavioral.
C. According to the hypothesis of natural selection, adaptations that involve no change in genetic information can be inherited.
D. An organism can correctly be thought of as a collection of adaptations that equip it to survive and reproduce in its usual environment.

Correction: _____

_____ 7. A. "The Origin of Species" was authored by Jean Baptiste Lamarck.
   B. The inability of worker and soldier castes to reproduce comprises evidence against the hypothesis of acquired traits.
   C. If acquired traits can be inherited, then this process would be most likely to operate in bacteria because the single cell of a bacterium is both the soma and the gamete.
   D. Some of the ossicles of the mammalian middle ear evolved from bones that, in reptiles, were part of the jaw.

   Correction: _____

_____ 8. A. The panda's thumb is not a finger, but an extension of a wrist bone instead.
   B. Warning coloration is used to escape predation.
   C. Cryptic coloration is used to escape predation.
   D. Warning coloration is used to warn other individuals of the presence of a predator.

   Correction: _____

_____ 9. A. Study of evolution at the molecular level involves techniques that determine the sequence of amino acids in proteins.
   B. Study of evolution at the molecular level involves techniques that determine the sequence of nucleotides in deoxyribonucleic acid.
   C. The "modern synthesis" refers to an approach to the study of evolution in which understanding of both genetics and natural history is used.
   D. Ontogenetic relationships can be determined on the basis of the degree of similarity in corresponding protein molecules in different species.

   Correction: _____

CHAPTER 36--ORIGIN OF SPECIES

CHAPTER OVERVIEW

In many chapters of your textbook and in this Study Guide, we have talked of diversity. In Chapter 33, we summarized the various ways by which variation is produced within the individual. Also in Chapter 33, we introduced various ways by which variation is produced within groups of individuals.

In this chapter, we examine patterns of variation between populations of the same species; several of the geographic patterns seen are expressed as ecogeographic rules. We conclude the chapter by showing various ways by which differences between populations become great enough to become distinct species.

TOPIC SUMMARIES

A. <u>Sources of Variation Within A Species</u>.--Despite our tendency to describe all individuals of a species in a particular way, variation exists among individuals of any population or species. In previous chapters, we have spoken of the critical role of variation: variation is the "stuff" on which natural selection operates to produce evolutionary change.

Recall that much of the diversity within a species is <u>not</u> inheritable. Such nongenetic variation includes seasonal effects, generational variation, social variation, ecological variation, and others. Only that variation that is due to changes in the genome can be inherited. Yet, non-heritable variation can influence whether an individual has an opportunity to pass on its genetic variation.

B. <u>Distribution of Variation in a Deme</u>.--Study of variation within a population generally involves use of <u>statistics</u>. Such studies involve quantification of the ways in which a trait is expressed in a sample of individuals from a population. Examples of traits are the length of the skull in a species of field mouse, or the number of scales on a fish. The sample can be characterized by a <u>frequency distribution</u> (Fig. 36.1) that depicts the number of individuals expressing the trait (e.g., number of seeds in a seed pod) in a particular way. A curve drawn through the data points might be <u>normal</u> ("bell-shaped"), with its most frequent value being the <u>mode</u> . . . or the curve might be <u>skewed</u>, with unbalanced distribution to either side of the mode. A distribution with two peaks is <u>bimodal</u> and frequently suggests two categories within the population; such categories (or <u>morphs</u>) might correspond to different sexes or age classes in the population.

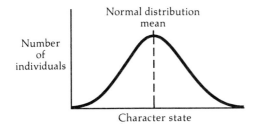 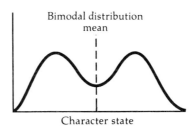

Fig. 36.1: Curves showing normal and bimodal distributions

C. <u>Differences Between Demes</u>.--Rarely, if ever, is one deme of a species genetically identical to any other deme of that species. Such differences occur because adjacent demes might be isolated from each other so that gene flow does not occur between them . . . or the selection pressures of the environments of the two demes might be slightly to greatly different. Yet, such differences are <u>not</u> so great that members of one deme cannot interbreed with members of another deme. Such interbreeding often produces intergradation in features of individuals found in the zone of overlap or interbreeding. Intergradation can be recognized by the presence of an intermediate character state (e.g., pink flowers from interbreeding of plants with white and red flowers) or by the presence of morphs in proportions intermediate to those found in either deme. If demes differ sufficiently from each other, then they might be recognized as separate <u>subspecies</u> or <u>races</u>. If demes differ so much that individuals from different demes can no longer interbreed and produce viable offspring, then each deme can be recognized as separate <u>species</u>.

Interdeme differences can take many forms. A pattern of <u>geographic variation</u> in which there is a gradual change in a trait over a considerable distance (e.g., hundreds or thousands of kilometers) is called a <u>cline</u>. Frequently, clinal variation corresponds to geographic trends in various variables (e.g., temperature, precipitation, elevation, etc.); this tendency suggests that there is an adaptive nature to geographic variation. Indeed, some of these patterns appear in so many species that they have been generalized into <u>ecogeographic rules</u>. <u>Bergmann's Rule</u>, for example, states that, for individuals of a particular species, individuals in colder parts of the species' range tend to have larger body sizes than found in individuals in warmer parts of the range. Why? <u>Allen's Rule</u> notes that appendages are longer in warmer climates and shorter in colder climates. <u>Gloger's Rule</u> addresses geographic trends in body surface coloration.

D. <u>Species</u>.--According to a widely used definition, a "species" is a "group of actually or potentially interbreeding natural populations that are reproductively isolated from other such groups." The concept of <u>reproductive isolation</u> of different species is key to this definition. Presence of hybrid individuals (F_1) does not necessarily mean that there is no reproductice isolation. The important question then becomes: can the hybrids breeding among themselves, or with the partental genotypes, produce viable offspring?

Yet, such isolation is difficult to demonstrate over time and space: How can we possibly ever know if a mouse from a 3 million-year-old fossil site could interbreed viably with modern mice? Even for organisms within the same time frame, how can we know if a certain type of bat from Florida can interbreed with a similar-looking bat from California? What about organisms that reproduce asexually, or that are self-fertile? In practice, taxonomists generally use the <u>morphological species concept</u> wherein organisms that have highly similar appearance are grouped into a species. Such a concept can be misleading: Suppose that individuals in two groups of organisms are extremely similar in external appearance, but that significant differences occurred in traits not visible externally (e.g., chromosome characters, presence or absence of certain enzymes).

E. <u>Speciation</u>.--Speciation is those events that produce new species. <u>Cladistic speciation</u> involves splitting or branching of a group from the main <u>lineage</u>. Such branching leaves two species--the old and the new. However, such splitting is not necessary for formation of new species: <u>Anagenetic speciation</u> occurs in continuous lineages, with one species grading into another; one species is replaced by another. Suppose that we have a particular kind of snake whose fossil history is recorded more-or-less continuously over the past million years. Suppose that the old specimens differ from the modern specimens only in size, being twice as long as the modern ones. Suppose, too, that a gradual trend of size decrease is recorded over the million years. The data might indicate to a taxonomist that the modern and ancient specimens should be recognized as separate species . . . but where along the lineage should the taxonomist draw the line? This sort of question is commonly encountered by the experts, and its resolution draws heavily upon familiarity with the particular group of organisms being studied.

In the context of these two types of speciation, the terms "clade" and "grade" are used. A clade includes those species descending (via branching) from a recent common ancestor. A grade can be thought of as a level of adaptation acheived, perhaps, in many independent lineages. In Fig. 36.2, each of the lettered segments of the phylogenetic diagram represents a separate clade. In this example, birds and dinosaurs can be thought of as members of the same grade because species in both groups are "egg-layers." Placental mammals occupy a different grade; their reproductive strategy has shifted from egg-laying to live-bearing.

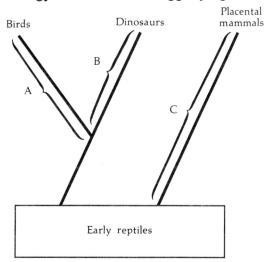

Fig. 36.2: A phylogenetic diagram showing relationships of three groups of vertebrates

F. <u>Rates and Patterns of Evolutionary Change</u>.--As suggested above, speciation can be gradual; considerable evidence support such processes in which character states shift slightly over time. However, the evidence for some lineages suggests that evolutionary change can be punctuated, via sporadic "leaps" in character states. Vigorous healthy debate goes on even now about which is the correct mechanism . . . if, indeed, there is only one correct answer!

The primary argument against punctuation is that a punctuated pattern might appear because the fossil record is incomplete; the intermediate fossils might not yet have been found. Gradualists feel that no genetic mechanism exists that can account for huge, rapid changes in the genome. However, recent advances in molecular genetics have identified plausible means of bringing about abrupt changes; these means involve introns, repetitive sequences, and "jumping genes."

Actually, at least one mechanism by which instantaneous speciation can occur is known. Occasionally, chromosome mutations and hybridization can produce polyploidy, situations in which one or more chromosomes are represented by extra copies. Such polyploid individuals cannot interbreed with their parental genotypes, but only among other individuals with the same chromosomal arrangement. This process is relatively common in plants, primarily because many plant species are self-fertile or can reproduce asexually.

G. Isolating Mechanisms.--In essence, speciation requires that genomes become isolated from other genomes. There are many such isolating mechanisms. Geographic isolation is, perhaps, the most straightforward. Individuals, no matter how similar their genomes, that cannot encounter each other clearly cannot mate. If the time of isolation is great enough, and if changes in the genomes are great enough, then bringing individuals of the formerly isolated genomes back together might demonstrate reproductive incompatibility. Such a scenario is an example of allopatric speciation.

The mechanism of sympatric speciation offers the apparent paradox that demes occupying the same area still can be isolated. Here, isolation is biological rather than geographic. The isolating mechanisms operating can be ecological (e.g., different microhabitats within the same habitat . . . different parts of the same tree), temporal (using the same microhabitats, but at different times of day), behavioral (each species having slight, but meaningful, differences in a courtship ritual), physiological (biochemical incompatibilities in the process of fertilization), or mechanical (anatomical incompatibilities of sex organs).

The action of an isolating mechanism might occur either before (prezygotic isolation preventing fertilization) or after (postzygotic isolation) fertilization. For species that are isolated by postzygotic means, the timing of exertion of isolation is quite variable. Effects could be as immediate as prevention of mitotic division of the zygote, or could be delayed until the hybrid individual mates and produces inviable offspring. An example of the latter is the infertile mule produced by mating of a horse and a donkey.

Parapatric speciation can occur in the apparently uncommon situation in which subgroups of a population diverge in characters while reproductive isolation is being established. This process occurs when genetically distinctive organisms in a population gain access to an unoccupied niche within the normal geographic range of the species. As a result, gene frequencies change abruptly over very short distances, and reproductive isolation occurs without the presence of a geographic barrier. Parapatric speciation probably produced much of the diversity of mole rats *(Spalax),* a group of burrowing rodents having limited dispersal ability.

H. Adaptive Radiation.--Rapid diversification of new species can occur when some combination of ecological opportunity, isolation, and novel characters occurs. "Darwin's finches" are a renowned example: A few individuals from the mainland of South America colonized the Galapagos Islands, an archipelago previously lacking birds with similar ecologies. Anatomically (with regard to form of beaks), these birds diversified along different feeding modes into many new species occupying vacant niches. Similar adaptive radiations produced the diversity of marsupial mammals in Australia and the diversity of bats worldwide; the novel feature in the ancestors of bats was wings. The invasion of land by certain fishes was followed by the adaptive radiations of land-dwelling vertebrates.

TERMS TO UNDERSTAND: Define and identify the following terms:

diversity

species

mode

bimodal

local Mendelian population

intergrade

subspecies

race

cline

Bergmann's rule

Allen's rule

Gloger's rule

hybrid

speciation

lineage

anagenetic speciation

cladistic speciation

ancestral species

descendant species

clade

grade

fossil record

stasis

punctuated equilibrium

polyploidy

instantaneous speciation

autopolyploidy

allopolyploidy

allopatric speciation

sympatric speciation

parapatric speciation

character displacement

prezygotic isolating mechanisms

postzygotic isolating mechanisms

interspecific selection

PEOPLE TO KNOW: What is the significant biological contribution of each of these individuals as indicated in Chapter 36?

R. C. Lewontin and J. L. Hubby

Julian Huxley

Hugo de Vries

P. G. Williamson

DISCUSSION QUESTIONS

1. Define <u>interspecific selection</u>. Discuss how interspecific selection can be interpreted as a version of natural selection.

2. It has been stated that isolated demes represent "incipient species." Elaborate on this statement.

3. Widths of zones of intergradation are quite varied. What can the distance over which intergradation occurs tell about how environmental factors vary?

4. In the context of Allen's Rule, explain the physiological rationale behind shifts in the lengths of various body appendages over the range of a species.

5. The concept of "species" seemingly is easily defined. How, then, is there a "species problem?"

6. How might the dispersal ability of organisms of a species affect the pattern of geographic variation among demes?

TESTING YOUR UNDERSTANDING

For each of the test items below, three of the lettered alternatives are true and the other is false. Determine which alternative is false and write its letter in the blank to the left of the question number. On the blank line below alternative D, write the corrected version of the false statement.

_____ 1. A. A species is a group of actually or potentially interbreeding populations that are reproductively isolated from other such groups.
   B. The pattern of character variation seen between demes often reflects variation in environmental factors.
   C. In a zone where interbreeding between demes occurs, some characters often are expressed in ways that are intermediate to the way those characters are expressed in either deme.
   D. Because nongenetic variation is not transmitted to subsequent generations, nongenetic variation does not have evolutionary importance.

   Correction: _____

_____ 2. A. Disruptive selection is suggested by a bimodal frequency distribution of phenotypes.
   B. When truncation selection is operating, intergradation between demes usually does not occur.
   C. Given the opportunity, individuals of different subspecies of the same species will interbreed freely and produce viable offspring.
   D. A sequence of demes with gradual change in traits from one area to another is called a cline.

   Correction: _____

_____ 3. A. Allen's Rule states that body size is greater in areas of warmer climates and lesser in regions of cooler climates.
   B. One aspect of the "species problem" is the inability to evaluate reproductive isolation in extinct organisms.
   C. Matings that produce hybrids demonstrate that the interbreeding forms are genetically related.
   D. Individuals of the same species tend to be similar phenotypically.

   Correction: _____

_____ 4. A. Each deme of a species evolves independently of other demes of that species.
   B. Cladistic speciation processes involve branching of a new species from a stem species.
   C. In anagenetic speciation, one species changes into another species without splitting the lineage.
   D. Macroevolution denotes the processes that produce genetic change and variation within a deme over two or more generations.

   Correction: _____

_____ 5. A. A clade can be defined as a unit of biological improvement.
   B. Species can belong to the same grade even if the species are not related to each other.
   C. According to cladistic theory, an ancestral species ceases to exist after it splits to produce new species.
   D. Evidence exists to support both the gradualist and the punctualist mechanisms of evolution.

   Correction: _____

_____ 6. A. The means by which abrupt evolutionary changes occur probably are related to introns and repetitive DNA sequences.
   B. Polyploidy is a means by which a new species can arise virtually instantaneously.
   C. Speciation by polyploidy is an example of allopatric speciation.
   D. Reproductive isolation can be acheived by geographic isolation.

   Correction: _____

_____ 7. A. Two species occupying the same geographic range are allopatric.
   B. Reproductive isolation of sympatric species can be achieved by ecological or temporal partitioning of resources in the habitat.
   C. Mating sometimes occurs among individuals that are not members of the same species.
   D. Prezygotic isolating mechanisms operate before fertilization of an ovum.

   Correction: _____

_____ 8. A. Postzygotic isolating mechanisms prevent development of an embryo into a viable individual.
   B. Parapatric speciation occurs in the absence of a geographic barrier.
   C. Parapatric speciation often produces abrupt changes in allelic frequencies over short distances.
   D. In allopatric speciation, divergence of traits occurs before reproductive isolation is achieved.

   Correction: _____

_____ 9. A. Independent evolution in demes that are geographically and reproductively isolated need not always result in genetic divergence.
   B. Industrial melanism in moths is an example of speciation despite the occurrence of gene flow.
   C. When two previously isolated demes resume gene flow, individuals in the zone of overlap tend to become more different from each other than do individuals not in the zone of overlap.
   D. Character displacement occurs when demes are isolated from each other.

   Correction: _____

_____ 10. A. The diversity of life on Earth was produced by evolutionary processes.
    B. For frequency distributions of characters in most demes, the values for the mean and the mode are usually the same.
    C. The distribution of character states in a deme having two morphs would be bimodal.
    D. A subspecies is a complex of demes that are more similar to each other than they are to other such deme complexes.

    Correction: _____

CHAPTER 37--PRINCIPLES OF CLASSIFICATION

CHAPTER OVERVIEW

Humans have a behavioral tendency to classify things found in our surroundings. A benefit of classification is the ordering of an array of diverse items. Some schools of biological thought think that classifications also should reflect the history of life. Approaches to biological classification have evolved for centuries.

In this chapter, we examine several approaches to classification and we study the history of biological classification. We consider the types of features that are most often used to develop classifications. We re-acquaint ourselves with basic features of organisms occurring within five kingdoms. Finally, we treat the concept of geological time and learn how to study past life.

TOPIC SUMMARIES

A. Purposes of Classifications.--If for no other reason, classifications are valuable because they help to impart order to the vast diversity of life existing on Earth. Such an organizational system facilitates retrieval of information and aids in communication. Depending upon how a classification is constructed, it may help us to understand the evolutionary relationships between different organisms.

B. Hierarchies.--In general, a classification is based upon similarities and differences of the entities being classified. In comparing appearances of organisms, it is clear that there are different degrees of similarities and differences. A common assumption is that those organisms that are more similar to each other are more closely related to each other than they are to less-similar organisms. The quandry raised by presence of degrees of difference is resolved by using a hierarchical classification system, an arrangement of smaller groups within larger groups. The level (within the hierarchy) that contains both of two organisms that are being classified indicates their degree of similarity. For example, cats and mice fit into Class Mammalia (a more-inclusive category, or taxon), whereas cats and dogs fit not only into Class Mammalia, but also into the same less-inclusive taxon (Order Carnivora). Hence, dogs and cats are more like each other than either is to mice.

C. The Linnaean Concept and its Successors.--Carolus Linnaeus published a hierarchical classification scheme, Systema Naturae, in 1758. The classification scheme now used worldwide by biologists is a modified version of Linnaeus' system. The major hold-over from his system is the binomial: Every species is identified uniquely by a two part name, genus and species. Both genus and species names of any organism always are italicized or underlined; the first letter of its genus is capitalized.

The Linnaean concept has undergone two revolutions since its origin. In Linnaeus' time, the number of species was thought to be fixed. In essence, his classification really was a catalogue of all kinds of organisms that existed. No mechanism was offerred by which new species could arise. All species had been created divinely, each according to its particular type. All organisms of a particular species supposedly came from the same mold; variation that did then (and does now) exist was dismissed as meaningless.

In the first revolution, systematists realized that species could and did change, and that species were in various ways related (by way of evolutionary descent) to each other. In the second revolution, the typological concept yielded to the population concept, wherein the existence of variation took on meaning. In this modern concept of systematics, then, a modified definition of species emerged. A species is all of the "populations of individuals of common ancestry that live together in similar environments in a particular region and tend to have similar ecological relationships and unified, distinctive, and continuing evolutionary roles."

D. <u>Modern Methods of Classification</u>.--Up until the middle of this century, none of the methods used for developing classifications were well-defined. Arbitrary use of different approaches in classifying the same organisms often produced vastly different results. Three practicing schools of classification, each with its specified procedures, are now recognized:

<u>Numerical phenetics</u> classifies entities purely (and objectively) on the basis of degree of similarity of appearance (phenotype). No effort is made (nor is the intent) to reflect evolutionary relationships in such a classification. Hence, species that evolved in parallel or via convergence are lumped together, thereby obscuring their different phylogenies.

<u>Cladistics</u> lies towards an end of the spectrum opposite that occupied by pheneticists. Cladistics (as you might recall from Chapter 35) is an objective approach that classifies taxa with regard to the pattern and sequence of branching by which the taxa evolved. Relationships are recognized on the basis of presence (or not) of <u>shared derived characters</u>.

The third school of classification, probably practiced by the majority of systematists, shares features of the other two. <u>Evolutionary systematics</u> evaluates both branching sequences and degree of similarity. Some subjective judgments must be made in this approach.

E. <u>Interpretation of Form and Descent</u>.--At least two of the three schools of systematics are concerned with knowing whether or not similarities are results of <u>common ancestry</u>. Determination of common ancestry requires understanding of several principles and processes.

(1) Homology, homoplasy, and analogy. Corresponding structures in organisms that are related by common descent are <u>homologous</u>. Homologues need not serve the same purpose in different species (Fig. 37.1). Indeed, designs of homologous structures might be different; in such cases, their embryological histories demonstrate the homology. The series of stages leading from ancestral condition to derived condition are referred to as a <u>transformation series</u> for those homologues.

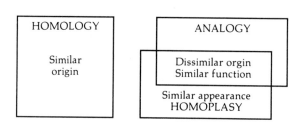

Fig. 37.1: Rectangles showing relationships of homology, analogy, and homoplasy

Structures that resemble each other, but that do not have share common ancestry, are <u>homoplastic</u>. Homoplastic structures that perform similar functions also are <u>analogous</u>; (e.g., wings of insects compared to wings of vertebrates). Some analogous structures, however, are not homoplastic because they are not anatomically similar (e.g., mammalian lungs compared to fish gills).

(2) Convergence, parallelism, and divergence. When populations of the same or different species are isolated, one of three evolutionary paths can be followed (assuming, of course, that

they do not go extinct). Unrelated and dissimilar species isolated in similar habitats and evolving under similar selection pressures often follow very similar evolutionary paths. The result is unrelated, derived species having similar form and function. The anteaters of South America, the numbats of Australia, and the pangolins of southeast Asia comprise such an example of convergence (Fig. 37.2). Parallelism is seen in groups of related organisms that evolve under a similar selection regime into derived forms that are similar in form and function. Two subfamilies of pocket gophers, a group of burrowing underground rodents, derived from a common ancestor a few million years ago and have evolved along similar courses to produce two groups of similar species. Parallelism and convergence are not always easy to distinguish.

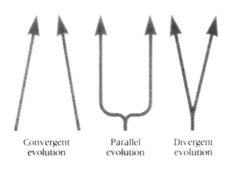

Fig. 37.2: Three patterns of evolution

Divergence is the third, and most frequently occurring, alternative of populations evolving in isolation. In previous chapters, we have addressed this idea, particularly in regard to allopatric speciation.

F. **The Practice of Classification**.--Classification involves three steps: describing and recognizing related groups of organisms, fitting these groups into a hierarchy, and then naming the taxa. Related groups of organisms are recognized by similarities of homologous characters. Such features might, for example, be anatomical structures, or related to methods of reproduction or development, or might pertain to sequences of amino acids in proteins or of nucleotides in nucleic acids. Because there is variation in a population, it is important to study a representative sample of organisms from that population in order to properly characterize that population.

G. **Classification Schemes**. The classification scheme now used by most biologists for most groups of organisms is a hierarchical scheme having the following categories in descending (less-inclusive) order: kingdom, phylum (= division for plants), class, order, family, genus, species. Interestingly, not all biologists agree on any one particular classification, even at the higher levels. Various arrangements have recognized from two to 27 kingdoms!

Prokaryotic organisms belong to Kingdom Monera. Eukaryotes compose the other four kingdoms, which are distinguished on the bases of the number of cells making up a body and the mode of nutrition. In the next several chapters, we'll study characteristics of these kingdoms.

H. **Studying the History of Life**.--The modern diversity of life on Earth evolved during a lengthy history of several billion years. Our estimates of the age of our planet derive from the techniques of radiometric dating, a method that applies known rates of decay of radioactive elements into their stable isotopes. Hence, the age of an object (e.g., a rock or a fossil) is related to the product of the ratio of the stable isotope (e.g., lead) to the radioactive form (e.g., uranium) and the half-life of the radioactive element. Other materials used for such dating include potassium-argon and carbon.

Radiometric dating has led to definition of four major time divisions. <u>Hadean time</u> preceeds 3.8 billion years ago; no rocks or evidence of life is known from this time. The <u>Archean eon</u>, ending about 2.5 billion years ago, is the earliest interval from which rocks and fossils (all monerans) are known. The <u>Proterozoic eons</u> ranged until 600 million years ago; during this time, eukaryotic organisms evolved. The <u>Phanerozoic eon</u> extends to the present; simple eukaryotic organisms radiated into the tremendous diversity seen in modern life.

Phanerozoic time is divided into three <u>eras</u>, which in turn are divided into <u>periods</u>, which in turn are divided into <u>epochs</u>, etc. Archaic plants and animals reigned during the <u>Paleozoic era</u>. The <u>Mesozoic era</u>, called the age of reptiles, marks the appearances of birds and mammals. We know the <u>Cenozoic era</u> as the age of mammals. The beginnings and endings of many time intervals are marked by <u>extinctions</u> of some groups and the appearances of others. You will find it useful to have a general grasp of the <u>geological time scale</u> (textbook Table 37-3).

I. <u>Paleontology</u>.--This field pertains to the study of ancient life as represented by <u>fossils</u>. Fossils tell us where an organism lived, when it lived, and other species with which it associated. Our abilities to glean such information from paleontology and geology rely upon certain principles, such as that of uniformitarianism. <u>Uniformitarianism</u> holds that the same rules (e.g., gravity, chemical activity, etc.) that govern the universe today always have been in effect. Fossils are our link to the past; they are indispensable elements in the search by systematists for the one true natural history of life on Earth.

TERMS TO UNDERSTAND: Define and identify the following terms:

system

systematics

taxonomy

nomenclature

taxon

hierarchy

habit

habitat

binomial

type

genus

numerical phenetics

cladism

evolutionary classification

cladogram

homology

homoplasy

analogous

transformation

convergence

parallelism

divergence

sample

population

Kingdom Monera

Kingdom Protista

Kingdom Plantae

Kingdom Animalia

Kingdom Fungi

radiometric dating

half-life

Hadean time

Archean eon

Proterozoic eon

Phanerozoic eon

era

epoch

period

geological time scale

fossil

paleontology

DISCUSSION QUESTIONS

1. The vast time during which Earth has existed has been divided into intervals. What are these major intervals? Discuss for each the particular factors that allow us to distinguish one such interval from another.

2. Compare and contrast the three philosophical approaches to classification of organisms. Which one of these is the best approach?

3. Discuss how the doctrine of uniformitarianism relates to study of the history of life.

4. In terms of the impact on use of a classification, why is it important that we be able to distinguish similarities resulting from convergence from those due to common ancestry?

5. Select a set of non-biological entities (e.g., cars of various designs, styles of houses, etc.). Develop three classification schemes for your selected set of entities. How do classifications that use cladistics differ from those using phenetics and from those using evolutionary classification?

CHART EXERCISE

Complete the chart below to summarize the major characteristics of organisms in each of the five currently recognized kingdoms. Are cells of these organisms prokaryotic or eukaryotic? Are the organisms unicellular or multicellular? Indicate the general mode of nutrition for each kingdom.

KINGDOM	CELL DESIGN	UNICELLULAR OR MULTICELLULAR	MODE OF NUTRITION
Monera			
Fungi			
Protista			
Plantae			
Animalia			

TESTING YOUR UNDERSTANDING

For each of the test items below, three of the lettered alternatives are true and the other is false. Determine which alternative is false and write its letter in the blank to the left of the question number. On the blank line below alternative D, write the corrected version of the false statement.

_____ 1. A. Animals, fungi, and many protists are heterotrophic.
　　　　　B. Animals, protists, and many fungi are unicellular.
　　　　　C. Plants are eukaryotic.
　　　　　D. Some monerans are autotrophic.

　　　　　Correction: _____

_____ 2. A. Classifications produced by the approaches of cladistics and of evolutionary classification show the evolutionary relationship of taxa.
B. Some phenetic classifications show evolutionary relationships.
C. The effects of convergence can obscure the evolutionary relationships of organisms contained in a cladogram.
D. One of the values of a classification scheme is that it imparts order to an assemblage of species.

Correction: _____

_____ 3. A. Taxonomy is the ordering of living organisms into groups on the basis of various relationships between those organisms.
B. Systematics is the scientific study of the kinds and diversity of organisms and of any and all relationships among those organisms.
C. The system of classification used in modern biology derived from the binomial system proposed by Linnaeus.
D. During the 18th century, species were viewed as fixed, unchanging units.

Correction: _____

_____ 4. A. The shift from the typological concept to the population concept of systematic units occurred when Darwin offerred an explanation for the role of variation seen in species.
B. Linnaeus was the author of Systema Naturae.
C. According to the typological concept, the blue and white forms of the goose *Chen caerulescens* should be considered to be members of the same species.
D. The typological concept held that each species was a divinely created, nonvarying form.

Correction: _____

_____ 5. A. Homologous structures can have the same function.
B. Analogous structures can have the same function.
C. Homoplastic structures have similar appearance.
D. Analogous structures can be homoplastic, and homoplastic structures also can be homologous.

Correction: _____

_____ 6. A. One of the roles of the malleus during its transformation into an ear ossicle was as a bone in the jaws of reptiles.
B. Evolutionary history is irrevocable and is irreversible.
C. Evolution in isolation can result in genetic divergence, by which new species might evolve.
D. Evolution in isolation can result in genetic convergence, by which related populations evolving under different selection regimes become similar.

Correction: _____

_____ 7. A. The first step in the practice of classification is applying names to groups of organisms.
   B. Allopatric speciation involves the process of genetic divergence.
   C. Sympatric speciation involves the process of genetic divergence.
   D. Most classifications are based on similarities of anatomical structures.

   Correction: _____

_____ 8. A. Order is a more-inclusive taxon than is class.
   B. Use of scientific names removes the possible ambiguities associated with use of common names.
   C. Botanists use "division" as a taxon that is equivalent to "phylum" of other kingdoms.
   D. Fungi is the kingdom recognized in the five-kingdom scheme that is not recognized in the four-kingdom scheme of classification.

   Correction: _____

_____ 9. A. The half-life of uranium is greater than the half-life of carbon-14.
   B. Carbon radiometric dating is useful for dating materials from the Mesozoic era.
   C. The time interval during which life originated on Earth is the Archaic eon.
   D. The Cenozoic era is known as the age of mammals.

   Correction: _____

_____ 10. A. Epochs are greater intervals of time than are periods.
   B. Paleontology is the study of ancient life, as based on fossils and other lines of evidence.
   C. Acceptance of the doctrine of uniformitarianism allows paleontologists to interpret events of the past in terms of processes that occur today.
   D. Usually, fossils represent only hard body parts, not soft tissues.

   Correction: _____

CHAPTER 38--KINGDOM OF THE SMALL: MONERA

CHAPTER OVERVIEW

The preceeding chapter introduced us to various concepts associated with the activity of classification. This chapter begins a six-chapter survey of the diversity of life. Within each of these six chapters, we will study classification, form, function, ecology, and distribution of many groups of organisms. In studying these organisms, you should focus on the features that are used to classify these organisms. Additionally, you should make the effort to understand the evolutionary relationships of these groups of species.

This chapter surveys the prokaryotic organisms--bacteria and their relatives. These species have profound implications for all life. Many are pathogens, and hence are of interest to aspiring health professionals. Many also play key roles in the cycling of materials (energy, nutrients, etc.) among species of all kingdoms of organisms. Without monerans, life as we know it probably would not exist.

TOPIC SUMMARIES

A. <u>Human Awareness of Monerans</u>.--Because of their small size (most visible only with magnification), the existence of monerans was unknown until technology advanced to the point that magnifying optics were available. With the discovery of bacteria and their relatives came understanding of many natural phenomena (e.g., many diseases, nutrient cycling in the biogeosphere, etc.) that previously were not understood.

B. <u>Classification of Monerans</u>.--Classification of these microorganisms has been more difficult than for most others for several reasons. The fossil record is poor. Being unicellular, there is no multicellular structure to examine for anatomical features useful in classification. Ontogeny involves changes in only one cell, so an embryological approach to identifying homologous structures is of little value.

Technological advances in <u>biochemistry</u> now supplement former reliance on <u>morphological features</u> (e.g., cell shape, size, color, motility, flagellar pattern, capsule, colonial morphology, staining properties) to establish classifications and to understand phylogeny. Other features now used in classification include pattern of nutrition, patterns of energy-yielding metabolism, production of characteristic chemical products, DNA composition, nucleotide sequences in DNA and in ribosomal RNA, cell-wall structure, immunological properties, and ecological relations.

C. <u>Early Evolution of Life</u>.--Since the discovery of monerans, the evolution of eukaryotes from early prokaryotes has been accepted widely. Recent discovery of the <u>archaebacteria</u>, however, suggests a modified view of the pattern of evolution of early prokaryotes and of eukaryotes from them. Fig. 38.1 shows major divisions of prokaryotes.

D. <u>General Features of Monerans</u>.--Now is the time to return to Chapter 5 to review the characteristics of prokaryotic cells. We may briefly summarize those features here: Genetic material is not organized into a nucleus. There are no membrane-bound organelles. DNA is not complexed into chromatin. If flagella are present, then they are not constructed on the eukaryotic 9+2 arrangement. Prokaryotes lack microtubules and tubulin. The cell wall is composed of amino sugars polymerized in cross-linking fashion.

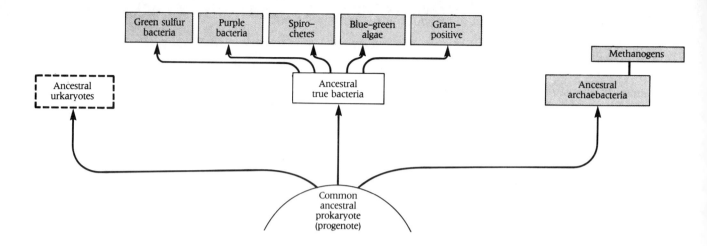

Fig. 38.1: Phylogeny of early prokaryotes

E. <u>Phylum Schizonta; Bacteria</u>. These are the familiar prokaryotes. They occur virtually everywhere in the biosphere. They are minute, only about a cubic micrometer in volume. Bacteria may be classified according to their <u>energy sources</u> (<u>saprobes</u> and other heterotrophs, autotrophs). Recall from Chapter 10 that autotrophic bacteria variously derive energy from light (photosynthetic autotrophs) or from simple chemicals (chemosynthetic autotrophs: from sulfur, hydrogen, simple nitrogen compounds). Various <u>oxygen requirements</u> allow recognition of aerobes, anaerobes, and facultative anaerobes. Some types of bacteria produce <u>spores</u> to enable survival of unfavorable conditions. Some are parasitic (includes disease-causing species). Some produce <u>toxins</u>.

(1) Class Eubacteria. Three cell morphologies occur among the "true bacteria." Spherical, rod-shaped, and spiral forms are called <u>cocci</u>, <u>bacilli</u>, and <u>spirilla</u>, respectively (Fig. 38.2). Eubacteria have thick, rigid cell walls. Generally, they are not motile, but a few propel themselves by flagella. Eubacterial species express an array of means of producing energy; many different carbon compounds can be used for sources of both carbon and energy.

Escherichia coli has been considered as the model species of eubacteria. As the fruit fly is to eukaryotic genetics, *E. coli* is a workhorse of biologists who are researching heredity, physiology, and other aspects of biology. Be sure to acquaint yourself with this coliform bacterium by studying the textbook section on *E. coli*.

(2) Class Myxobacteria. Cells of myxobacteria are gram-negative rods that generally occur in soils. They secrete a slime track upon which they glide about. Some forms can produce fruiting bodies, structures that contain spores that allow endurance of adverse conditions. Some aggregate into colonies that exhibit behavior.

(3) Class Actinobacteria. These unicells often link into mat-like mycelial masses, a characteristic that led to their previous classification as a group of fungi. They do not produce internal spores. Some are pathogens, such as those that cause leprosy and tuberculosis, whereas others are producers of important antibiotics.

(4) Class Spirochaeta. Characteristic features of spirochaetes include coiled form and axial filaments that resemble internal flagella. The presence of filaments in some species has led to the suggestion that the flagellar microtubules of eukaryotes might have resulted from the invasion of eukaryotic cells by spirochaetes.

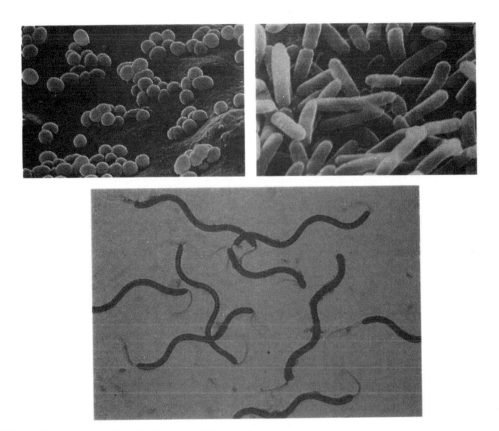

Fig. 38.2: Diversity of cell shapes of bacteria

(5) Mycoplasmas and Rickettsiae. These endoparasitic forms apparently derived from generalized bacteria via loss of various internal structures and functions. Some forms cause diseases, such as Rocky Mountain spotted fever and typhus. These are the smallest of all prokaryotic species.

F. Phylum Cyanonta.--Although not all members of this phylum are blue-green in color as the phylum name suggests, all are photosynthetic oxygen producers. All have thylakoids, membranous structures associated with the cell membrane, that contain molecules of various photosynthetic pigments, including chlorophyll a, phycocyanin, phycoerythrin, and others. Sexual processes are unknown in this group.

Some cyanonts exist as free-living solitary cells whereas others form clumps or filaments. Spirolina, the blue-green alga that is sold in many health-food stores, is helically shaped. Within some filamentous forms, a division of labor occurs among individual cells of the colony. Nitrogen fixation, the conversion of atmospheric nitrogen gas into ammonia, is the role of certain specialized cells (heterocysts) within the filament. Cyanonts are virtually independent nutritionally--they require (beyond a few minerals) only water, nitrogen, carbon dioxide, and light. These meager requirements enable them to pioneer seemingly sterile situations, such as surfaces of rocks and mineralized waters.

G. Phylum Prochloronta.--This recently discovered group is represented by only one genus, Prochloron. These photosynthetic oxygen producers possess a suite of photosynthetic pigments found in green plants: chlorophylls a and b, carotenoids, and xanthophylls. They differ from cyanonts in the absence of phycocyanin and phycoerythrin. These organisms are of key interest because they occupy the gap between cyanophytes and green algae. Also, various features suggest that the chloroplasts of plants might have evolved from ancient prochloronts via an <u>endosymbiont association</u>.

TERMS TO UNDERSTAND: Define and identify the following terms:

Monera

Phylum Schizonta

Phylum Cyanonta

Phylum Prochloronta

Bergey's Manual

Gram stain

bacterium

progenote

urkaryotes

endosymbionts

saprobes

spores

sporulation

decay

parasite

Class Eubacteria

coccus

bacillus

spirillum

Class Myxobacteria

fruiting body

Class Actinobacteria

mycelial mass

Class Spirochaetae

axial filament

mycoplasma

Rickettsiae

green sulfur bacteria

purple sulfur bacteria

purple nonsulfur bacteria

Escherichia coli

conjugation

phycocyanin

phycoerythrin

heterocyst

nitrogen fixation

MATCHING EXERCISES

Complete the following matching exercise to check your familiarity with characteristics and examples of the various higher categories of Kingdom Monera. Correct answers must include all characteristics and examples from the right column that match the taxon in the left column. Some characteristics might apply to more than one taxon.

TAXON   CHARACTERISTICS & EXAMPLES

____ 1. Myxobacteria   A. cocci

____ 2. Rickettsiae   B. form fruiting bodies

____ 3. Cyanonta   C. archaebacteria

____ 4. Eubacteria   D. cell shape is helical

____ 5. Prochloronta   E. *Escherichia coli*

____ 6. Spirochaeta   F. heterotrophic

____ 7. Actinobacteria   G. form mycelial masses

                                     H. are intracellular parasites

                                     I. photosynthetic

DISCUSSION QUESTIONS

1. For what reason(s) is it important for biologists to learn the phylogeny of monerans?

2. Which of the modern groups of prokaryotes is (are) the most likely candidate(s) for the role of endosymbionts that evolved into various organelles of eukaryotic cells? Why?

3. Discuss the nutritional strategies used by various monerans. How do their photosynthetic strategies differ from those used by plants?

4. Summarize the reproductive strategies found among members of the Kingdom Monera.

5. Outline the economic importance(s) of each moneran phylum.

TESTING YOUR UNDERSTANDING

For each of the test items below, three of the lettered alternatives are true and the other is false. Determine which alternative is false and write its letter in the blank to the left of the question number. On the blank line below alternative D, write the corrected version of the false statement.

    _____ 1. A. The smallest members of Phylum Schizonta belong to classes Mycoplasma and Rickettsiae.
            B. The blue-green algae are members of Phylum Prochloronta.
            C. All members of Phylum Cyanonta are autotrophs.
            D. Prochloronta appears to represent an early stage in the evolution of eukaryotes from prokaryotes.

            Correction: _____

_____ 2. A. The type of chlorophyll found in all monerans that possess chlorphyll is chlorophyll *b*.
   B. Some mycoplasmas are internal parasites, whereas others occur in soils.
   C. Rod-shaped cells characterize at least some species in classes Eubacteria, Myxobacteria, and Actinobacteria.
   D. Motility of myxobacteria is via gliding.

   Correction: _____

_____ 3. A. A primary ecological role of many bacteria is decomposition of organic material.
   B. Cyanonta frequently occur in ecologically barren situations.
   C. Many species of autotrophic bacteria cause diseases in humans.
   D. Production of spores enables some kinds of bacteria to endure harsh conditions.

   Correction: _____

_____ 4. A. The heterocysts of blue-green algae conduct the process of nitrogen fixation.
   B. Cyanonta previously was classified as a group of plants because of their ability to photosynthesize.
   C. In Phylum Prochloronta, molecules of photosynthetic pigment, such as chlorophyll, are contained within chloroplasts.
   D. Kingdom Monera is one of several kingdoms that include unicellular species.

   Correction: _____

_____ 5. A. The scanty fossil record of monerans contributes to the difficulty of understanding the phylogeny of monerans.
   B. Morphological features are the primary characteristics used in the classification of monerans.
   C. The Gram-staining characteristic of a bacterium is determined by the composition of its cell wall.
   D. Among early bacteria, anaerobic metabolism apparently arose several times from aerobic ancestors.

   Correction: _____

_____ 6. A. The endosymbiont hypothesis for the origin of eukaryotic cells involves invasion of cells by various prokaryotes to produce cytoplasmic structures including mitochondria, chloroplasts, and flagellar microtubules.
   B. Some photosynthetic bacteria are anaerobic.
   C. Photosynthetic sulfur bacteria assimilate carbon dioxide and use water as a reducing agent.
   D. The reducing agents used by some of the anaerobic, nonsulfur, photosynthetic bacteria include simple organic compounds, such as lactic acid and ethanol.

   Correction: _____

_____ 7. A. *Escherichia coli* is a gram-negative bacillus.
B. The normally used way to determine if a bacterial cell is alive is to examine it by using a microscope.
C. One of the normal habitats of *E. coli* is the human large intestine.
D. The generation time for *E. coli* is about twenty minutes.

Correction: _____

_____ 8. A. Although *E. coli* normally reproduces via conjugation, occasionally some cells undergo fission.
B. The ribosomes of prokaryotic cells usually are associated closely with the DNA of the cell.
C. Bacterial cells do not have chromatin.
D. Prokaryotic cells have cell walls composed of polymerized amino sugars.

Correction: _____

_____ 9. A. Saprobes are organisms that derive their energy and nutrient needs from dead organic material.
B. Many bacterial species can survive within a broader range of temperatures than can eukaryotic cells.
C. The chemosynthetic autotrophs convert light energy into chemical energy that is stored as carbohydrates.
D. The chains of cells formed by many bacterial species are produced by simple adhesion of cells that were produced by cell division.

Correction: _____

_____ 10. A. Decomposition is the principle ecological role of myxobacteria.
B. The cellular structure of prochlorophytes closely resembles the structure of chloroplasts in green algae.
C. "Bergey's Manual of Determinative Bacteriology" uses a classification scheme based on descriptive morphology.
D. For each species of bacteria, the content of guanosine and cytosine in the cell's proteins is constant.

Correction: _____

CHAPTER 39--PROTISTS

CHAPTER OVERVIEW

In the previous chapter, we surveyed the diversity of monerans, a kingdom of primarily unicellular (but also some colonial and multicellular forms) prokaryotes. In this chapter, we are introduced to the first of four kingdoms of eukaryotes. Two major evolutionary advances occurred within this kingdom: both multicellularity and sexuality evolved for the first time. At least one view of the evolutionary history of life holds that all other eukaryotic organisms (fungi, plants, animals) arose from ancestral protists. The major subdivisions of Protista corresponds largely to mode of nutrition--autotrophy (subkingdom Algae) compared to heterotrophy (subkingdom Protozoa).

TOPIC SUMMARIES

A. <u>Position of Protista in Evolutionary History</u>.--The earliest eukaryotic cells probably were protists. These presumably evolved from prokaryotes whose cells were invaded by other prokaryotic cells (see endosymbiont hypothesis in Chapters 5, 9, and 38). Although many protists are unicellular, mitotic division in some species produces colonial forms; whether such colonies represent merely a union of individual protists or an interactive group of cells is uncertain. Nevertheless, the existence of some degree of apparent "multicellularity" places protists on the path towards evolution of higher, multicellular organisms (i.e., plants, fungi, and animals).

Protista is a diverse kingdom. In part, this results from its role as a "catch-all" kingdom. Upon recognition of fungi, plants, and animals as kingdoms, many systematists arbitrarily placed all remaining eukaryotes into kingdom Protista. Therefore, Protista might not be a natural group having a monophyletic origin. Subdivisions of Protista reflect similarities to plants (subkingdom Algae) and to animals (subkingdom Protozoa).

B. <u>Subkingdom Algae: Autotrophic Protists</u>.--About 30,000 species of eukaryotic algae have been described. Many are unicellular, but most exhibit some degree of multicellularity and some degree of specialized roles for individual cells. Almost all are adapted to life in water, whether marine or fresh. Most algae are photosynthetic; indeed, an estimated 90% of all photosynthesis occurring on Earth is conducted by marine algae. Classification of Algae is based on features including the identity of photosynthetic pigments, the number of flagella, chemical composition of the cell wall, and the nature of the food reserves produced.

Let's briefly examine six major phyla of Algae. You will find textbook Table 39-1 helpful in organizing your study efforts. Also, you should review the appropriate life cycles (Chapter 21) to refresh your understanding of reproductive modes of these algae.

(1) Phylum Euglenophyta. Euglenoid flagellates, as they are called, have <u>flagella</u> that are used in locomotion. Most species are unicellular, aquatic, and photosynthetic. The photosynthetic storage product is <u>paramylon</u>. Other species are colonial or marine or parasitic. As in all other photosynthetic algae, one of the pigments present is <u>chlorophyll *a*.</u>

Euglena is the exemplary form (Fig. 39.1): Like many other euglenoids, it has features rendering it both animal-like and plant-like. Motility via flagella could ally it with animals, yet, when grown in light, presence of chloroplasts and photosynthetic pigments suggests it to be a plant. *Euglena* placed in darkness lose photosynthetic structures. Reproduction is by mitosis, an asexual process.

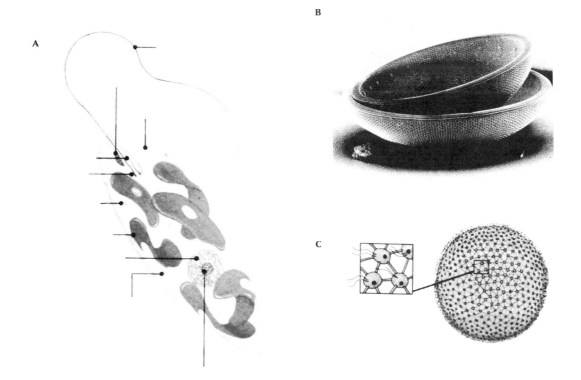

Fig. 39.1: Sketches of *Euglena* (A), a diatom (B), and *Volvox* (C)

(2) Phylum Pyrrophyta. Most dinoflagellates are unicellular, although a few are colonial. Most are marine forms abundant in warm seas. Their contribution to marine productivity is second only to that of cyanobacteria and diatoms (Phylum Chrysophyta, below). Photosynthetic pigments present include chlorophylls *a* and *c* and, unique to Pyrrophyta, peridinin. The photosynthetic storage product is starch. Some species, however, are heterotrophic. A few forms, the zooxanthellae, live symbiotically within tissues of various coral reef animals. These symbiotic forms lose the external cellulose plates that encase free-living forms. Sexual reproduction occurs rarely. Some species are responsible for red tides (textbook Box 39-1).

Another distinctive feature of dinoflagellates is substitution of hydroxymethyluracil for some of the thymine in DNA. Additionally, their nuclei, chromatin, and DNA fibrils have an unusual arrangement that places these organisms intermediate to prokaryotes and eukaryotes.

(3) Phylum Chrysophyta. The color of the golden algae and diatoms results from the presence of various golden-yellow carotenoids, including fucoxanthin. Also present are chlorophylls *a* and *c*. Chrysolaminarin is their storage product. These organisms, along with the dinoflagellates, comprise a large portion of photosynthetic plankton, and, thus, form part of the broad base of many marine food chains.

Diatoms are the better studied of the chrysophytes (Fig. 39.1). These usually unicellular organisms generally produce a skeleton containing silica; this two-part skeleton is arranged as a pillbox. Body symmetry is either bilateral (as in humans) or radial. Unlike most Chrysophyta, diatoms occasionally reproduce sexually.

(4) Phylum Chlorophyta. The over 7,000 species of green algae (family Chlorophycae) exhibit an array of body forms and modes of life. Within this family, a progression from unicellular life to colonial life is apparent: This series extends from the one-celled *Chlamydomonas* bto simple colonial forms *(Oedogonium, Pandorina, Eudorina)* to the complex colonial form, *Volvox* (Fig. 39.1). Additionally, the life cycle such as that of *Chlamydomonas* probably represents the first instance of sexual reproduction in a unicellular organism. The suite of photosynthetic pigments (chlorophylls *a* and *b* and various carotenoids) present in Chlorophyta matches those found in plants; hence, green algae are considered the stem group of plants.

The stoneworts (Class Charophyceae) are a less-diverse group of this phylum. Different body regions are differentiated into structures resembling (and having similar functions as) roots, stems, leaves, and seeds of plants. The cell walls of stoneworts contain calcium carbonate.

(5) Phylum Phaeophyta. All brown algae are multicellular, showing regional differentiation of the body (thallus) into "stems," "leaves," and flotation bladders. Most species are marine. The combination of several photosynthetic pigments (fucoxanthin, chlorophylls *a* and *c)* produces the characteristic brown color. Cell walls are composed of cellulose and of alginic acid. The kelps (a group of seaweeds) belong to this phylum.

Reproduction is either sexual or asexual. Several versions of alternation of generations between haploid gametophyte and diploid sporophyte are seen in this group. In *Ectocarpus,* neither generation dominates, whereas the sporophyte dominates in *Laminaria.*

(6) Phylum Rhodophyta. Most red algae are marine seaweeds. Carbohydrate reserves are stored as granules of floridean starch. Cell walls are made of cellulose and other polysaccharides that are used in production of agar (a common medium for growing microbes) and other gels. The pigment phycoerythrin contributes to the red coloration of some species. Calcium carbonate deposited within their cell walls contributes to building of coral reefs.

Red algae occupy an interesting phylogenetic position. They share some features of other algae and some features of blue-green "algae" (Cyanonta of Kingdom Monera). Features linking red algae with cyanonts include accessory pigments (phycoerythrin and phycocyanin) and the absence of flagellated cells. Yet, red algae are indeed eukaryotic. Presence of chlorophyll *a* and carotenoids allies red algae with higher plants.

C. Subkingdom Protozoa: Heterotrophic Protists.--Five phyla compose the animal-like protists. "Animality" of these protozoans is suggested by ingestion of food (all are heterotrophs), mobility in their surroundings (with associated behavior), storage of energy reserves in animal-like forms (glycogen and lipids), and absence of a rigid cell wall.

(1) Phylum Zoomastigina. The zooflagellates are unicellular organisms that possess one to many flagella or cilia. Sexual and asexual reproductive modes exist. Some are free-living, but many are parasitic within bodies of plants and animals. Some diseases (e.g., elephantiasis, Chaga's disease, African sleeping sickness) are caused by zooflagellates of genus *Trypanosoma*. Others are involved in symbiotic associations: termites can live on a diet of wood because certain zooflagellates living within the gut of the termite can digest cellulose and, thereby, release nutrients to the host.

Zooflagellates probably arose from euglenoids that lost their chloroplasts. These primitive zooflagellates then probably evolved into various other protozoans and into plants and animals (textbook Fig. 39-15).

(2) Phylum Sarcodina. The amoebas (Fig. 39.2) and their relatives comprise this group. Sacrodines possess pseudopods, extensions of the cell membrane and associated cytoplasm. Pseudopods enable both locomotion and capture of food (via phagocytosis). Whereas amoebas lack shells, other members of the phylum (foraminiferans and radiolarians) are encased in hard shells of calcium carbonate or silicate. Major portions of the sediments of ocean bottoms are formed by accumulation of these shells; the white cliffs of Dover represent such sediments that have been uplifted. Reproduction may be asexual (by mitosis) or sexual, a process th involves meiosis. A few species are parasitic.

Fig. 39.2: Sketches *Amoeba* (left) and *Paramecium* (right)

(3) Phylum Sporozoa. These parasitic forms have complex life cycles that involve both sexual and asexual phases. Adult forms lack flagella. Body form is amoeboid. Malaria is caused by infection by sporozoans of genus *Plasmodium*. Certain mosquitos *(Anopheles)* are the hosts in which the sexual phase of the parasite's life cycle occurs. Asexual multiplication of sporozoites into merozoites occurs in an alternate host, humans.

(4) Phylum Ciliophora. Numerous cilia, used for propulsion, arise from the cell membranes of ciliates. Paramecium (Fig. 392) represents this group, the most diverse and complex group of protists. Although unicellular, ciliates possess organelles whose functions correspond to the functions of entire organ systems of higher organisms. For example, there are "systems" for locomotion, feeding and digestion, support, and others. These complex levels of organization probably enabled this group to trace an evolutionary trend of increasing cell size. Reproduction can be by both self-fertilization or by conjugation. The ciliates probably evolved from ancestral zooflagellates.

(5) Phylum Myxomycota. The slime molds are among the strangest of all life forms. Their taxonomic placement is unsettled, being placed variously in kingdoms Fungi, Plantae, or Protista. Two major groups, differing in their life cycles, are recognized:

The true slime molds progress through a life cycle with the following steps: You most likely are familiar (if at all) with slime molds in their multinucleate form, the plasmodium, that slides about like a giant amoeba. Fruiting bodies develop from the plasmodium, the basal cells of which degenerate. Within the fruiting bodies (sporangia), pairs of haploid nuclei fuse to produce diploid nuclei. Meiosis follows. Haploid spores are released. These germinate into flagellated unicells (actually gametes). The flagellae are lost, the unicells reproduce by fission, and these now amoeboid individuals aggregate into the large plasmodium.

The life cycle of cellular slime molds is asexual, all nuclei being haploid. The slug-like aggregated form, the <u>pseudoplasmodium</u>, is both multinucleate and multicellular. A stalked <u>fruiting body</u> forms. Non-flagellated spores are released from the fruiting body. These spores germinate into the amoeboid cells that eventually form the pseudoplasmodium.

<u>TERMS TO UNDERSTAND</u>: Define and identify the following terms:

symbiosis

algae

euglenoid flagellates

paramylon

locomotor flagellum

internal flagellum

cytopharynx

contractile vacuole

stigma

photoreceptor

dinoflagellates

plankton

diatoms

peridinin

zooxanthellae

hydroxymethyluracil

chrysophytes

silica

swarmer cells

diatomaceous earth

Chlorophycae

Charophycae

Phaeophyta

fucoxanthin

alginic acid

antheridia

Rhodophyta

floridean starch

phycoerythrin

chromatic adaptation

Protozoa

Zoomastigina

Sarcodina

pseudopodia

foraminiferans

Sporozoa

Ciliophora

macronucleus

micronuclei

Myxomycota

plasmodium

fruiting bodies

pseudoplasmodium

MATCHING EXERCISE

Complete the following matching exercise to check your familiarity with characteristics and examples of the various higher categories of Kingdom Protista. Correct answers must include all characteristics and examples from the right column that match the taxon in the left column. Some characteristics might apply to more than one taxon.

TAXON   CHARACTERISTICS & EXAMPLES

____ 1. Euglenophyta   A. autotrophy

____ 2. Myxomycota   B. includes some multicellular forms

____ 3. Ciliophora   C. contains pigment chlorophyll *b*

____ 4. Sarcodina   D. possess flagella or cilia

____ 5. Chlorophyta   E. energy storage material is glycogen

____ 6. Sporozoa   F. parasitic

____ 7. Rhodophyta   G. includes some forms that reproduce sexually

                              H. none of the above

DISCUSSION QUESTIONS

1. Discuss the evidence that supports multiple, independent origins for chloroplasts.

2. Label as many structures as possible in the accompanying drawing of a protist. What are the functions of the structures? With which phylum of protists is this organism most likely classified?

3. Select one phylum of protists and present the evidence supporting the evolution of multicellularity from unicells within that phylum.

4. Construct a phylogenetic tree that depicts how the three kingdoms of multicellular organisms evolved from Kingdom Protista. What features support this alignment of these kingdoms?

5. Summarize the symbiotic associations in which various protists are found.

TESTING YOUR UNDERSTANDING

For each of the test items below, three of the lettered alternatives are true and the other is false. Determine which alternative is false and write its letter in the blank to the left of the question number. On the blank line below alternative D, write the corrected version of the false statement.

_____ 1. A. All protists are prokaryotic.
B. Many protists are multicellular.
C. Some protists are colonial.
D. Three kingdoms of multicellular organisms probably evolved from Protista.

Correction: _____

_____ 2. A. Asexual reproduction in protists is by mitosis.
B. Organization of individual cells into colonies might have been a step along the evolutuionary pathway towards metazoans.
C. Protists have been variously classified into kingdoms Plantae, Fungi, and Animalia.
D. The various species of protists are either autotrophic or heterotrophic, but no individual species is both autotrophic and heterotrophic.

Correction: _____

_____ 3. A. A mutually-beneficial relationship between two species is called a symbiosis.
B. The ability of some animals to derive energy from a diet of wood results from an association with certain types of protists.
C. Chloroplasts probably arose independently several times during evolution.
D. Mitochondria probably evolved via invasion of prokaryotic cells by certain eukaryotic cells.

Correction: _____

_____ 4. A. The primary ecological role of organisms in subkingdom Algae is production of food via photosynthesis.
B. The blue-green algae are properly classified into Kingdom Monera.
C. Almost all species of algae are unicellular.
D. Most algal species occur in aquatic habitats.

Correction: _____

_____ 5. A. Flagella are found in all known species of subkingdom Algae.
B. Chlorophyll *a* occurs in chloroplasts of some species of all phyla of Algae.
C. Carotenoid pigments occur in some species of all phyla of subkingdom Algae.
D. Rhodophyta is the group of Algae that is most closely related to Cyanonta.

Correction: _____

_____ 6. A. *Euglena* is used in the laboratory to assay vitamin B_{12} content of solutions.
B. All of the following structures are found in *Euglena:* macronucleus, contractile vacuole, and photoreceptor.
C. Bioluminescence is a feature of many species of dinoflagellates.
D. Zooxanthellae are dinoflagellates that have lost their armor coats and have taken up residence within tissues of certain coral reef animals.

Correction: _____

_____ 7. A. The term mesokaryotes is sometimes applied to species of dinoflagellates.
B. Diatoms and some members of Phylum Sarcodina have the ability to deposit calcium-carbonate skeletons around themselves.
C. The chemical form in which energy is stored in golden algae is chrysolaminarin.
D. In some Chrysophyta, new colonies are established by swarmer cells that swim away from existing colonies.

Correction: _____

_____ 8. A. Charophycae is a group of green algae that made an unsuccessful attempt to evolve into higher plants.
B. The volvicine series of green algae reflects an evolutionary trend extending from unicellularity to colonial form.
C. Most green algae are marine species.
D. The life cycle of *Chlamydomonas* represents probably the first instance of sexual reproduction in a unicellular organism.

Correction: _____

_____ 9. A. Antheridia and oogonia are multicellular sex organs found in Rhodophyta.
B. All of the following are materials found in the cell walls of various algae: alginic acid, cellulose, and various mucilaginous polysaccharides.
C. All of the following are chemical forms in which various algae store their energy reserves: paramylon floridean starch, and leucosin.
D. The presence of phycocyanin and phycoerythrin in the chloroplasts of Rhodophyta is evidence suggesting a close relationship with Cyanonta.

Correction: _____

_____ 10.   A. The hallmark of animality is the ingestion of particulate food.
B. Pathogenic species of Protozoa occur in phyla Zoomastigina, Sarcodina, and Sporozoa.
C. All phyla of Protozoa include at least some members that possess cilia or flagella.
D. The primary food reserve of protozoans is glycogen.

Correction: _____

_____ 11.   A. The phenomenon of variation in pigmentation resulting from variation in the color of incident light is called chromatic adaptation.
B. Mobility and heterotrophy among protozoans sets these organisms apart from the other subkingdom of Protista.
C. Zoomastigina includes some species that are symbiotic and some species that are parasitic.
D. Sexual reproduction is not known to occur in two of the phyla of Protozoa.

Correction: _____

_____ 12.   A. Diseases caused by various species of sarcodines include amoeboid dysentery, malaria, and Chagas' disease.
B. The pseudopods of sarcodines are used in locomotion and in food procurement.
C. Contractile vacuoles are structures usually found in protozoans living in a hypotonic environment.
D. Foraminiferans and radiolarians are sacrodines that secrete hard shells of calcium carbonate or silicates.

Correction: _____

_____ 13.   A. Two prominent features of sporozoans are a paratisitic existence and a complicated life cycle that includes both sexual and asexual phases.
B. *Plasmodium* is a genus of true slime molds.
C. *Paramecium* is a multinucleate ciliate.
D. Merozoites are haploid cells in the life cycle of a sporozoan.

Correction: _____

_____ 14.   A. The ciliates represent the greatest structural complexity known for any unicellular species.
B. The two major groups of slime molds can be distinguished by the presence or absence of sexual phases in their life cycles.
C. The two major groups of slime molds can be distinguished by the presence or absence of cell membranes between nuclei of the plasmodium or pseudoplasmodium.
D. The spores of cellular slime molds are flagellated.

Correction: _____

CHAPTER 40--FUNGI

CHAPTER OVERVIEW

We continue our survey of the diversity of life with study of fungi, a group of organisms for which few of us have adequate appreciation until such a survey as this. Well over 100,000 species are known. All are heterotrophic. Many are parasites and agents of diseases in many species of plants and animals. Others are decomposers that play a critical role in cycling of nutrients through the biogeosphere. Still others enable many species of plants to obtain nutrients from soil. The phyla of fungi are classified largely according to their modes of reproduction.

TOPIC SUMMARIES

A. General Characteristics.--A feature shared by all fungi is heterotrophy, as evidenced by the absence of chlorophyll. Whether subsisting as saprobes, symbionts, or parasites, all fungi engage in extracellular digestion of food. They secrete digestive enzymes into a mass of food and then absorb the simpler products of extracellular digestion.

Some fungi, particularly the yeasts, are unicellular. Most, however, are multicellular species whose bodies consist of masses (mycelium) of threadlike filaments (hyphae). In most fungi (excluding Zygomycota and Oomycota that have aseptate hyphae) adjacent cells in a hypha are separated by cross walls (septa). Each cell in a hypha contains two or more nuclei. Cell membranes are surrounded by cell walls of cellulose or chitin.

Classification into phyla is based largely on structures and methods associated with reproduction. The life cycles of fungi often are complex and may include both sexual and asexual phases. Fig. 40.1 depicts the life cycle of a sac fungus; you can modify this scheme to reflect the life cycles of species in the other sexually reproducing phyla of Fungi.

Meiosis in most fungi is zygotic, with meiosis occurring immediately after formation of the zygote. Hence, the only diploid cell in the life cycle is the zygote. Both sexual and asexual modes produce spores, cells that can develop into an adult organism without fusing with another cell. Sexually produced spores are nonmotile, whereas asexually produced spores of most fungi are dispersed by wind (asexual spores of aquatic fungi are motile via flagella). Asexual spores are produced by or contained within sporangia. Gametes are produced by or contained within gametangia.

B. Importance of Fungi to Other Species.--At least a few repesentatives of each phylum of fungi are known agents of disease in humans and/or other organisms. Examples include athlete's foot, coccidiodomycosis, late blight of potatoes, downy mildew of grapes, histoplasmosis, ergotism, and *Amanita* poisoning. Hence, as with monerans, there is practical value in learning the biology of fungi. Conversely, many benefits accrue from fungi: antibiotics produced by *Penicillium,* cheeses (textbook Box 40-1), ethyl alcohol produced by yeasts, leavened bread, mychorrhizal associations, etc. Certain fungi contribute directly to the diets of many animal species: lichens eaten by reindeer, mushrooms on pizzas consumed by humans. As decomposers, many species of fungi bear great responsibility in nutrient recycling.

C. Phylum Oomycetes.--The water molds, white rusts, and downy mildews are filamentous or unicellular species having cell walls made mainly of cellulose. Asexual reproduction produces zoospores having two flagella of unequal length; zoospores germinate into new individuals (textbook Fig. 40-7). In sexual reproduction, gametes are produced by gametangia called oogonia and antheridia; fusion of eggs and sperm yields thick-walled cells, oospores.

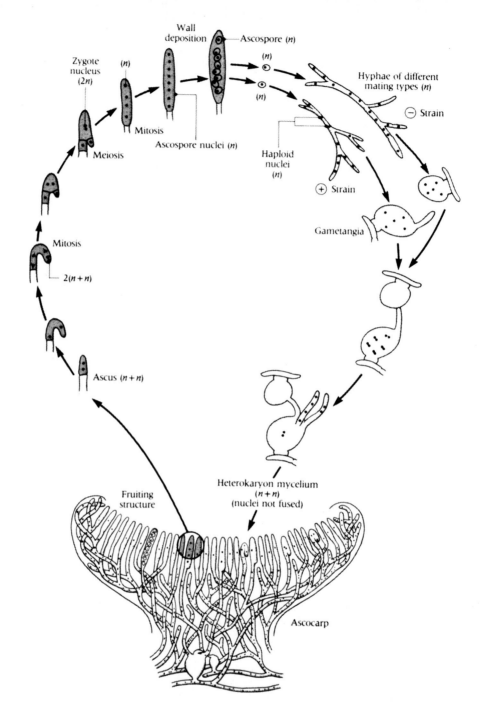

Fig. 40.1: Diagram of life cycle of an ascomycete fungus

D. <u>Phylum Zygomycetes</u>.--The <u>bread molds</u> are terrestrial filamentous species having walls of <u>chitin</u> (in what other organisms have we found this material?). Cells are not defined by septa except to separate sporangia and gametangia from the rest of the mycelium. Sexual reproduction of species in this group produces resting spores (<u>zygospores</u>) that uniquely develop from the zygote (text Fig. 40-5). No flagellated spores are produced in the life cycle.

E. <u>Phylum Ascomycetes</u>.--The <u>sac fungi</u> (yeasts, powdery mildews, red bread molds, etc.) constitute the largest phylum of this kingdom. The group is named for the sac-like structure (a sporangium called an ascus) that contains the products (eight ascospores) of sexual reproduction. Ascomycetes may be unicellular (e.g., yeasts) or multicellular. Division of the ascomycetes into two major groups is on the basis of the presence (euascomycetes) or absence (hemiascomycetes, e.g., yeasts) of a fruiting structure.

In multicellular forms, hyphae are divided by septa, with each compartment containing one or more nuclei. The asexual spores, conidia, are borne at the tips of specialized hyphae called conidiophores (Fig. 40.1). Sexual reproduction involves production of a heterokaryon, a mycelial stage formed when haploid nuclei are passed from hyphae of one individual to hyphae of another individual. The newly-formed heterokaryon then includes nuclei from different parents that divide in synchrony, but that do not fuse until an ascus forms. If the nuclei are of the same genetic type, then the heterokaryon is homokaryotic; if the nuclei are different genetically, then the mycelium is heterokaryotic. In the ascus, two of the haploid nuclei fuse into a zygote that quickly undergoes meiosis and a mitotic division to produce eight haploid ascospores. How does sexual reproduction in hemiascomycetes differ from the process just described for euascomycetes?

F. Phylum Basidiomycetes.--The sexual spores (basidiospores) of club fungi are borne exogenously on many basidia. This group includes the mushrooms and bracket fungi.

Three developmental stages characterize most species of club fungi (textbook Fig. 40-15). First, germination of a basidiospore forms the primary mycelium. Septa partition the mycelium into units, each containing a nucleus like one or the other of the original haploid partners. A heterokaryon forms when cells containing genetically different nuclei fuse. In the second stage (dikaryotic stage), a secondary mycelium grows and produces a basidiocarp, a fruiting body we know as a mushroom. Nuclear fusion eventually occurs to produce a diploid nucleus. Finally, these dikaryotic mycelia grow while meiosis occurs to produce four haploid basidiospores attached to basidia that line the surfaces of the gills of the mushroom.

G. Phylum Deuteromycetes.--This large phylum includes many species of presumably unrelated fungi; their shared feature is the absence of sexual stages. Whether these species have evolutionarily lost their sexual modes of reproduction or biologists simply have not discovered sexual modes in these species is uncertain. Some deuteromycetes, however, do exhibit parasexuality--wherein fusion of haploid nuclei of a heterokaryon occurs, but without the involvement of specialized mycelia.

H. Symbiotic Relationships Involving Fungi.--Recall that symbiosis is a mutually beneficial relationship between two or more species. Many such relationships evolve to the point of being obligatory--neither participant being able to survive alone.

The body (thallus) of a lichen is a union of filaments of fungi within which reside algae, either cyanobacteria or chlorophytes (textbook Fig. 40-17). The alga is autotrophic, supplying food to the fungus, and the fungus in turn supplies water and dissolved salts to the alga. Fungal participants may be from phyla Ascomycetes, Basidiomycetes, or Deuteromycetes. Lichens inhabit harsh environments and often are the first stage of life occupying otherwise sterile situations (e.g., bare rock). Also, lichens are a primary food source of animals such as reindeer and caribou.

Mycorrhizae are a second type of symbiotic association in which fungi form an association with roots of a plant. The fungus obtains organic carbon compounds from the plant, but in turn supplies the plant with soil nutrients such as phosphorus, zinc, and copper. Approximately 80% of all vascular plants require this ancient symbiotic association, that dates to at least the Carboniferous. Two mycorrhizal arrangements are known: Endomycorrhizal associations involve hyphae (usually of Zygomycetes) that penetrate the outer cells of the root (textbook Fig. 40-20). <u>Ectomycorrhizal</u> associations involve basidiomycetes whose hyphae surround but do not invade the root (textbook Fig. 40-21). Such associations might have been instrumental in enabling plants to invade land.

TERMS TO UNDERSTAND: Define and identify the following terms:

yeast

hypha

septum

mycelium

chitin

saprobe

symbiont

mold

bracket fungi

antibiosis

spore

zoospore

zoosporangium

oospore

zygospore

ascospore

basidiospore

sporangiophore

gametangium

oogonium

antheridium

water mold

rust

mildew

sac fungus

conidium

conidiophore

ascus

ascocarp

homokaryotic

heterokaryotic

monokaryotic

dikaryotic

euascomycete

hemiascomycete

basidium

basidiocarp

gill

parasexuality

thallus

ascolichen

basidiolichen

deuterolichen

endomycorrhiza

ectomycorrhiza

haustorium

rhizoid

DISCUSSION QUESTIONS

1. Discuss the various modes of nutrition demonstrated in the various phyla of Fungi. Provide examples as part of your answer.

2. Compare and contrast the processes by which Basidiomycetes and Ascomycetes produce sexual spores.

3. What are the major characteristics by which the phyla of Fungi are distinguished?

4. Using examples, discuss several ways in which various fungi have had significant effects on the course of human history.

5. *Neurospora* is a euascomycete that has been used extensively in genetics research. What features of this bread mold make it a suitable model species for such research?

6. How does <u>sexuality</u> differ from <u>parasexuality</u>? In which phyla do these reproductive modes occur?

MATCHING EXERCISE

Complete the following matching exercise to check your familiarity with characteristics and examples of the various higher categories of Kingdom Fungi. Correct answers must include all characteristics and examples from the right column that match the taxon in the left column. Some characteristics might apply to more than one taxon.

TAXON

_____ 1. Euascomycetes

_____ 2. Zygomycetes

_____ 3. Basidiomycetes

_____ 4. Hemiascomycetes

_____ 5. Deuteromycetes

_____ 6. Oomycetes

CHARACTERISTICS & EXAMPLES

A. includes at least some pathogenic or parasitic species
B. does not exhibit sexual reproduction
C. sexual reproduction produces eight spores within a sporangium
D. sexual reproduction produces four spores within a sporangium.
E. exhibits asexual reproduction
F. life cycle includes flagellated cells
G. mushrooms
H. yeasts

TESTING YOUR UNDERSTANDING

For each of the test items below, three of the lettered alternatives are true and the other is false. Determine which alternative is false and write its letter in the blank to the left of the question number. On the blank line below alternative D, write the corrected version of the false statement.

_____ 1. A. Although most species of fungi lack chlorophyll, a few species are autotrophic.
B. A primary ecological role of fungi is decomposition of dead organic matter.
C. The mass of hyphae that constitutes the body of many species of fungi is called a mycelium.
D. The cell walls of many kinds of fungi are made largely of the same material that is found in the exoskeletons of insects.

Correction: _____

_____ 2. A. Yeasts are single-celled fungi.
B. Yeasts are euascomycetes.
C. In sexual reproduction, yeasts produce asci usually containing eight spores.
D. All kinds of fungi absorb needed sugars and other preformed organic molecules from their surroundings.

Correction: _____

_____ 3. A. The mushrooms that were a topping on the pizza you ate last week are classified into Phylum Basidiomycetes.
B. Athlete's foot, also known as trichophytosis, is a condition caused by infection by a fungus classified into Phylum Deuteromycetes.
C. Histoplasmosis is a respiratory disease caused by an deuteromycete.
D. Ergotism is a fungus-produced disease that humans can contract by ingestion of beef.

Correction: _____

_____ 4. A. Conidia are sexual spores produces by certain oomycetes.
B. An ascus is a sac that contains the sexual spores of ascomycetes.
C. A hypha that contains nuclei of the same genetic constitution is referred to as homokaryotic.
D. Any monokaryotic hypha is homokaryotic.

Correction: _____

_____ 5. A. Species of sac fungi that lack fruiting structures are called hemiascomycetes.
B. During sexual reproduction in ascomycetes, a crozier forms at the tip of the heterokaryotic hypha.
C. The fruiting body of zygomycetes is called a basidiocarp.
D. Fungi Imperfecta is a synomym for Deuteromycetes.

Correction: _____

_____ 6. A. Sexual reproduction is unknown in any species of deuteromycetes.
B. The parasexual process of deuteromycetes involves fusion of diploid nuclei.
C. Antibiotics are derived from substances produced by some fungi of Phylum Deuteromycetes.
D. The type of meiosis that occurs in fungi that reproduce sexually is zygotic meiosis.

Correction: _____

_____ 7. A. Zoospores are motile, flagellated spores produced by asexual processes in Oomycetes.
B. The sexual spores produced by some types of fungi are flagellated.
C. Both spores and gametes can develop into an adult organism.
D. All of the following are types of spores produced by sexual processes in various groups of fungi: zygospores, ascospores, basidiospores, oospores.

Correction: _____

_____ 8. A. The water molds are classified into Phylum Oomycetes.
B. The sac fungi are classified into Phylum Ascomycetes.
C. The bracket fungi are classified into Phylum Basidiomycetes.
D. The lichens are classified into Phylum Zygomycetes.

Correction: _____

_____ 9. A. Most species of plants participate in mycorrhizal associations.
   B. The fungal component of an endomycorrhizal association is a zygomycete.
   C. The fungus of an ectomycorrhizal association benefits by receiving water and minerals from the plant.
   D. The possession of mycorrhizal roots may have enabled the earliest plants to invade lands with soils that lacked organic matter.

   Correction: _____

_____ 10. A. Lichens are autotrophic organisms.
   B. The photosynthetic partner in all known types of lichens is a cyanobacterium.
   C. Lichens are ecological pioneers that often play an important role in the first steps of breaking down rocks into soil.
   D. Lichens form a significant part of the diets of somes species of animals.

   Correction: _____

CHAPTER 41--PLANTS

CHAPTER OVERVIEW

Plants constitute a tremendously diverse group of photosynthetic autotrophs that evolved from green algae, over 400 million years ago. Plants are multicellular organisms that generally inhabit land. Invasion of land by ancestral plants, and achievement of true independence from aquatic environments, involved evolution of supporting tissues and of reproductive modes that do not require free water. Supporting (vascular) tissues also have the critical role of transporting materials (water, food) within the plant body.

Tracheophyta, the major group that possesses vascular tissues, is divided further according to several features including the presence, absence, or degree of development of roots, stems, and leaves. Other key characters are the presence or absence of seeds, whether the seeds are "naked" or covered, and the presence or absence of flowers.

Plants are the principal producers of the chemical energy consumed by animals and by many organisms of other kingdoms. Plants are sources of various medicines and of building materials. Hence, a general understanding of botany is important for all biologists and for all humans in general.

TOPIC SUMMARIES

A. <u>General Features of Plants</u>.--Plants are multicellular eukaryotes that are autotrophic due to their ability to photosynthesize. Plant cells possess plastids (such as chloroplasts) containing chlorophylls *a* and *b* and various carotenoids. Plant cells have cell walls containing cellulose. Most species of plants are terrestrial.

Although some systematists include various of the algae among Kingdom Plantae, we will classify the autotrophic protists into Kingdom Protista. Our more narrowly defined plant kingdom includes two broad groups: The <u>bryophytes</u> are generally nonvascular. The <u>tracheophytes</u>, however, have vascular tissues; phloem and xylem were discussed in Chapters 7 and 24.

B. <u>Early Evolution of Plants</u>.--Early plants probably evolved from green-algae (Chlorophyta) ancestors. Invasion of land by primitive plants probably occurred during the Silurian Period, about 415 million years ago. Presumably, the earliest land plants were similar to modern bryophytes in the absence of vascular tissues. However, the earliest fossils of land plants known to us (Zosterophyllophyta, Rhiniophyta, and Trimerophytophyta of Devonian times) have vascular tissues.

The <u>transition of plants from water to land</u> required resolution of at least four major problems. Water is required for biochemical processes. Unicellular ancestors living in water had a ready supply of water, but early land plants were exposed to the <u>dehydrating effects of air</u>. Also, land plants needed ways of transporting water from areas of water availability to cells needing water; vascular tissues evolved in response to this need to <u>transport water throughout the body</u>. <u>Specialized organs</u> (roots to collect water and nutrients; stems to support the plant body; leaves as the primary sites of photosynthesis) evolved to conduct specific functions. <u>Changes in reproductive strategy</u> also were required; whereas water plants could merely shed gametes into water, land plants required alternate means of getting gametes together. Pollen, flowers, and seeds represent some of these reproductive adaptations. Various <u>other physiological adaptations</u> were needed to respond to environmental challenges such as gas exchange across moist surfaces, and means of withstanding extreme temperatures, ultraviolet radiation, and other hazards not encountered in aquatic settings.

C. <u>Subkingdom Bryophyta: Nonvascular Plants</u>.--The liverworts, hornworts, and mosses generally lack vascular tissues; if present, such tissues are very rudimentary. Bryophytes occupy an intermediate position in the spectrum of nonvascular-to-vascular plants. Lacking the transport ability and the support offered by vascular tissues, these plants necessarily are small--generally only a few centimeters in height. They occur in low, moist situations.

The reproductive cycle of bryophytes emphasizes the gametophyte, upon which the simple sporophyte lives. You should review this life cycle (Chapter 21). Note the importance of water in reproduction: Motile sperm cells swim through a thin film of water coating the plant surface.

Mosses (Class Muscopsida) make up the majority of the approximately 24,000 extant bryophyte species. The shape of the thallus is the source of the common name of the liverworts (Class Hepaticopsida), a group of about 9,000 species. Only about 100 species of hornworts (Class Anthocerotopsida) exist; in this group the sporophyte is not nutritionally dependent on the gametophyte.

D. <u>Subkingdom Tracheophyta: Vascular Plants</u>.--One of the major trends in the evolution of vascular plants is the increasing dominance of the sporophyte generation (compare this with the bryophytes). In the following survey of five major groups of tracheophytes, we'll study the position of each group along the gametophyte-to-sporophyte dominance continuum.

E. <u>Division Psilophyta: Whisk Ferns</u>.--Various psilopsids have been around for at least 400 million years; indeed, early in the history of land plants they were the dominant land plants. The modern genera *Psilotum* and *Tmesipteris* are the sole survivors of Psilophyta. These plants lack roots. Only a few species possessed leaves, and these leaves were small and simple. The sporophyte is the dominant generation. Gametophytes are small, colorless (i.e., not photosynthetic), and subterranean. No seeds are produced. The psilopsid genus *Rhynia* is among the earliest known land-plant fossils.

F. <u>Division Lycopodiopsida: Club Mosses</u>.--This once-diverse group now is represented by five extant genera: *Lycopodium, Selaginella, Phylloglossum, Isoetes,* and *Stylites*. Although modern forms generally are low-lying herbs or vines, many extinct species were great trees that contributed greatly to deposits of coal. True (although simple) roots, stems, and leaves characterize the club mosses. As another advance, two types of spores (microspores and macrospores) are produced; these develop into tiny, free-living male and female gametophytes, respectively. Additionally, some club mosses exhibit the first evolutionary indication of seeds: Although not true seeds, cases containing embryonic sporophytes are borne within the female gametophyte.

G. <u>Division Sphenophyta: Horsetails</u>.--Though many species once flourished long ago, only about 15 species (all in genus *Equisetum)* remain. They occur rather commonly in moist situations. The bodies of these <u>horsetails</u> consist mainly of a hollow stem bearing whorls of small leaves and joints in the stem. Seeds are unknown in sphenodsids. Spores generally do not differentiate into microspores and megaspores. The presence of silica in the stems makes them useful in cleaning, hence another common name, "scouring rushes."

H. <u>Division Pteridophyta: Ferns</u>.--The <u>ferns</u> were the first plants to possess leaves (<u>megaphylls</u>) with complex venous systems; leaves (<u>microphylls</u>) in the preceeding groups have only one vascular channel. As with whisk ferns, club mosses, and horsetails, ferns were more diverse and adundant during the Carboniferous Period. However, unlike these other groups, ferns have maintained considerable (although reduced) diversity.

Temperate-zone species are usually small, reaching heights of a couple of meters or less, whereas some tropical forms grow as trees. The leaves (fronds) arise from horizontal stems (rhizomes) or from trunks. True roots usually are present. The vascular system is well-developed.

The sporophyte dominates in the life cycle. Some leaves (sporophylls) bear sporangia wherein meiosis produces spores that are released to germinate into a free-living gametophyte. On the gametophyte develop archegonia and antheridia, structures that produce egg and sperm cells, respectively. Flagellated sperm cells swim through the surface film of water to reach the archegonium wherein fertilization occurs (note the dependence of reproduction in ferns on presence of external water). The developing sporophyte grows directly upon the gametophyte. Refer to Chapter 21 for details on fern life cycles.

I. Division Spermophyta: Seed Plants.--Seed plants successfully completed the evolutionary transition to life on land: Male gametes do not require a surface film of water to enable transport to ova. Indeed, the male gametophyte (i.e., the germinated pollen grain together with the pollen tube) is resistant to dessication. The female gametophyte lives its whole life, protected from dehydration, within the tissues of the parental sporophyte. The seed, first appearing in Spermophyta, encloses the embryonic sporophyte with an independent supply of nutrients adequate to carry the new plant through germination when it then can produce its own food.

Subdivision of Division Spermophyta is based on the presence or absence of flowers and on the degree of protection provided to seeds by the parental sporophyte:

J. Subdivision Gymnospermae: Nonflowering Plants.--Four of the five extant classes of seed plants are referred to as gymnosperms. Seeds of gymnosperms are not enclosed by protective ovarian walls and they are borne on exposed surfaces of sporophylls.

The ancestral seed plants arose during the Carboniferous and radiated during the latter Paleozoic. Several groups--seed ferns [Class Pteridospermopsida], cycadeoids [Class Cycadeoidopsida], cordaites [Class Cordaitopsida]--are extinct. The ginkgos [Class Ginkgopsida], cycads [Class Cycadopsida], and joint firs [Class Gnetopsida] remain today in lesser diversity within subsets of the geographic ranges occupied by their ancestors. The most abundant and widespread modern gymnosperms are the conifers [Class Coniferopsida]; these include the pines, cedars, and many other species. Be sure to review the life cycle of conifers as presented in Fig. 41.1 below.

At least two trends are evident in the evolution of gymnosperms. Early forms (e.g., seed ferns) bore seeds individually along the frond or leaf stalk, whereas derived forms (e.g., cycadeoids, cycads, and later forms) bear seeds in cones (strobili). For those groups having strobili, the trend has been from one kind of strobilus that bears both pollen and egg cells to separate strobili, each bearing gametes of only sex.

K. Subkingdom Angiospermae: Flowering Plants.--Angiosperms differ from gymnosperms not only in the presence of flowers, but in the enclosure of the seed by a protective ovarian wall. Flowering plants are the predominant group of land plants today, numbering over a quarter of a million species. The earliest fossil record of angiosperms is from the early Cretaceous, about 135 million years ago.

The apparently greater success of angiosperms over gymnosperms is directly related to the major differences between the groups. Flowers evolved in ways that attracted various animals

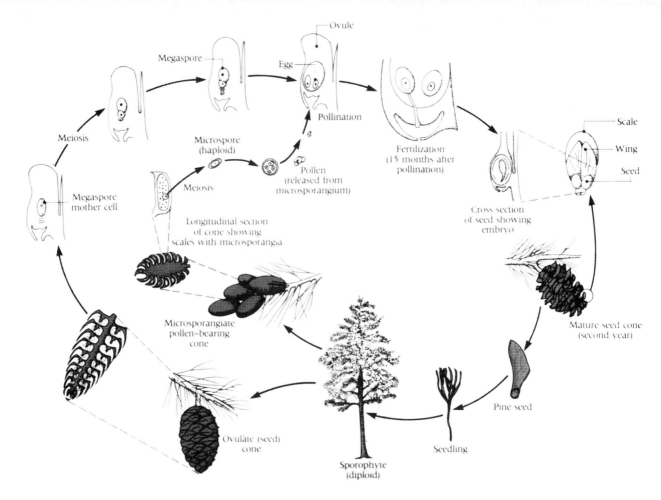

Fig. 41.1: Life cycle of a conifer

that serve as <u>pollinators</u>. These pollinators carry pollen to other flowers of the same plant species; such insect-pollinated species waste fewer gametes than do plants that depend on haphazard dispersal of pollen by wind. Genetic diversity within populations is promoted further by genetic <u>self-incompatibility</u> in many species, wherein a plant cannot fertilize itself. Dispersal of individuals from parents is enhanced by attractive features (color, taste, odor, wings, etc.) of the ovarian cover of the seed; this, too, decreases inbreeding. It seems that differences in reproductive strategies explain the decline of gymnosperms and the simultaneous rise of angiosperms.

Differences in early development of embryos of flowing plants allow recognition of two groups of angiosperms. The nutrients (especially starch) of an embryo are stored in leaf-like structures called <u>cotyledons</u>. Those species having one cotyledon are called <u>monocotyledons</u>, whereas those with two seed leaves are <u>dicotyledons</u>. With further growth beyond the embryo, other anatomical features discriminate between monocots and dicots; the chart exercise in this chapter gives you the opportunity to compare these groups.

TERMS TO UNDERSTAND: Define and identify the following terms:

Bryophyta

Tracheophyta

Zosterophyllophyta

Rhinophyta

roots

leaves

stem

Hepaticopsida

Anthocerotopsida

Muscopsida

Psilotopsida

Lycopodiopsida

Sphenophyta

Pteridophyta

Spermophyta

Equisetum

microphylls

megaphylls

fronds

pinnae

pinnules

rhizoids

archegonia

antheridia

homosporous

heterosporous

epiphytes

seeds

pollen grain

carpels

sporophylls

gymnosperms

angiosperms

seed ferns

cycadeoids

cordaites

ginkgos

cycads

joint firs

conifers

strobili

evergreens

deciduous

flowers

fruits

perfect vs. imperfect

androecium

gynoecium

carpellate

pistillate

staminate

genetic self-incompatibility

monocotyledons

dicotyledons

CHART EXERCISE

Complete the following chart to compare the two major groups of flowering plants, the monocotyledons and the dicotyledons. You can expand this exercise by adding columns for other groups of plants.

CHARACTER	MONOCOTYLEDONS	DICOTYLEDONS
Number of seed leaves		
Numbers of flower parts		
Arrangement of vascular bundles		
Venation of leaves		
Presence of seeds		
Presence of flowers		
Examples		

DISCUSSION QUESTIONS

1. Describe the problems faced by the plants that left aquatic habitats and invaded land. What structural features enabled these plants to cope successfully with these problems? Provide examples.

2. Reproductive strategies of the earliest land plants differ significantly from those of modern land plants. Briefly summarize the evolutionary trends related to differences in reproduction.

3. In what ways are the reproductive strategies of angiosperms superior to those of gymnosperms in achieving greater genetic diversity?

4. The fossil record has not yet yielded fossils of the earliest bryophytes. On the basis of what you know about modern bryophytes and about early tracheophytes, describe the features that you would expect to find in the earliest (as yet unknown) bryophytes.

MATCHING EXERCISE

Complete the following matching exercise to check your familiarity with characteristics and examples of the various higher categories of Kingdom Plantae. Correct answers must include all characteristics and examples from the right column that match the taxon in the left column. Some characteristics might apply to more than one taxon.

TAXON	CHARACTERISTICS & EXAMPLES
_____ 1. Coniferopsida	A. has motile sperm
_____ 2. Ginkgopsida	B. has seeds
_____ 3. Pteridophyta	C. has true roots
_____ 4. Hepaticopsida	D. gametophyte dominates life cycle
_____ 5. Monocotyledonae	E. lacks vascular tissues
_____ 6. Sphenophyta	F. has flowers
_____ 7. Angiospermae	G. all species of taxon are extinct

TESTING YOUR UNDERSTANDING

For each of the test items below, three of the lettered alternatives are true and the other is false. Determine which alternative is false and write its letter in the blank to the left of the question number. On the blank line below alternative D, write the corrected version of the false statement.

_____ 1. A. The earliest-known land plants, Zosterophyllophyta and Rhiniophyta, are bryophytes.
B. Plants invaded land approximately 415 million years ago.
C. Some classification schemes include algae within Kingdom Plantae.
D. Plants are multicellular, photosynthetic, eukaryotic organisms.

Correction: _____

_____ 2. A. The plastids of plants contain chlorophylls *a* and *b*.
B. The cell walls of plant cells contain cellulose, a polysaccharide.
C. One of the problems that had to be solved during the invasion of land by plants was dehydration of plant tissues.
D. Ancient aquatic plants required free water through which sperm could swim to fertilize ova, but modern bryophytes do not rely on external water for reproduction.

Correction: _____

_____ 3. A. Vascular tissues serve the dual roles of supporting the plant body and of conducting food and water throughout the plant.
B. The waxy cuticle on many plant surfaces retards loss of water from tissues.
C. Some modern plants shed gametes into the atmosphere.
D. The invasion of land by plants probably occurred after land was invaded by insects and vertebrates.

Correction: _____

_____ 4. A. In the less-advanced plants, the sporophyte is the predominant phase of the life cycle.
B. In ferns, the sporophyte develops directly upon the gametophyte.
C. In all of the following groups, motile sperm swim to sites in which ova are located: mosses, ferns, lycopsids, ginkgos.
D. Bryophytes never achieved large body sizes, probably because they lack vascular tissues.

Correction: _____

_____ 5. A. In advanced tracheophytes, the gametophyte is not photosynthetic.
B. *Psilotum* is a modern representative of a group (Psilophyta) that dominated the landscape during Silurian and Devonian times.
C. The "coal forests" of Carboniferous times were composed mainly of gymnosperms.
D. Lycopsids have true roots, stems, and leaves.

Correction: _____

_____ 6. A. Some of the extinct species of all of the following groups grew as large trees: Sphenophyta, Lycopodiopsida, Bryophyta.
B. A microphyll of a sphenopsid has a single vascular channel.
C. Modern flowering plants have megaphylls with complex vein systems.
D. Sporangia are borne on the fronds of the sporophytes of ferns.

Correction: _____

_____ 7. A. Rhizoids are the root-like filaments of the gametophytes of pteridopsids.
B. Separate archegonia and antheridia are borne on the gametophytes of ferns.
C. Flagellated ova of ferns swim through an external watery medium to fertilize sperm cells located in antheridia.
D. Some species of ferns grow as trees.

Correction: _____

_____ 8. A. In seed plants, the germinated pollen grain (including the pollen tube) represents the male gametophyte.
B. Male gametophytes of spermopsids are highly resistant to dehydration, a problem that has limited most other tracheophytes to moist habitats.
C. The female gametophytes of spermopsids are photosynthetic.
D. The embryonic sporophytes of spermopsids are contained within seeds.

Correction: _____

_____ 9. A. Seeds evolved earlier in geological time than did flowers.
B. Seeds are borne on carpels.
C. Plants with flowers and with seeds encased in an ovary are angiosperms.
D. Seeds of gymnosperms are surrounded by ovarian tissues that are actually tissues of the parental plant.

Correction: _____

_____ 10. A. The seed ferns, cycadeoids, and cordaites are extinct groups of gymnosperms.
B. Gingkos and cycads are extant groups of gymnosperms.
C. Cycads and conifers are similar in that both groups have two types of strobili, one that produces pollen and another that produces ova.
D. Conifers probably evolved from the ancient group Ginkgopsida.

Correction: _____

_____ 11. A. The only group of Spermophyta possessing motile sperm is Class Cordaitopsida.
B. Most conifers are evergreens.
C. Cones are clusters of modified leaves borne on a short stem.
D. The male gametophyte of a conifer consists of only a pollen grain and pollen tube containing a total of six nuclei.

Correction: _____

_____ 12. A. Gymnosperms are a more successful group of land plants than are angiosperms.
   B. Angiosperms possess both flowers and seeds.
   C. The distribution of pollen by gymnosperms is haphazard and wasteful in comparison to means by which pollen is distributed by most angiosperms.
   D. The tendency of many angiosperm species to be self-incompatible has enhanced genetic diversity.

   Correction: _____

_____ 13. A. An imperfect flower is one that contains only male or female structures.
   B. A staminate flower contains only female structures.
   C. The expansion of the diversity of angiosperms coincided with the decline of gymnosperms.
   D. Monocotyledonous species have longitudinal, parallel venation.

   Correction: _____

_____ 14. A. Vascular tissues of dicots are arranged in cylindrical bundles.
   B. In monocots, floral parts are arranged in multiples of three.
   C. Cotyledons are seed leaves in which the nutrient supply for the embryo is stored.
   D. Dicots are one of the two major groups of gymnosperms.

   Correction: _____

_____ 15. A. Xylem transports water and salts.
   B. Roots serve to anchor plants into the ground and to obtain water and nutrients from the soil.
   C. Sexual reproduction is known for all major plant groups except Pteridospermopsida.
   D. In most bryophyte species, the sporophyte is essentially a parasite of the gametophyte.

   Correction: _____

CHAPTER 42--INVERTEBRATE ANIMALS

CHAPTER OVERVIEW

In the next two chapters, we will survey the diversity, adaptations, and evolutionary relationships of species in Kingdom Animalia. We will examine the characteristics that define an organism as an animal; in doing so, we will see that distinguishing between animal and plant is not always a simple matter.

Chapter 42 focusses on animals that lack a vertebral column, the invertebrates. Such invertebrates comprise the vast majority of animal species. Chordates and other vertebrates that make up the balance are treated in Chapter 43.

TOPIC SUMMARIES

A. <u>What Is An Animal</u>?--Animals are <u>multicellular</u>. Although some animals are sessile (attached to substrate) during much of their lives (e.g., sponges), most are <u>mobile</u> for their entire lives. All species of animals are mobile during at least some phase of their lives.

Animals are <u>heterotrophs</u>. For most species, ingested food is digested within an internal cavity. When compared to plants, <u>growth</u> in animals generally is restricted to early phases of life. Animal cells <u>lack rigid cell walls</u> and usually <u>lack vacuoles</u>. Many animals possess specialized structures and systems (e.g., kidneys) for <u>regulating the internal environment</u>. Presence of muscles and nerves allows animals to be much more <u>responsive</u> to environmental stimuli than are plants. With only a few exceptions (especially among insects), <u>gametes represent the only haploid phase</u> of the life cycle in animals.

B. <u>Diversity of Animals</u>.--Over two million species of living animals have been described. Certainly, many more remain to be discovered, and many more are known only as fossils. A true appreciation and understanding of the diversity and relationships of animals comes with many years of study; the same can be said for organisms of other kingdoms. Perhaps the best way to begin such a study of diversity is to examine various <u>trends</u> that have marked the evolution of the group. The ways in which certain characters are expressed in a particular group indicates that group's position along the evolutionary continuum.

C. <u>Evolutionary Trends in Kingdom Animalia</u>. Along the ancestral-to-derived continuum, animals have become <u>structurally more complex</u>. A <u>division of labors</u> arose, with different types of cells (e.g., muscle cells, neurons, etc.) becoming specialized to carry out different functions. <u>Tissues</u> (groups of cells of mixed type that work toward the same function) evolved as functions became more complex. Eventually, <u>organs and organ systems</u> arose. The <u>complexity of embryonic tissues</u> increased, from ill-defined cell layers to forms with two and three distinct cell layers. The ancestral <u>digestive system</u> is merely a simple pouch with only one opening, whereas derived digestive systems have openings at both ends and contain various chambers with specialized digestive functions. Body forms lacking a true <u>body cavity</u> (acoelomates) evolved into forms with lined <u>coeloms</u> (coelomates). <u>Support of the body</u> may not involve a skeleton (ancestral) or might involve an exoskeleton or endoskeleton (derived). As body size increased (and surface-to-volume ratio became unfavorable), systems (e.g., circulatory, respiratory) evolved <u>to deliver materials</u> to all body cells. Derived body designs exhibit <u>segmentation</u> and/or <u>appendages</u>. <u>Symmetry</u> shifted from radial (ancestral) to bilateral (derived). Systems for perceiving stimuli (<u>sensory structures</u>) and for processing information (<u>nervous systems</u>) evolved with increasing body complexity and increasingly active lifestyles.

D. <u>Survey of Selected Phyla</u>.--In the following section, we'll survey some of the major features of structure and natural history of the more diverse phyla of Kingdom Animalia. You will find textbook Table 42-1 to be a useful summary of additional features of these phyla.

(1) Phylum Porifera. <u>Sponges</u> are aquatic (mainly marine) animals that spend their lives attached to the substrate. Their strategy of feeding is to bring food (water-borne organic materials; plankton) to the animal rather than to go after food. Flagellated choanocytes lining the body cavity establish a current that flows through the body. Division of labors among cells is slight; the body is little more than an aggregation of relatively independent cells that function in a somewhat coordinated fashion. These and other features suggest that sponges represent a lineage (Parazoa) that long ago split away from the rest of Animalia (Fig. 42.1).

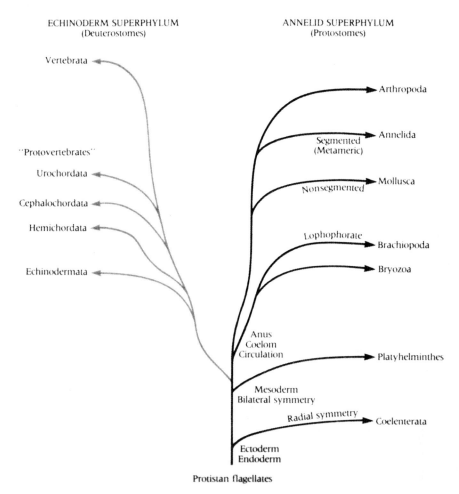

Fig. 42.1: The general phylogeny of animals

(2) Phylum Coelenterata. <u>Jellyfishes</u> (Class Scyphozoa), <u>corals</u> (Class Anthozoa), and <u>hydras</u> (Class Hydrozoa) are aquatic (mainly marine) species having two embryonic germ layers and a digestive system consisting of a blind sac, the <u>gastrovascular cavity</u>. Specialized tissues and organs are defined. Food-gathering involves capture of prey by stinging cells (<u>nematocytes</u>) that discharge small "harpoons" (nematocysts). The life cycles of coelenterates entail alternation between a sessile <u>polyp</u> and the motile <u>medusa</u>.

(3) Phylum Platyhelminthes. The ribbon-like body form of <u>flatworms</u> represents an evolutionary response to the need to provide adequate surface area to enable exchange of

materials between the body and environment. Flatworms are bilaterally symmetric and lack a body cavity (i.e., are acoelomate). Classes Cestoda (tapeworms) and Trematoda (flukes) include species that are parasitic.

(4) Phylum Aschelminthes. The roundworms (Class Nematoda) and their relatives differ from flatworms in their rounder shape, in the presence of a complete digestive tract (having both anterior and posterior openings), and a pseudocoel, a type of body cavity delimited by mesoderm of the body wall and endoderm of the gut. Many species are parasites of humans, producing diseases including trichinosis.

(5) Phyla Ectoprocta and Entoprocta. Formerly, these phyla were grouped as bryozoans ("moss animals") because of their superficial resemblance to mosses. Ectoprocts and entoprocts occur in mats encrusting shells and rocks along the bottom of shallow seas and other aquatic situations. Individuals live in cup-like depressions of a limy skeletal framework. Ectoprocts have tentacles (comprising the lophophore) bearing cilia that beat rhythmically to establish a current that carries food into the stomach. In entoprocts, the cilia beat in uncoordinated fashion. The digestive system has two openings. Simple muscular and nervous systems are present. Most species have a true coelom, a cavity lined by mesoderm.

(6) Phylum Brachiopoda. The brachiopods are another phylum of lophophorate animals. These live individually in marine habitats. Only about 300 species remain from a diverse heritage that included over 30,000 species. Brachiopods possess a two-valved shell, one above and one below. They attach to the substrate by a fleshy stalk. Their body organization is more complex than for the phyla considered thus far in this chapter. A true coelom and well-developed nervous, muscular, digestive, circulatory, excretory, and reproductive systems are present.

(7) Phylum Mollusca. Including over 100,000 species, this phylum is the second most diverse group of invertebrates. Their diversity is apparent in the following list of classes: Gastropoda (snails and relatives), Bivalvia (clams and relatives), Cephalopoda (octopus, squids, and relatives), Amphineura (chitons), Aplacophora (solenogasters; see footnote 7 in textbook), Monoplacophora (monoplacophorans), and Scaphopoda (tusk shells). Refer to the textbook for distinguishing features of species of these seven classes of molluscs.

Despite the structural diversity of molluscs, certain features are shared throughout the phylum: The body is not segmented, and usually there is an anterior head, a ventral foot, and a dorsal visceral mass (including well-developed organ systems found in other animals). The visceral mass is covered by a mantle that often includes glands that secrete a shell. Gills, for respiration, lie within the mantle cavity.

(8) Phylum Annelida. The annelids include the segmented worms and leeches. The bodies of annelids consist of repeated ring-like segments to which setae (bristles) are attached. Organs of various systems (e.g., circular blood vessels, nephridia) are repeated in each segment, whereas those of other systems (nerve cords, some blood vessels) run the length of the body without evidence of segmentation. A complete digestive system courses from the anterior mouth to the posterior anus. Annelids probably evolved from flatworm ancestors.

Class Oligochaeta includes terrestrial forms (e.g., earthworms) and fresh-water forms. Leeches and various other semiparasitic species are grouped into Class Hirudinea. The majority (about 6,000 species) of annelids belongs to Class Polychaeta; polychaetes are marine forms, some of which are free-swimming with others residing in burrows or tubes.

(9) Phylum Arthropoda. The most diverse of all animal phyla (perhaps 10 million species, most yet to be described!), the arthropods probably evolved from ancestral annelid

worms. Arthropods occur in every imaginable habitat in the biosphere. Crustaceans, spiders, and insects are united into this phylum by the presence of jointed, movable legs (hence, the name of the phylum) and an exoskeleton. Other improvements on the annelid body plan include defined muscle groups mechanically related to movement of specific body parts, development of jaws, and nervous systems and sensory structures (e.g., eyes, antennae) of greater complexity.

Thirteen classes of arthropods are recognized. The trilobites (Class Trilobita) and giant water scorpions (Class Eurypterida) are extinct. Centipedes (Class Chilopoda) and millipedes (Class Diplodopa) are collectively referred to as myriapods; two other minor classes (Pauropoda and Symphyla) include centipedelike animals with fewer legs. Crabs, lobsters, shrimps, and other crustaceans belong to Class Crustacea. Crustaceans and insects (Class Insecta) comprise the subphylum Mandibulata (distinguished from other arthropods by the presence of mandibles and sensory antennae). Spiders, ticks, and their relatives form Class Arachnida. Arachnids plus Class Merostomata (horseshoe crabs) and two other classes make up subphylum Chelicarata (arthropods lacking antennae, presence of chelicarae, and division of body into cephalothorax and abdomen). This brief summary cannot do justice to arthropods; study the diversity of and adaptations in this group as presented in your textbook.

(10) Phylum Echinodermata. All echinoderms occupy marine habitats. Among the most highly evolved invertebrates, they possess some features suggesting a possible role in the evolution of chordates (includes vertebrates; see next chapter). Although their larvae are bilaterally symmetric, adults possess radial symmetry. From the outer surfaces of echinoderms protrude spines, hence the name of the phylum. Echinoderms possess a complete digestive tube, a coelom, and specialized excretory, reproductive, nervous, and circulatory systems. A water-vascular system, unique to echinoderms, courses throughout the body. Bulbed structures (tube feet) at the ends of this system operate hydraulically and by muscle action in locomotion and in grasping (e.g., supplying force needed by sea stars to open bivalve shells).

Five classes of extant echinoderms are recognized. Class Crinoidea includes the sea lilies and feather stars; crinoids may be sessile or free-swimming, they lack tube feet, and are housed in a calcareous box. Sea stars (Class Asteroidea) are perhaps the best-known echinoderms. Brittle stars (Class Ophiuroidea) differ from sea stars in having long, slender arms. Class Echinoidea includes sea urchins and sand dollars. Sea cucumbers comprise Class Holothuroidea. Refer to the textbook for more details distinguishing these classes.

TERMS TO UNDERSTAND: Define and identify the following terms:

choanocytes

amoeboid cells

spongin

Parazoa

nematocytes

polyp

medusa

hydrozoans

scyphozoans

anthozoans

acoelomate

turbellarians

planarians

tapeworms

flukes

nematodes

pseudocoelom

bryozoans

lophophore

radula

ctenidia

mantle

shell

mantle cavity

chitons

tusk shells

operculum

anopedal flexure

nudibranchs

prosobranchs

pulmonates

chromatophores

setae

cuticle

trilobites

myriapods

antennae

mandibles

maxillae

molting

barnacles

thorax

sclerotin

epicuticle

apterygotes

hemimetabolous

chelicerae

pedipalps

cephalothorax

abdomen

book gills

eurypterids

water-vascular system

sieve plate

tube feet

DISCUSSION QUESTIONS

1. Using specific examples, describe the evolutionary trends seen among phyla of invertebrates in the degree of division of labors among cells and tissues.

2. What evolutionary trends are evident in the evolution of invertebrates for methods of food gathering? Use specific examples.

3. Compare the ecological diversity of species in Class Insecta with that of organisms in Phylum Ectoprocta.

4. Review the anatomy and natural history of sponges. In what ways are sponges like plants? In what ways are they like animals? In which kingdom are sponges most appropriately classified?

5. Compare and contrast the structure and natural history of brachiopods and of bivalves.

6. Adults of some animal species are sessile, whereas others are motile. Examine the advantages conferred by sessile and motile lifestyles. Use specific examples.

MATCHING EXERCISE

Complete the following matching exercise to check your familiarity with characteristics and examples of the various invertebrate phyla of Kingdom Animalia. Correct answers must include all characteristics and examples from the right column that match the taxon in the left column. Some characteristics might apply to more than one taxon.

TAXON   CHARACTERISTICS & EXAMPLES

_____ 1. Aschelminthes   A. coelomate

_____ 2. Mollusca   B. pseudocoelomate

_____ 3. Echinodermata   C. digestive system with anterior and posterior openings

_____ 4. Coelenterata   D. segmented

_____ 5. Entoprocta   E. adults with radial symmetry

_____ 6. Brachiopoda   F. adults of at least some species of taxon are sessile

_____ 7. Platyhelminthes   G. all species of taxon are aquatic

_____ 8. Porifera   H. at least some species of taxon are terrestrial

_____ 9. Annelida   I. all species of taxon are extinct

TESTING YOUR UNDERSTANDING

For each of the test items below, three of the lettered alternatives are true and the other is false. Determine which alternative is false and write its letter in the blank to the left of the question number. On the blank line below alternative D, write the corrected version of the false statement.

_____ 1. A. In flatworms, respiratory gases reach body cells by diffusion across the body wall, whereas in segmented worms such gases are transported in a closed circulatory system.
B. All animals are heterotrophs.
C. All plants are photosynthetic.
D. Most plants are sessile, whereas all animals are motile during at least some phase of their lives.

Correction: _____

_____ 2. A. Because of the presence of special sensory structures and muscle tissue in animals, animals are more responsive to environmental stimuli than are plants.
B. The life cycles of most animals include both haploid and diploid phases.
C. The predominant phase of the life cycle in advanced animals is the diploid phase.
D. The predominant phase of the life cycle in advanced plants is the haploid phase.

Correction: _____

_____ 3. A. The most diverse phylum of invertebrate animals is Arthropoda.
   B. Species in the following phyla occupy aquatic habitats: Porifera, Ectoprocta, Echinodermata, Aschelminthes.
   C. One of the least diverse phyla of invertebrates is Entoprocta.
   D. All species of phylum Trilobita are extinct.

   Correction: _____

_____ 4. A. Mesozoans are characterized by unusually small bodies, consisting of only about 30 cells.
   B. A trend evident in evolution of animals from ancestral protozoans is from unicellularity to multicellularity.
   C. Most species of terrestrial invertebrates possess skeletons, whereas most species of aquatic invertebrates lack skeletons.
   D. A major trend seen in evolution of digestive tracts of invertebrates is from a tract with two openings to a tract with one opening.

   Correction: _____

_____ 5. A. The presence of radial symmetry in echinoderms indicates that these species are primitive invertebrates.
   B. The presence of segmentation in some animals is a derived feature.
   C. Materials produced by amoeboid cells of sponges include spongin and spicules made of calcium carbonate or silica.
   D. Sponges seem to have evolved from protists independently of other animals.

   Correction: _____

_____ 6. A. Unlike most coelenterate species, *Hydra* resides in fresh water.
   B. Nematocytes are the harpoon-like structures housed in nematocysts of coelenterates.
   C. Both coelenterates and poriferans are radially symmetric.
   D. Nervous tissue is present in species of these phyla: Coelenterata, Mollusca, and Arthropoda.

   Correction: _____

_____ 7. A. All of the following are classes of phylum Coelenterata: Anthozoa, Scyphozoa, Monoplacophora, Hydrozoa.
   B. Some species of phylum Aschelminthes are parasitic to humans.
   C. Turbellarians are acoelous, whereas nematodes are pseudocoelous.
   D. Ectoprocts and entoprocts sometimes are grouped as the bryozoans.

   Correction: _____

_____ 8. A. A lophophore is an organ that bears numerous tentacles used in food gathering.
   B. Both corals and ectoprocts produce calcareous exoskeletons that accumulate to form reefs.
   C. Both bivalves and brachiopods possess shells consisting of two parts.
   D. Like bivalves, most brachiopods are sessile, attached to the substrate by a fleshy stalk.

   Correction: _____

_____ 9. A. The following are anatomical features of many species of molluscs: mantle, visceral mass, radula, foot.
B. Body segmentation is apparent in some species of molluscs.
C. The operculum is a structure that prevents dehydration of tissues in many species of molluscs.
D. Squids and octopuses feed on plankton that they filter from the surrounding water.

Correction: _____

_____ 10. A. Choanocytes are specialized cells that contain pigments that color the skin.
B. Nervous and sensory structures are well developed in certain species of cephalopods.
C. Unlike shells of snails, the space in the shells of nautiluses is partitioned into many chambers.
D. Annelids possess complete digestive tracts, and their excretory structures are repeated in each body segment.

Correction: _____

_____ 11. A. Phylum Annelida includes both terrestrial and aquatic species.
B. Locomotion in oligochaetes is assisted by setae.
C. Leeches are parasites classified into Class Hirudinea.
D. Annelids probably evolved from ancestral arthropods.

Correction: _____

_____ 12. A. Two features that contributed to the tremendous success of arthropods are jointed legs and a durable exoskeleton.
B. Subphylum Mandibulata consists of arthropods belonging to classes Insecta and Crustacea.
C. Arachnids characteristically possess two pairs of antennae, a pair of mandibles, and two pairs of maxillae.
D. The bodies of insects consist of three segments, whereas bodies of chelicarates consist of a cephalothorax and an abdomen.

Correction: _____

_____ 13. A. Insects occupy every imaginable habitat on land and in fresh water.
B. In insects, most respiratory gases reach body cells via the circulatory system.
C. The cuticle of insects is composed of chitin and a protein called sclerotin.
D. Apterygotes are more primitive than are the winged insects.

Correction: _____

_____ 14. A. Holometabolous insects have a life cycle in which an individual passes through only a few of the stages of metamorphosis.
B. Arachnids possess four pairs of legs, whereas insects have three pairs of legs.
C. All species of spiders are carnivorous.
D. Respiratory gas exchange in horseshoe crabs is by book gills.

Correction: _____

CHAPTER 43--CHORDATES AND ANIMAL PHYLOGENY

CHAPTER OVERVIEW

This chapter is our second installment examining the diversity of animals. Here, we focus on Phylum Chordata, the phylum into which we are classified. Most chordates have vertebral columns and, hence, are called vertebrates. A few chordates, lack this feature and, therefore, are referred to as the invertebrate chordates. Though adults of the vertebrate and invertebrate chordates bear little resemblance to each other, their evolutionary relationships become apparent through study of their embryos. We'll spend much of this chapter surveying the diversity and adaptations of chordates.

Chapter 43 concludes by summarizing the major trends seen in the evolution of Kingdom Animalia. These trends are placed into the "bigger picture" of how multicellular organisms might have arisen from unicellular protists. The evolutionary relationships among animal phyla are presented and depicted in a phylogenetic tree.

TOPIC SUMMARIES

A. <u>Phylum Chordata</u>.--This phylum includes seven extant classes of animals (the <u>vertebrates</u>) that are familiar to most of us and three not-so-familiar groups, the <u>nonvertebrate chordates</u>. Union of these superficially dissimilar groups is based on three essential features that are present during some, if not all, stages of life: All chordates possess a <u>notochord</u>, a flexible, yet stiff rod for supporting an elongated body. All chordates possess a <u>hollow nerve cord located dorsally</u>. All chordates possess <u>gill slits</u> in the walls of the pharynx (throat region).

B. <u>Invertebrate Chordates</u>.--The <u>sea squirts</u> or <u>tunicates</u> (subphylum <u>Urochordata</u>) form one of the subphyla of chordates that lack a vertebral column. Adults, which are sessile blob-like forms, bear little resemblance to vertebrates. However, <u>tunicate larvae</u> possess tails supported by a notochord.

Adult <u>lancets</u> (subphylum <u>Cephalochordata</u>) have a notochord running virtually the full length of the body. <u>Amphioxus</u> (genus *Branchiostoma*) is the common representative form. Cephalochordates are more similar to vertebrates than are the other invertebrate chordates.

Some textbooks (such as this text) recognize a third subphylum of invertebrate chordates, <u>Hemichordata</u>. The <u>acorn worms</u> and their relatives possess a rod-like, supporting structure in the <u>proboscis</u>; this structure is viewed by some systematists as homologous to the notochord, whereas others do not recognize this apparent homology and, therefore, place acorn worms into a phylum of their own. These and the other invertebrate chordates occupy aquatic (primarily marine) habitats and are filter feeders.

C. <u>Subphylum Vertebrata</u>.--Vertebrates are, perhaps, the most progressive of all animals; generally, they (we) lead highly mobile lifestyles and have highly developed sensory structures and nervous systems. Vertebrates have achieved a pinnacle in <u>cephalization</u>, the positioning of sensory structures and brain towards the anterior of the body. Vertebrates possess closed circulatory systems, and they have segmented bodies.

In the sections below, we'll survey selected classes of vertebrates, which can be grouped into two categories (Fig. 43.1). <u>Pisces</u> includes the fishes, classes Agnatha, Placodermi, Chondrichthyes, and Osteichthyes, groups whose members generally are finned and breathe via gills. The four-limbed lung-breathing mainly terrestrial vertebrates are grouped as <u>Tetrapoda</u>.

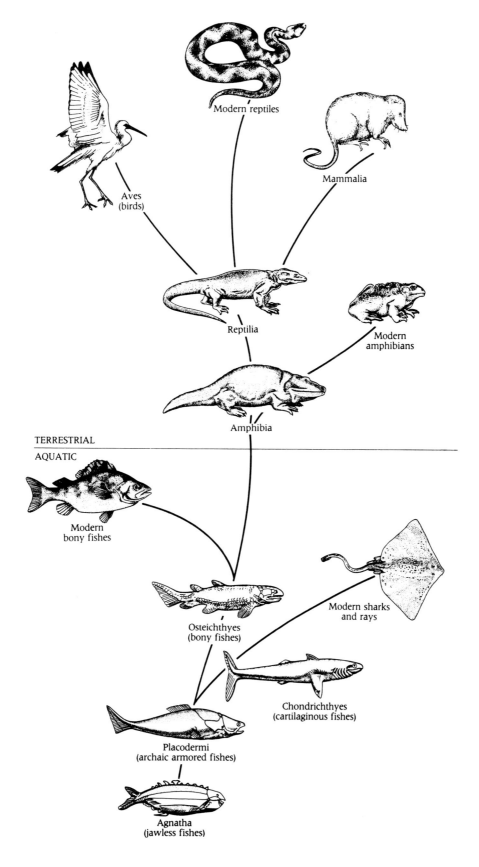

Fig. 43.1: The general phylogeny of vertebrate animals

D. <u>Class Agnatha</u>.--Appearing in Ordovician times, the first vertebrates lacked jaws and lacked paired fins. Though much more diverse in the past, a few forms survive today. Modern forms include <u>lampreys</u>, that feed parasitically on the body fluids of living fishes, and <u>hagfishes</u>, that scavenge for food. Ancestral forms probably were filter feeders. The larva, called <u>ammocoetes</u>, has anatomical similarities with the cephalochordates.

E. <u>Class Placodermi</u>.--This extinct group is one of several classes of armored fishes. Placoderms may have been the first group of vertebrates to evolve jaws. Their skeletons were made of bone. Refer back to Chapter 34 for an account of how jaws probably evolved from a pair of gill arches.

F. <u>Class Chondrichthyes</u>.--Two successful groups of fishes arose from the placoderms, the bony fishes (see G below) and the cartilaginous fishes, class <u>Chondrichthyes</u>. During evolution of this group, <u>cartilaginous skeletons</u> (a primary feature of this group) supplanted bony skeletons. <u>Sharks</u>, <u>skates</u>, and <u>rays</u> are the commonly known members of this class; less familiar are the <u>chimaeras</u>. Chondrichthyans possess jaws and paired appendages (fins). Teeth of these fishes probably evolved from a peculiar type of scale (dermal <u>denticles</u>) located near the mouth opening. These fishes sequester urea in their tissues to bring body fluids into osmotic equilibrium with the surrounding seawater. They lack lungs or swim bladders.

G. <u>Class Osteichthyes</u>.--Bony fishes (<u>Osteichthyes</u>) make up nearly half of the over 40,000 living species of vertebrates. Like arthropods and molluscs, osteichthyans occur in a tremendous array of sizes, shapes, and lifestyles. They occupy seemingly the full spectrum of available aquatic habitats. Bony fishes possess a feature absent in chondrichthyans, an <u>operculum</u>, a bony flap covering the external openings of the gill slits. Most possess either lungs or a swim bladder.

Two subclasses are recognized on the basis of structure of the paired fins. <u>Sarcopterygians</u> have fleshy fins; these fins probably were the evolutionary forerunners of limbs of terrestrial vertebrates. The coelacanth, *Latimeria,* is the only known living species of <u>crossopterygian</u>, the group of sarcopterygians that gave rise to land-dwelling vertebrates. Other living sarcopterygians include the <u>lungfishes</u> (order <u>Dipnoi</u>). The vast majority of bony fishes are <u>actinopterygians</u>, the ray-finned fishes. These fishes are divided into subclasses according to the degree of ossification of the skeleton.

H. <u>Class Amphibia</u>.--Amphibians lead a double-life; their reproduction and, to some extent, respiration are tied to water, although adults may lead a terrestrial existence. <u>Frogs and toads</u> (order Anura), <u>salamanders</u> (order Urodela), and <u>caecilians</u> (order Apoda) evolved from crossopterygian ancestors that invaded land early in osteichthyan history. The invaders of land crawled on fleshy fins and breathed with lungs.

I. <u>Class Reptilia</u>.--Evolving from early amphibians over 300 million years ago, <u>reptiles</u> became the first truly terrestrial vertebrates. Neither their respiration nor reproduction required them to occupy aquatic habitats. A key feature freeing reptiles from water was the <u>amniotic egg</u>, an egg having a membrane (<u>amnion</u>) that encloses the embryo within a miniature aquatic environment.

Only four of the 16 recognized orders of reptiles have living representatives. Lizards and snakes comprise order <u>Squamata</u>. <u>Chelonia</u> includes turtles and tortoises. Crocodiles, alligators, and their relatives make up order <u>Crocodilia</u>. The tuatara is the only extant species of order <u>Rhynchocephalia</u>. Two other orders included the ancestors of birds (<u>Thecodontia</u>) and of mammals (<u>Therapsida</u>).

J. <u>Class Aves</u>.--<u>Feathers</u> are the essential feature distinguishing birds from reptiles. Feathers function not only as flight surfaces, but also in insulation. The source of body heat in birds (and in mammals) is internal metabolism; birds and mammals are <u>homeotherms</u>. Rapid metabolism is required for the energetic demands of flight; feathers provide insulation to conserve body heat. Even the earliest known birds, *Archaeopteryx,* possessed feathers. Homeothermy is promoted further by a double-circuit circulatory system; separation of oxygenated from deoxygenated blood is facilitated by a four-chambered heart. Over 8,000 species of birds live today.

K. <u>Class Mammalia</u>.--The class that includes humans is named for a diagnostic feature, the <u>mammary glands</u> that produce milk to nourish young. Insulation in mammals, another class of homeotherms, is achieved by <u>hair</u>. Division of Mammalia is based on reproductive features. The <u>monotremes</u> (e.g., duck-billed platypus; <u>Prototheria</u>) lay eggs rather than give live birth; they also possess other reptilian characters. The remaining mammals are viviparous. Pouched mammals (<u>marsupials; Metatheria</u>) give birth to young that are not well-developed. "Placental" mammals (<u>Eutheria</u>) have longer gestation periods and give birth to well-developed young. The greater success of placentals over marsupials is attributed largely to differences in reproductive strategies.

The approximately 4,500 species of living mammals are divided into 16 orders. Over 40% of modern mammals are <u>rodents</u> (order Rodentia); <u>bats</u> (Chiroptera) are the next most diverse order. Humans are classified into order Primates.

L. <u>Basic Relationships Among Animal Phyla</u>.--Probably all phyla of animals arose in aquatic environments. All animal phyla have aquatic species. Only three animal phyla (Mollusca, Arthropoda, Chordata) contain species that are fully terrestrial; interestingly these are the three most diverse phyla. The evolutionary relationships among these phyla are manifest in various anatomical features. Let's review the major trends in certain characters:

<u>Tissues</u> are ill-defined in sponges. Two relatively distinct layers are apparent in coelenterates. Species of other phyla have embryos with three embryonic germs layers that undergo further differentiation during development.

No special <u>digestive structures</u> are present in Porifera. Coelenterates have a blind gastrovascular cavity with one opening. Most other major phyla have complete digestive systems with two openings.

Acoelomate species have no <u>body cavity</u> between the digestive cavity and the body wall. Aschelminths are pseudocoelomate, having a body cavity lacking a cellular lining. Most other phyla possess true coeloms, cavities with mesodermal linings.

<u>Circulatory systems</u> are absent in sponges, coelenterates, flatworms, aschelminths, and bryozoans. Most other animal taxa have closed circulatory systems in which blood flows through vessels.

Although <u>body segmentation</u> is apparent in three phyla, such segmentation probably arose independently two times. Annelids, in which organs are repeated in each segment, probably gave rise to arthropods, in which fewer segments are present and in which there is less repetition of organs by segments. Segmentation in chordates probably arose in relation to locomotion.

Generally, radial <u>body symmetry</u> is ancestral to bilateral symmetry. Presence of radial symmetry in adult echinoderms probably represents an evolutionary return to an ancestral condition; this is suggested by the bilateral symmetry of echinoderm larvae.

M. <u>Origins of Multicellular Organisms</u>.--Evolution of multicellular organisms (<u>metazoans</u>) from unicellular protistans probably occurred over a billion years ago. The unicellular ancestors of metazoans probably was a <u>flagellated protozoan</u>, much like modern zooflagellates. The method of locomotion seen in such flagellates (cells propelled by flagella) also is seen in male gametes (and perhaps other cell types) of virtually all animal species and in the more-primitive plants. The tendency of some flagellates to form colonies hints at multicellularity. After multicellular aggregates of organisms occurred, probably there was differentiation into various tissues that conferred various sensory and behavioral advantages.

Sponges, coelenterates, and ctenophores diverged from hypothetical colonial flagellates along a dead-end course as the <u>parazoans</u>. The other animal phyla comprise the Eumetazoa. Three hypotheses are offerred for the origin of <u>Eumetazoa</u>: One hypothesis suggests that metazoans, like sponges, evolved from colonial flagellates, the main difference being a greater degree of differentiation of individuals in the colony. A series of changes in shape and tissue organization (<u>blastaea</u> and <u>gastraea</u> stages of development) occurred to yield radial, and later bilateral, symmetry. A revision of this hypothesis was proposed when it was learned that gastrulation in coelenterates occurs not by invagination, but by inward wandering of cells to form a planula larva; hence, a <u>planuloid</u> stage was substituted for the gastraea stage.

The third hypothesis holds that metazoans evolved from unicellular protozoans having multiple nuclei. Multicellularity would result merely by partitioning the <u>syncytium</u> into individual cells. The greater complexity of more-advanced eumetazoans involved formation of a circulatory system, an anus, and a coelom.

N. <u>Protostomes and Deuterostomes</u>.--Study of larval features and of developmental mechanisms (particularly regarding whether the embryonic blastopore becomes the anus or mouth of the adult) has produced an understanding of the relationships of Eumetazoans (textbook Fig. 43-28; Study Guide Fig. 42.1).

The <u>Protostomia</u> includes, among others, phyla Ectoprocta, Brachiopoda, Mollusca, and Annelida. All of these phyla have the <u>trochophore</u> larva (Fig. 43.2). Larvae of flatworms are similar to trochophores. Although arthropods lack the trophophore larva, they are linked with the protostomes by other features. These related phyla comprise the <u>annelid superphylum</u>.

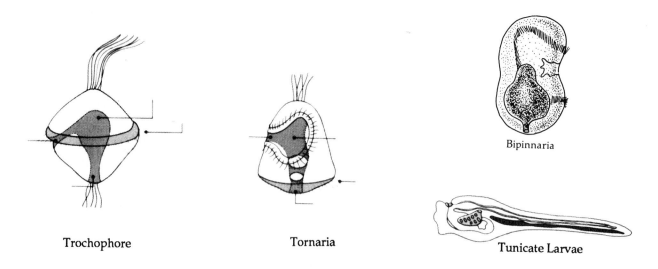

Trochophore   Tornaria   Bipinnaria

Tunicate Larvae

Fig. 43.2: Larvae of several groups of invertebrate animals

The echinoderms, chordates, and protochordates (Hemichordata, Cephalochordata, and Urochordata) form the echinoderm superphylum (= Deuterostomia). The tornaria larva (Fig. 43.2) of hemichordates is quite similar to the pluteus or bipinnaria larva (Fig. 43.2) of echinoderms. Generally, vertebrates lack free-swimming larvae (some fishes and amphibians have re-evolved larvae), hence their relationships with other deuterostomes is based on other features.

TERMS TO UNDERSTAND: Define and identify the following terms:

chordate

hemichordate

tunicate

cephalochordate

vertebrate

Pisces

Tetrapoda

ammocoete

placoderm

Chondrichthyes

denticles

chimaera

Osteichthyes

teleosts

Sarcopterygii

Actinopterygii

crossopterygians

coelacanth

swim bladder

chondrostean

Amphibia

anurans

caecilians

urodeles

Reptilia

chorion

allantois

amnion

dinosaurs

Rhynchocephalia

Chelonia

Crocodilia

Squamata

therapsids

thecodonts

Aves

feathers

Archaeopteryx

Mammalia

Theria

Prototheria

monotreme

marsupials

placentals

rodents

primates

blastaea

gastraea

planuloid

protostomes

deuterostomes

Eumetazoa

trochophore

tornaria

pluteus (= bipinnaria)

MATCHING EXERCISE

Complete the following matching exercise to check your familiarity with characteristics and examples of the various chordate phyla of Kingdom Animalia. Correct answers must include all characteristics and examples from the right column that match the taxon in the left column. Some characteristics might apply to more than one taxon.

TAXON

_____ 1. Mammalia

_____ 2. Amphibia

_____ 3. Cephalochordata

_____ 4. Hemichordata

_____ 5. Vertebrata

_____ 6. Urochordata

_____ 7. Aves

CHARACTERISTICS & EXAMPLES

A. pseudocoelomate

B. homeothermic

C. blastopore of embryo becomes anus of adult

D. taxon includes some fully-terrestrial species

E. display some degree of body segmentation

F. taxon includes some species having amniotic eggs

G. taxon includes some species that fly

DISCUSSION QUESTIONS

1. The ancestry of chordates is not certain. Summarize the evidence suggesting origin from (a) ancestors of annelids and (b) ancestors of echinoderms. Which of these two options do you think is the more likely ancestor of chordates? Why?

2. How is embryological evidence (including features of larvae) useful in understanding the relationships among animal phyla? Include specific examples in your answer.

3. The last several chapters have provided a survey of the diversity of life on Earth. With this overview in mind, reconsider the validity of viewing the diversity of life in terms of a "Ladder of Life."

4. Phylum Chordata is one of only three animal phyla including fully terrestrial species. What problems did the land-invading chordates face as they left water for land? How were these problems resolved? Who were the chordates that invaded land?

TESTING YOUR UNDERSTANDING

For each of the test items below, three of the lettered alternatives are true and the other is false. Determine which alternative is false and write its letter in the blank to the left of the question number. On the blank line below alternative D, write the corrected version of the false statement.

_____ 1. A. The notochord is a flexible stiffening rod located on the ventral surface of the body cavity of all chordates.
   B. Chordates possess a hollow, dorsal nerve cord.
   C. At some time during the life history of every chordate, pharyngeal gill slits are present.
   D. Chordates are coelomate animals with complete digestive tracts.

   Correction: _____

_____ 2. A. Tunicates, acorn worms, and lancelets occupy marine habitats.
B. The notochord of urochordates is present in adults but not in larvae.
C. Subphylum Vertebrata is the most speciose of all subphyla of the chordates.
D. The bodies of chordates are segmented.

Correction: _____

_____ 3. A. The following taxa are classified into superclass Tetrapoda: Aves, Reptilia, Osteichthyes, Amphibia.
B. The earliest vertebrates lacked jaws.
C. The earliest vertebrates lacked paired fins.
D. Placoderms possessed jaws.

Correction: _____

_____ 4. A. An evolutionary link between cephalochordates and agnathans is apparent in the similarity of adult amphioxus and ammocoetes.
B. Chondrichthyes is a group of extinct fishes.
C. The teeth of sharks probably evolved from scales called denticles.
D. Species having cartilaginous skeletons are included in Osteichthyes and in Chondrichthyes.

Correction: _____

_____ 5. A. Sharks use urea to maintain an osmotic balance between their tissues and the surrounding saltwater.
B. More species of vertebrates are classified as bony fishes than into any other vertebrate class.
C. Many of the earliest fishes were armored; a few modern species of fishes, such as gars, still possess bony armor.
D. The vertebrates that invaded land belonged to the taxon Actinopterygii.

Correction: _____

_____ 6. A. Some fishes breathe by using lungs rather than gills.
B. The operculum is the bony flap that covers the external openings of gills in bony fishes.
C. There are more species of terrestrial vertebrates than there are aquatic vertebrates.
D. Amphibians are not truly terrestrial because they must return to water to reproduce.

Correction: _____

_____ 7. A. The legless vertebrate species include snakes (Reptilia) and caecelians (Amphibia).
B. The invasion of land was facilitated by the evolution of the amniotic egg.
C. Reptiles and mammals evolved from different groups of amphibians.
D. Reptiles, birds, and mammals are referred to collectively as the amniotes.

Correction: _____

_____ 8. A. Species of flying vertebrates are included in classes Reptilia, Aves, and Mammalia.
B. Turtles are classified into order Chelonia, and lizards are classified into order Crocodilia.
C. Feathers, present only in birds, function to insulate the body and as flight surfaces.
D. Homeothermic physiology of birds and mammals is accompanied by a four-chambered design of the heart.

Correction: _____

_____ 9. A. Although ancestral birds possessed teeth, modern birds lack teeth.
B. Features present in mammals, but not in any other vertebrates, include hair and mammae.
C. Mammals evolved from thecodont reptiles.
D. Respiratory efficiency in mammals is enhanced by the diaphragm.

Correction: _____

_____ 10. A. Although most mammals are viviparous, a few species of mammals lay eggs.
B. The "pouched mammals" (marsupials) occur in Australia and in Africa.
C. There are more species of rodents than of any other order of mammals.
D. Humans are classified as humanoid primates.

Correction: _____

_____ 11. A. In terms of tissue differentiation, phylum Porifera is less advanced than is phylum Annelida.
B. In terms of complexity of the digestive tract, phylum Platyhelminthes is less advanced than is phylum Echinodermata.
C. In terms of type of body cavity, species in phylum Echinodermata and phylum Chordata are grouped together.
D. Presence of body segmentation is a less-advanced feature than is absence of body segmentation.

Correction: _____

_____ 12. A. The circulatory systems of mammals are closed.
B. The closed circulatory system evolved as body size and complexity increased.
C. The presence of radial symmetry in echinoderms and in poriferans implies a close evolutionary relationship between these two phyla.
D. The presence of body segmentation in annelids and in arthropods implies a close evolutionary relationship between these two phyla.

Correction: _____

_____ 13. A. Multicellular plants probably evolved from flagellated protozoans.
B. Multicellular animals probably evolved from flagellated protozoans.
C. Eumetazoans evolved from advanced sponges.
D. The protozoans from which metazoans evolved probably were colonial.

Correction: _____

_____ 14. A. Evolution of a coelom, an anus, and a more-complex circulatory system were important innovations in the evolution of eumetazoans.
B. Deuterostome phyla include Mollusca and Annelida.
C. One of the types of larvae found among protostome phyla is the trochophore.
D. The similarity of the tornaria and bipinnaria larvae suggests common ancestry of hemichordates and echinoderms.

Correction: _____

_____ 15. A. The close relationship of vertebrates with cephalochordates is based on similarity of the tunicate larvae with larvae of vertebrates.
B. Vertebrates probably evolved as neotenic tunicates.
C. Adult tunicates are sessile.
D. Dispersal of many kinds of kinds of invertebrates whose adults are sessile is by means of motile larvae.

Correction: _____

CHAPTER 44--HOW DIVERSITY AROSE: THE HISTORY OF LIFE

CHAPTER OVERVIEW

Chapter 44 concludes our study of the diversity of life. In the previous six chapters we have surveyed structure and natural history of major groups of organisms. Here, we roll together organisms of all five kingdoms with abiotic events in Earth history to develop a broad view of how the modern diversity of life arose.

It is a long, complex story that we must, of necessity, treat superficially. While developing this scenario, we recognize major features, tendencies, and processes that have operated during this history. We end the chapter by placing humans into this evolutionary perspective.

TOPIC SUMMARIES

A. <u>Earliest Life</u>.--In Chapter 4, we saw that organic molecules characteristic of life can be produced from inorganic precursors by abiotic means. We learned, too, that some such molecular systems are capable of self-reproduction. Eventually, one or more aggregates of such molecules became a true multimolecular organism, having the qualities of "life." We estimate that life was present on Earth by about 3.5 billion years ago.

A <u>key question</u> at this point is whether the earliest forms of life were autotrophic or heterotrophic. It has been assumed generally that <u>autotrophy</u> evolved first. After all, if there were no other organisms to consume, then what would a heterotroph do for sources of nutrients and energy? However, more recent thinking supports first appearance of <u>heterotrophs</u>: autotrophy requires cellular organization and metabolic capabilities that are much more complex than required for heterotrophy. Such early heterotrophs fed either on each other or, more likely, on abiotically formed molecules in their surroundings. Such heterotrophs were a factor that modified the molecular composition of the atmosphere: they consumed (or otherwise used) simple carbon compounds (methane, etc.) and produced carbon dioxide.

Apparently some of these heterotrophic forms mutated such that they could synthesize required complex compounds instead of having to find preformed compounds in the surrounding medium. This lead to evolution of <u>photosynthesis</u> (about 2.8 billion years ago) and progress toward nutritional self-sufficiency. Eventually, <u>carbon dioxide</u> became the primary carbon source. Atmospheric carbon dioxide content dropped as atmospheric <u>oxygen</u> content increased. Some of this oxygen was irradiated by <u>ultraviolet radiation</u> to form atmospheric <u>ozone</u>, a layer of which blocked ultraviolet radiation from impinging on Earth's surface. The ozone layer effectively prevented further abiotic synthesis of organic molecules, with the effect of <u>halting further spontaneous origin</u> of molecular systems that might again evolve into life. Hence, modern life on Earth stems from evolution from ancient unicellular autotrophs and heterotrophs. Eukaryotic cells probably appeared some time after 2 billion years ago.

B. <u>Precambrian and Cambrian Life</u>.--For the time before the Cambrian (which began about 600 million years ago; refer now and throughout this chapter to Fig. 44.1 to refresh the geological time table in your mind), fossils are rare. Those that have been found indicate that prokaryotes (bacteria) were present about 3.2 billion years ago. The earliest known occurrence of blue-green algae was about 2 billion years ago. Trace fossils of simple invertebrate animals are known from about 850 million years ago. The Cambrian is marked by abundant fossils of marine plants and marine invertebrates.

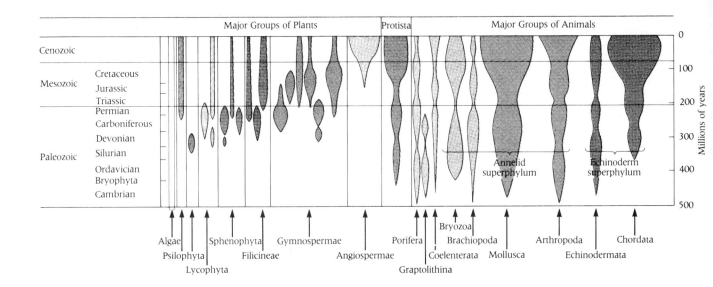

Fig. 44.1: Geological time table showing broad features of phylogeny of life

C. <u>Tendencies Characterizing the History of Life</u>.--One of the generalizations apparent from survey of the history of life is that <u>life has expanded</u>--in total number of organisms, in bulk of biomass, and in diversity of kinds. Expansion has not been continuous; <u>adaptive radiations</u> produced rapid expansions, whereas <u>extinctions</u> produced rapid decreases in rate of expansion.

The <u>availability of new habitats</u> and the occupation of new environments by organisms is a major reason for general expansion. Perhaps the greatest expansions accompanied the invasions of land by many groups of organisms on different occasions. With increasing length of time that a new environment was occupied, <u>niches became more finely subdivided</u> and <u>organisms became increasingly specialized</u> to occupy those niches. On many occasions when organisms have comfortably become stuck in an "ecological rut," some other species that is better adapted to that rut <u>replaces</u> the current occupant. Another trend that is sometimes seen as a factor in expansion is <u>increasing structural complication and functional improvement</u>.

D. <u>Ancient Seas and the Conquest of the Land</u>.--Before the invasion of land by metazoans, a striking diversity of plant and animal life existed in the seas. Major animals of <u>Cambrian</u> seas were trilobites, brachiopods, monoplacophorans, and eocrinoids. Major <u>Ordovician</u> groups inlcuded protists (flagellates, foraminiferans, radiolarians) and animals (sponges, coelenterates, graptolites, ectoproctans, brachiopods, mollusks, arthropods, echinoderms, agnathan fishes).

The <u>Silurian and Devonian</u> brought great diversification of fishes, not only agnathans (armored ostracoderms), but various jawed fishes (armored placoderms, chondrichthyans, osteichthyans) as well. Even in the Devonian, bony fishes were differentiated into three groups: lung fishes, paleoniscids, and crossopterygians. Among invertebrates, ammonites diversified during the Devonian.

<u>Invasion of land</u> occurred successfully during the Silurian. Plants, mainly small leafless psilopsids, were the earliest vascular plants. Psilopsids flourished, but gave way to lycopsids, sphenopsids, and ferns by late Devonian. Various invertebrate animals (arthropods and probably some wormlike phyla) invaded land probably soon after plants were established. During the Devonian, arthropods became better established on land and crossopterygian fishes invaded to produce the first nonfish vertebrates, amphibians.

The <u>Carboniferous and Permian</u> periods of the late Paleozoic were times of giant swamps and forests composed mainly of lycopsids, seed ferns, cordaites, and conifers. The remains of these plants were preserved as various fossil fuels (peat, petroleum, natural gas). These forests and swamps teemed with terrestrial invertebrates, amphibians, and early reptiles. The end of this era was marked by the Permo-Triassic crisis, when many groups declined (brachiopods, early tracheophytes) and many went extinct (trilobites). Among plants, replacement groups that expanded into the Mesozoic included ammonites, cycadeoids, cycads, ginkgos, and conifers.

The <u>Mesozoic Era</u> was a time of tremendous diversification of reptiles; this was the age of reptiles, primarily dinosaurs. Although birds (from thecodont reptiles) and mammals (from therapsid reptiles) were present since early Mesozoic, their diversification was suppressed until the end of the Mesozoic when the "<u>great dying</u>" occurred. Dinosaurs, certain other reptiles, various marine invertebrates, and certain plants suffered mass extinctions. Much scientific speculation has been directed towards explaining the cause(s) of this Cretaceous-Tertiary extinction.

E. <u>Modernization of the Living World</u>.--Most of the major groups of <u>aquatic plants, protists, and prokaryotes</u> probably have been present for several hundreds of millions of years. The main recent change is the appearance of diatoms in the Jurassic; diatoms are major components of aquatic floras today.

Only a few of our modern <u>invertebrate groups</u> were present in the Paleozoic; these include corals, ectoprocts, and clams. Most modern invertebrate groups represent replacements for extinct forms. The successful modern <u>fishes</u> (mainly teleosts and chondrichthyans) largely represent offshoots of archaic Paleozoic groups that either are extinct or are present as relicts. Many groups of archaic <u>amphibians</u> were lost in the Permo-Triassic crisis. Several groups of <u>marine reptiles</u> (mosasaurs, plesiosaurs, ichthyosaurs) were diverse during the Mesozoic; these interestingly were lost without replacement by other reptiles. Several groups of <u>mammals</u> (whales, seals, sea cows) re-invaded water during the early Tertiary.

Modernization occurred on land as well as in water. A nearly modern flora largely of conifers and flowering plants existed by the end of the Cretaceous. Likewise, most major groups of insects in today's fauna were present and in modern form by late Cretaceous. Birds of essentially modern structure evolved by the end of the Cretaceous. Diversification of birds has continued during the Cenozoic, which could as reasonably be called the "age of birds" as the "age of mammals." The early Cenozoic was a time of rapid radiation of mammals, a group that seems to have been held in check (perhaps by competively superior reptiles) until the "great dying."

F. <u>History of Mammals</u>.--The Mesozoic represents the first two-thirds of mammalian history; it is the remaining one-third of mammalian history (Cenozoic) about which we know most. Placental mammals rapidly radiated beginning in the Paleocene, about 70 million years ago. The modern orders of mammals had evolved by the Eocene, beginning about 58 million years ago. Modern families of mammals appeared by the end of the Oligocene, which ended about 25 million years ago. The Eocene diversity of mammals (at the level of order) was greater than before or after because both declining orders and new orders were present.

Much of what we know of the mechanisms of evolution is based on changes recorded in the fossil record. The evolution of horses is unusually well documented. Examine textbook Figs. 44-30 and 44-31 to see how various anatomical features changed over time. What do such changes tell us about lifestyles of these horses?

G. <u>Phylogeny of Primates</u>.--It is natural for us as humans to be interested in the events that resulted in *Homo sapiens*. We are members of order <u>Primates</u>. Hence, our history is part of the history of this order. The earliest primates known are of Cretaceous age; these departed from the rest of the mammals by becoming <u>arboreal</u> (tree-dwelling). Early primates probably more-closely resembled <u>prosimians</u> than they did <u>anthropoids</u>. The only remaining prosimians are the tarsiers, lemurs, and a few others largely restricted to Madagascar. The more-familiar primates, the anthropoids, diverged from ancestral prosimians in the Oligocene. Family <u>Hominidae</u>, of which humans are the only living representative, appeared in the Miocene as an offshoot of earlier Old World hominoids. The family having closest affiliation with hominids is <u>Pongidae</u>, which includes gibbons, chimpanzees, gorillas, and orangutans; interestingly, some classifications place humans into Pongidae and do not even recognize Hominidae!

Study of human phylogeny is being pursued actively by many researchers in many locations. With such intense activity, it is not surprising that new information is continually coming to light and that not all interpretations of the evidence agree with each other. Therefore, the textbook presents more than one scenario for human origin (summarized in Fig. 44.2).

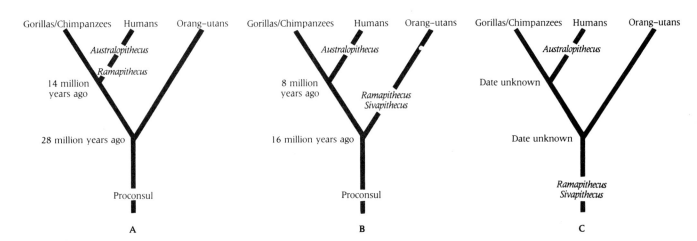

Fig. 44.2: Three phylogenies of hominid evolution

Let's develop an overview of hominoid evolution from the Miocene forward: Early hominoids *(Sivapithecus, Ramapithecus, Gigantopithecus,* and others) probably occupied open woodland habitats from 14 to 8 million years ago. *Ramapithecus* is placed differently in the phylogenies shown in Fig. 44.2--as an early homonid directly in line towards humans, as an early ape on line to orangutan, or as a common ancestor to both pongids and hominids.

The hominid fossil record is largely nonexistennt before 4 million years ago. Various species of hominids (<u>australopithecines</u> in genus *Australopithecus)* occupy the gap between divergence of families Pongidae and Hominidae (somewhere between 28 and 16 million years ago) and the appearance of genus *Homo* (around 2 million years ago). The group of australopithecines from which *Homo* arose is not certain. However, anthropologists generally agree on the sequence of appearance of species in genus *Homo*: *H. habilis* (around 2 million years) to *H. erectus* (around 1.5 million years) to *H. sapiens* (around 200,000 years ago). By about 35,000 years ago, modern *Homo sapiens* were the only remaining hominid species. Entry of modern humans into the New World probably occurred only about 10,000 years ago. Variation in modern humans is described by recognition of from six to 30 races.

TERMS TO UNDERSTAND: Define and identify the following terms:

biopoiesis

RNA world

ribonucleoprotein

RNP world

DNA world

Precambrian

expansion

replacement

Age of Fishes

paleoniscids

ammonites

fossil fuels

labyrinthodonts

pelycosaurs

therapsids

Age of Reptiles

mass extinctions

crisis

thecodonts

The Great Dying

iridium

Age of Mammals

prosimians

hominoids

australopithecines

brachiation

bipedalism

neanderthals

Homo sapiens

DISCUSSION QUESTIONS

1. Select a time interval (e.g., an epoch) in Earth history and indicate the status of ten major groups of organisms during that interval. (This is an excellent study excerise. You can modify this question to allow you to check your grasp of the status of each major group of organisms during each time interval.)

2. What major tendencies recur during evolution of life during Earth history? Provide at least one example illustrating each tendency.

3. Summarize the various explanations for the cause(s) of the "great dying." Is there one explanation that you think is the most likely? Provide evidence to support your choice.

4. What problems did organisms face as they invaded land? Were the same problems important to both plants and animals?

5. Present a summary of the evolutionary history of primates, with emphasis on the events that led to humans.

MATCHING EXERCISE

Complete the following matching exercise to check your familiarity with characteristics of, and various events that occurred during, selected time intervals during Earth history. Some descriptions might apply to more than one interval.

INTERVAL | DESCRIPTION

_____ 1. Mesozoic   A. interval when trilobites were present

_____ 2. Plio-Pleistocene   B. interval when plants invaded land

_____ 3. Cretaceous   C. Age of Reptiles

_____ 4. Silurian   D. interval when all extant orders of vertebrates are present

_____ 5. Cenozoic   E. Age of Fishes

_____ 6. Cambrian   F. interval whose end is called the "great dying"

_____ 7. Devonian   G. Age of Mammals

_____ 8. Precambrian   H. interval when only organisms present were prokaryotes and protistans

I. interval when genus *Homo* appeared

TESTING YOUR UNDERSTANDING

For each of the test items below, three of the lettered alternatives are true and the other is false. Determine which alternative is false and write its letter in the blank to the left of the question number. On the blank line below alternative D, write the corrected version of the false statement.

_____ 1. A. Earliest Earth atmosphere consisted of many gases, including oxygen as a major component.
B. Heterotrophic organisms probably evolved before autotrophic organisms.
C. Early in Earth history, ultraviolet radiation penetrating the atmosphere was an effective energy source for the abiotic synthesis of organic molecules.
D. Ozone does not absorb significant amounts of visible light.

Correction: _____

_____ 2. A. Eukaryotic cells appeared about 2 billion years ago.
B. Prokaryotic cells appeared by about 3.5 billion years ago.
C. Photosynthesis first appeared in eukaryotic cells.
D. Photosynthesis first appeared in unicellular organisms.

Correction: _____

_____ 3. A. The tendency of life to expand refers to an increase in numbers of species within a phylum.
   B. The tendency of life to expand refers to an increase of biomass in a phylum.
   C. Trilobites were diverse during the Cambrian, but were extinct in the Mesozoic.
   D. Amphibians were diverse during the Triassic, but were extinct in the Cenozoic.

   Correction: _____

_____ 4. A. The vast coal forests of the Cretaceous were made mainly of lycopsids, sphenopsids, and ferns.
   B. The earliest fishes lacked jaws.
   C. The earliest fishes lacked paired fins.
   D. Some of the earliest fishes were armored.

   Correction: _____

_____ 5. A. Land was invaded by fishes classified as paleoniscids.
   B. Ammonites were dominant marine mollusks of the Mesozoic.
   C. The coelacanth Latimeria is the only crossopterygian known to be living today.
   D. Classes of fishes with representative species living during the Mesozoic include Agnatha, Osteichthyes, and Chondrichthyes.

   Correction: _____

_____ 6. A. Some of the problems faced by aquatic organisms invading land included dehydration, support of the body, and temperature extremes.
   B. The first vascular land plants were the psilopsids.
   C. According to the fossil record, the earliest land animals were mollusks.
   D. Fossils fuels represent the remains of vegetation that grew during the Carboniferous.

   Correction: _____

_____ 7. A. The earliest reptiles were the cotylosaurs.
   B. Mammals evolved from a group of reptiles called the thecodonts.
   C. Reptiles acheived their greatest diversity during the Mesozoic.
   D. Among the groups of organisms that went extinct during the "great dying" were dinosaurs, ammonites, and ichthyosaurs.

   Correction: _____

_____ 8. A. In efforts to explain the cause of the mass extinction at the end of the late Cretaceous, some scientists are studying the amount of iridium present in rock layers.
   B. In aquatic environments, modernization of plant life occurred millions of years before modernization of animal life.
   C. Diatoms are an ancient component of the aquatic flora.
   D. Modern classes of fishes appeared before modern class of terrestrial vertebrates appeared.

   Correction: _____

_____ 9. A. By the late Cretaceous, the flora of land environments was modernized.
B. Most, if not all, of the main groups of insects were present in modern form in the late Cretaceous.
C. Adaptive radiation of birds and mammals occurred during the Cenozoic.
D. Amphibians are more diverse now than they were during the Mesozoic.

Correction: _____

_____ 10. A. The diversification of mammals was suppressed during the Mesozoic because of competition with birds.
B. Most modern orders of mammals were present by the Eocene.
C. Most modern families of mammals were present by the Oligocene.
D. The fauna of mammals today is less diverse than the mammalian fauna of the Eocene.

Correction: _____

_____ 11. A. All of the following are groups of reptiles: mosasaurs, rhynchocephalians, therapods.
B. All of the following are groups of amphibians: labyrinthodonts, pterosaurs, salamanders.
C. All of the following are groups of fishes: crossopterygians, actinopterygians, agnathans.
D. All of the following are groups of mammals: rodents, primates, oreodonts.

Correction: _____

_____ 12. A. All of the following are features of some primates: opposable first digit, stereoscopic vision, ability to brachiate.
B. The early evolution of hominids occurred in the Old World tropics.
C. Bipedalism is a feature of hominids.
D. *Homo erectus* is the ancestor of *Homo habilis*.

Correction: _____

CHAPTER 45--POPULATIONS

CHAPTER OVERVIEW

This chapter begins a new unit in your textbook. In this and the following chapters, we will examine a variety of ecological topics--topics that address the ways by which organisms interact with each other and with their physical surroundings. We'll see that interspecific relationships are numerous and intricate. We'll see, too, that humans are part of the ecological linkage of species and environment on our planet.

Let's begin our study of ecology by looking at ways that individuals of the same species and of different species interact with each other. We'll examine mathmatical models that predict rates of increase in population size. Rates of population increase are affected by a variety of limiting factors. Interspecific interactions include competition for resources and predator-prey relationships.

TOPIC SUMMARIES

A. <u>Ecology: Definition and Levels of Organization Involved</u>.--<u>Ecology</u> is the study of the interactions between organisms and their surroundings. Organisms interact with their physical surroundings as well as with other organisms of the same and other species. We can study ecology at the levels of <u>populations</u> (among individuals of the same species), <u>communities</u> (among individuals of one or more other species living in a given area), and <u>ecosystems</u> (interaction with individuals of the same and other species and with the surrounding physical environment). The most inclusive level of ecological study is the <u>biosphere</u>, the thin shell of air, earth, and water that encases our planet.

B. <u>Population Dynamics</u>.--One of the attributes of organisms is reproduction. As new individuals are added to a population (via birth or immigration), the size of the population changes. As individuals leave the population (via death or emigration), the size of the population changes. The study of changes in populations is <u>population dynamics</u>.

If we eliminate <u>immigration</u> and <u>emigration</u>, the rate at which populations change is determined by rates of birth and death. If all needed materials (food, shelter, water, etc.) are available in unlimited supply, then any population of any species will multiply rapidly in <u>exponential</u> fashion. In exponential growth (Fig. 45.1), rate of increase in population *(I)* equals the difference *(r)* in average birth rate *(b)* and average death rate *(d)* times the number of individuals in the population *(N)*. In equation form, the relationship is:

$$I = (b - d) \times N.$$

The difference between *b* and *d* is *r*, the <u>intrinsic rate of increase</u>. Every species has a particular value for *r* under a particular set of conditions. A population increasing at its maximum potential for many generations eventually would exceed available resources (e.g., food, space, etc.) and would "crash." This arrangement of unrestricted population growth is seen occasionally in some species (e.g., bacterial growth in a culture, diatoms in oceans). [Read in your textbook about the furor caused in 18th-century England when <u>Thomas Malthus</u> included humans among the species to which this growth curve applies.]

Much more common, however, is a pattern of restricted population growth described by the <u>logistic growth curve</u> (Fig. 45.1). Growth in more realistic settings where one or many factors are limiting is exponential only for a while; growth slows when the carrying capacity of the surroundings is approached. <u>Carrying capacity</u> *(K)* is defined as the maximum number of individuals of a species that a given area can support indefinitely. The J-shaped exponential curve can be changed to the sigmoid (S-shaped) logistic curve by modifying the equation to account for carrying capacity as follows:

$$I = r \times (K - N)/(K) \times N.$$

The quantity of $(K - N)/K$ allows nearly exponential (rapid) growth when N is low; it reflects slow growth when N is large and virtually no growth when N approaches K. For a particular species under a particular set of conditions, K is a constant value. The slope of the growth curve indicates the rate of increase for a particular N; this rate of increase is greatest at the point of inflection (square in Fig. 45.1) on the curve. The point of inflection represents the maximum sustainable yield, the ideal N for harvesting of individuals from the population.

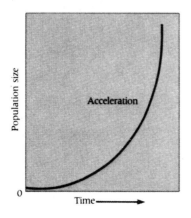
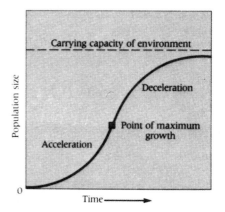

Fig. 45.1: Exponential (left) and logistic (right) growth curves

C. Survivorship.--Not every individual in a population lives as long as its possible lifespan; some die from causes other than old age (e.g., predation, starvation, illness). So, the potential value for d in the equations above is rarely acheived. Survivorship accounts for the loss of individuals along the way. Fig. 45.2 shows four commonly seen types of survivorship curves.

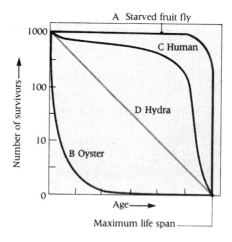

Fig. 45.2: Four common patterns of survivorship curves

For a few species, such as *Hydra,* individuals have about the same probability of dying at any point during their lives. Many other species, such as oysters, endure great mortality during early stages of the lifespan; an individual that survives for a few days or weeks likely will live to near the maximum lifespan. Humans are an example of species that suffer minor mortality just after birth and a slightly increasing mortality rate over most of the rest of the potential lifespan; near the potential lifespan, mortality rate increases dramatically.

D. Factors that Limit Population Growth.--Birth rate, death rate, and survivorship are not the only factors that affect population growth; there are other limiting factors as well. Operation of these factors might be dependent on population size (density-dependent) or not dependent on population size (density-independent). Often, catastrophic environmental agents are density-independent: a flood or fire or ice storm generally will kill most or all individuals present, regardless of how many individuals are present. However, the effectiveness of many biotic factors (e.g., parasitism, predation, competition) depends on population density.

(1) Food chains. Individuals of most species obtain energy and materials of many kinds from individuals occupying preceding links of the food chain. Trophic relationships such as parasitism and predation are situations in which we can examine a type of interspecific interaction.

Usually, a change in population density of a prey species (e.g., deer) will affect population density of a predator species (e.g., cougar). Your textbook examines the predator-prey relationship of *Paramecium caudatum* (prey) and *Didinium nasutum* (predator). One type of interaction, necessarily short-lived, is for the predator to consume all (or nearly all) prey, with the result of extinction of the predator (or both) populations. "Boom and bust" cycling of predator and prey can occur. A third approach that can be maintained indefinitely is stability achieved by a predator population of appropriate size that is harvesting a prey population near carrying capacity (this is the ideal interaction between deer and deer hunters). History teaches us that humans have been the predators that caused the extinction of many species (e.g., dodo bird, passenger pigeon, sea cow, etc.).

(2) Competition. In real life, one or often many resources are available in limited amounts . . . hence, the logistic growth curve. Certainly, individuals of the same species compete (intraspecific competition) with each other for resources. These are but a few of the resources for which humans compete: employment, grades, mates, food, etc. One of the results of intraspecific competition is the establishment (by individuals or groups of individuals) of territories, areas within the normal (larger) area of activity that are defended against other individuals of the same species (see textbook Box 45-2). Territories often include home (burrow, nest, etc.), feeding sites (a squirrel might prevent any others from feeding in a particular tree), areas occupied by mates, a watering hole, or any of myriad other limiting resources.

Interspecific competition also occurs. The closer the requirements of interacting species are, the greater the potential for there to be competition between the species for those resources. Intensity of competition varies inversely with the availability of shared resources and directly with population densities of the competing species. Much of our understanding of interspecific competition stems from study of ecologically-similar species of *Paramecium*.

E. Other Types of Intraspecific Interactions.--Not all intraspecific interactions are adversarial. Many interactions enhance likelihood of survival. A simple form of interaction is aggregation, in which individuals operate in groups but in which there is no differentiation of roles. Aggregations offer opportunities for selection of mates. They may offer "safety in numbers" via the dilution effect: a predator may have difficulty targeting an individual among a moving herd or flock.

A more complex form of interaction is <u>cooperation</u>. Predators may be more effective if they cooperate in the hunt: one lion might not be able to subdue a wildebeest, but several working together can. Prey might be better able to fend away predators if several prey individuals work together. Individuals might communicate to other individuals of the group information about danger or about location of food. <u>Societies</u>, such as those we studied in Chapter 33, might well have evolved from cooperative interactions in ancestral species.

F. <u>Reproductive Strategies</u>.--Each species has a particular ecological approach to continuation of its species. Species can be positioned along a continuum according to the proportion of their energy that they expend on reproduction and according to how that energy is expended. <u>K-selected</u> species maintain populations near the carrying capacity. They usually are large-bodied, grow slowly, and have few young to whom considerable parental care is provided. They generally occur in environments which are stable and/or predictable. Large mammals (including humans) are examples of <u>K-strategists</u>.

Conversely, <u>r-selected</u> species are capable of rapid population growth. They live in unstable and/or unpredictable environments. A high rate of increase enables recovery of populations from castastrophic events. Body size is small, offspring mature to reproductive age rapidly, and there is little parental care. Among the mammals, mice and rabbits could be considered <u>r-strategists</u>.

TERMS TO UNDERSTAND: Define and identify the following terms:

ecology

populations

communities

ecosystems

biosphere

birth rate

death rate

intrinsic rate of increase

exponential growth curve

logistic growth curve

carrying capacity

maximum sustained yield

population dynamics

survivorship curve

limiting factors

food chain

parasitism

predation

interspecific competition

intraspecific competition

population periodicity

territoriality

aggregation

dilution effect

conspecific

K-selection

r-selection

DISCUSSION QUESTIONS

1. Both Darwin and Malthus realized that organisms produce more organisms than can survive. Both saw implications of excessive production of offspring. Summarize the conclusions that Darwin drew from these observations and the different conclusions that Malthus drew from these observations. Is there a logical connection between their conclusions?

2. Earth's human population seems to be tracking an exponential growth curve (textbook Fig. 45-4). How long can such accelerating increase continue? What factors, if any, will decrease our rate of increase? What role has technology and medicine had in our pattern of population growth?

3. Compare and contrast the logistic and exponential growth curves. How do differences in their formulae yield curves of different shapes? You can best understand the behavior of these curves and formulae by working the same data set through the formulae. Calculate I and plot population growth for 10 generations beginning with two individuals (of opposite sex) and assuming $r = 2$. (You can vary I and r to see how this effects population growth.)

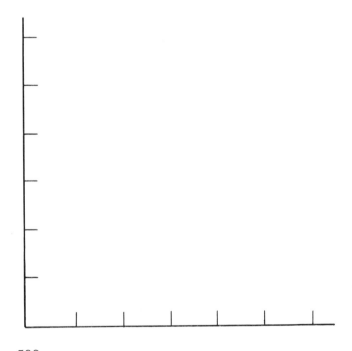

4. Compare the implications of cannabalism and of altruism for a population.

TESTING YOUR UNDERSTANDING

For each of the test items below, three of the lettered alternatives are true and the other is false. Determine which alternative is false and write its letter in the blank to the left of the question number. On the blank line below alternative D, write the corrected version of the false statement.

_____ 1. A. The logistic growth curve takes into account the reality of limiting factors.
B. The exponential growth curve levels off when the population size reaches carrying capacity.
C. The logistic growth curve is S-shaped.
D. The exponential growth curve is J-shaped.

Correction: _____

_____ 2. A. A community includes all individuals of a species within a given area, whereas a population includes individuals of several species within a given area.
B. The intrinsic rate of increase (r) is computed as the difference between birth rate and death rate for a population.
C. For any particular species, the intrinsic rate of increase varies according to environmental conditions.
D. Birth rate is often measured as number of live births per 1,000 individuals in a population.

Correction: _____

_____ 3. A. Carrying capacity is the maximum number of individuals that a given area can support indefinitely.
B. According to the logistic curve, growth slows as K nears N.
C. According to the logistic curve, growth is rapid at intermediate values of N.
D. Malthus realized that the number of offspring produced by individuals of a population usually exceeded the number of individuals that the resources in the area occupied by that population could support.

Correction: _____

_____ 4. A. Survivorship curves depict the schedule of mortality for a group of offspring produced by a population.
B. *Hydra* is an example of a species that suffers equivalent mortality over the entire potential lifespan.
C. Oysters and humans are examples of species that suffer great mortality shortly after hatching or birth.
D. An effect of medical care has been that more individuals will more nearly acheive their potential lifespans.

Correction: _____

_____ 5. A. Natural selection is the process that culls the poorly-adapted excess progeny of a population.
B. The effects of a prolonged drought on a population probably are density-independent.
C. The detrimental effects of a parasite on a population of hosts probably operate independently of population density.
D. The ultimate source of energy for life on Earth is sunlight.

Correction: _____

_____ 6. A. If a parasite is to be successful, it must not kill its host.
B. Gause demonstrated that an excessive level of predation could exterminate a prey population and, eventually, cause extinction of the predator population.
C. If two ecologically similar species are placed into direct competition in a situation of limited resources, they acheive equilibrium population levels at which they can coexist.
D. Territories are established as a means of withstanding competition from other individuals of the same species or of ecologically similar species.

Correction: _____

_____ 7. A. Cooperation can occur between individuals of the same or of different species.
B. Predator-prey relationships can result in oscillations in populations of both prey and predator species.
C. Territoriality generally is not so intense that all other individuals of the species are excluded; oftentimes, territories are shared with mates or family members.
D. In intraspecific interactions such as aggregation, participants usually carry out specific activities for the benefit of the group.

Correction: _____

_____ 8. A. The dilution effects refers to the increased safety that an individual gains by associating with a larger group of individuals of its species.
B. *K*-strategists usually produce large numbers of young in each litter.
C. *K*-strategists usually occur in environments that are predictable and stable.
D. *K*-strategists usually provide large amounts of parental care to their young.

Correction: _____

_____ 9. A. Examples of *r*-strategist species include mice, rabbits, and elephants.
   B. *r*-strategists usually produce large numbers of young in each litter.
   C. *r*-strategists usually occur in environments that are unpredictable.
   D. *r*-strategists usually provide little or no parental care to their young.

   Correction: _____

_____ 10. A. An *r*-strategy of reproduction enhances opportunity for a population to quickly recover from a catastrophe.
   B. Territories of individuals in a population are non-overlapping areas.
   C. High population densities can decrease population size due to exhaustion of food supply.
   D. High population densities can decrease population size by promoting spread of disease.

   Correction: _____

CHAPTER 46--ECOSYSTEMS

CHAPTER OVERVIEW

An ecosystem includes all living organisms within an area as well as all abiotic components of that area. Study of ecology at this level allows us to follow cycling of materials between the living and non-living world. It also allows us to trace the flow of energy from its abiotic source (usually the Sun), through one or more biotic trophic levels, and back to the physical world. In this context, we will examine the processes of biogeochemical cycling, the new phenomenon of pollutant cycling, and energy transfers.

TOPIC SUMMARIES

A. <u>Some Definitions</u>.--An <u>ecosystem</u> is a fundamental functional unit having biotic and abiotic components. These components include all organisms living within a defined physical environment. Examples include a pond and all organisms living within the pond and interacting with it . . . or tall-grass prairie and its inhabitants . . . or a coniferous forest and its inhabitants. Approaches to study of the functioning of an ecosystem include documenting <u>flow of energy</u> or <u>flow of materials</u> through all living and nonliving components of the ecosystem.

The totality of all living and nonliving components (that is, the sum of all ecosystems) in the biosphere represents the <u>environment</u>. Let's examine various ecological topics separately for the abiotic environment and then for the biotic environment.

B. <u>Abiotic Environment</u>.--Numerous physical factors affect the well-being of organisms. These factors include solar radiation, climate, minerals, water, and others.

(1) Solar radiation. Earth is showered with all wavelengths of electromagnetic radiation produced by the Sun. This <u>insolation</u> is the ultimate source of virtually all of the energy required for life processes of virtually all organisms on Earth. Plants and other photosynthetic organisms directly utilize solar energy. <u>Heterotrophs</u>, as consumers of autotrophic tissues, indirectly rely on solar energy.

Much of the electromagnetic energy from the Sun is converted into heat energy. Whether a place is suitable for occupation by an organism is, in part, determined by the <u>heat content</u> of the physical environment.

(2) Water. All forms of life require <u>water</u>. Many organisms live in water. Although new water molecules are being produced constantly by respiration and existing water molecules are being broken down constantly by hydrolysis (such as in photosynthesis and in various digestive processes), the global volume of water seems to have been constant in the recent geological past. Water cycles (<u>hydrologic cycle</u>, Fig. 46.1) between <u>oceanic</u>, <u>atmospheric</u>, and <u>terrestrial reservoirs</u>. Solar energy is the primary force that drives this cycle. Water evaporates from oceans and other bodies of water, water falls as various forms of precipitation onto land, water passes through terrestrial ecosystems, and water eventually courses back to the seas. The winds required to bring rainclouds over land also are a result of insolation.

(3) Microclimate, niche, and substratum. The amount and pattern of solar radiation and the pattern of precipitation are major factors that produce short-term <u>weather</u> and long-term <u>climate</u> of a particular area. Within a region having a particular climate, there is much localized variation such that smaller areas have <u>microclimates</u>. For example, the peninsula of Florida is described as having a subtropical climate; yet, individual areas have microclimates such as

those prevailing in swamps, in pine-flatwoods forests, and salt-water marshes. It is the local conditions that determine whether an area is suitable for occupation by a particular species.

The environmental setting in which an organism occurs is its habitat. For example, the habitat in which the eastern flying squirrel occurs is eastern deciduous and/or coniferous forests. Habitat is described in terms of both biotic and abiotic characteristics. Because organisms, for at least part of their lives, come into contact with solid ground (either on land or at the bottom of bodies of water), substratum is an important contributor to habitat. The relationship with substratum is particularly evident for most plants and for animals that spend considerable time working the soil. For example, soil quality is a prime determinant of suitability of a habitat for use by pocket gophers or moles. A niche can be thought of as the relationship of an organism with its biotic and abiotic microhabitat.

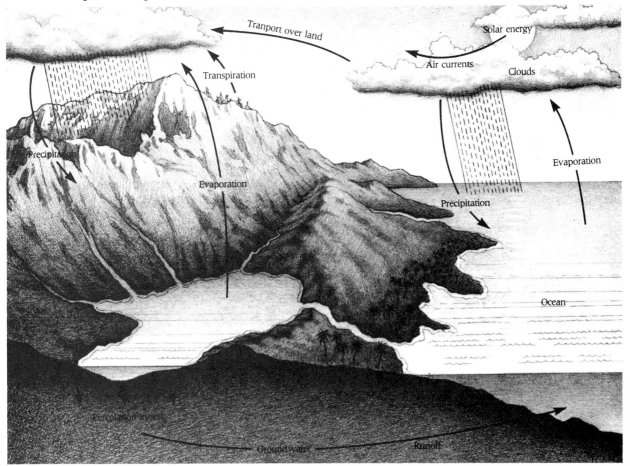

Fig. 46.1: Diagram of the hydrologic cycle

(4) Chemical environment. The chemical quality of the environment also influences habitat selection and use by organisms. Globally, the atmospheric gas composition is relatively constant: nitrogen, 78%, oxygen, 21%, and carbon dioxide, 0.4%. Variation in gaseous composition includes water-vapor concentration (usually measured in terms of relative humidity) and various gaseous pollutants (carbon monoxide, oxides of sulfur and nitrogen, hydrogen sulfide, and hydrogen fluoride). Pore spaces between particles of soil also contain water vapor and atmospheric gases.

As already mentioned, chemical composition and structural features of soils are important components of habitats. The organisms that occupy soils often contribute to changing the character of the soils. Excreta and tissues of organisms contribute organic material (humus) to the soil. Various bacteria convert various molecules (see D2 below) into forms usable by other organisms.

Concentrations of molecules and ions in <u>water</u> have marked osmotic effects on tissues; some aquatic species are faced with the problem of dehydration (e.g., a fish living in hypertonic ocean water). Other aquatic species would become waterlogged if they were unable to unload the continuing influx of water from hypotonic fresh-water surroundings.

(5) Tolerances and optima. Rarely, if ever, is any aspect of any environment constant. Hence, if an organism is to occupy an area over an interval of time, it must be able to tolerate a range of conditions. Ranges of tolerance have upper and lower limits. Species that can operate within a broad range of environmental parameters are labelled <u>euryecious</u>, whereas those with relatively narrow zones of tolerance being <u>stenoecious</u>. Interestingly, a species might be euryecious for one parameter but stenoecious for another. The prefixes <u>eury-</u> and <u>steno-</u> can be joined with appropriate suffixes to describe ecological tolerances of the species, with the suffix of the term indicating the parameter. For example, a species that tolerates a wide range of temperature is called <u>eurythermic</u>, whereas a species that requires little variation in temperature is <u>stenothermic</u>.

Usually, for each species there is an <u>optimum</u> value within tolerance ranges for each parameter. We sometimes say that a species "<u>prefers</u>" such an optimal situation. Unless we have evidence that the organisms being considered have cognitive abilities, we should avoid use of preference. How do the suffixes -<u>phile</u> and -<u>phobe</u> facilitate description of optimum conditions?

C. <u>Biotic Environment</u>.--We may define biotic environment of a species to include all living things that directly or indirectly affect the species. We have addressed many aspects of the biotic environment in earlier chapters; these include parasitism, predation, competition, and others.

D. <u>Cycling of Materials</u>.--As regards energy, Earth is an <u>open system</u>; solar energy enters the system, and heat and light energy leave Earth. Earth, however, is a <u>closed system</u> with respect to materials. The limited amounts of elementary materials (carbon, nitrogen, various minerals; perhaps as many as 40 elements) are cycled between biotic and abiotic segments of the planet through various <u>biogeochemical cycles</u>. Study of a few cycles here indicates the principles of biogeochemical cycling.

(1) Carbon cycle. <u>Carbon</u> is a component of all organic molecules. <u>Photosynthesis</u> is the primary means of transferring inorganic carbon (mainly in the form of carbon-dioxide gas in the atmosphere) into organic carbon (initially as simple carbohydrates in <u>autotrophs</u>). Such <u>producer</u> organisms are eaten by <u>consumers</u>; organisms that consume plant materials are <u>herbivores</u> and those consuming herbivores are <u>carnivores</u>. Food chains and webs often consist of many such <u>trophic levels</u>.

At each trophic level, some carbon returns to the atmospheric reservoir as a waste product (carbon dioxide) of <u>respiration</u>. Some carbon (in organic form) is passed on to the next trophic level. The rest of the carbon (mainly as dead tissues or other products) passes to the <u>decomposers</u>, fungi and bacteria (and some other organisms) that oxidize organic compounds into carbon dioxide and other products. <u>Fossil fuels</u> represent great stores of carbon that have temporarily been taken out of the cycle; every time you consume gasoline while driving, or burn coal, oil, or gas for heating, you return carbon to the cycle. How does the <u>greenhouse effect</u> fit into the carbon cycle?

(2) Nitrogen cycle. Nitrogen, too, is required for life. Nitrogen is a component of proteins and of nucleic acids (such as RNA and DNA). A difference between this and the carbon cycle is that nitrogen must pass through a series of microorganisms to completely pass through the cycle.

Although gaseous nitrogen makes up nearly 80% of the atmosphere, this form of nitrogen is useless to all organisms except for the nitrogen-fixing microbes. Nitrogen-fixing bacteria (e.g., *Rhizobium, Frankia*) and cyanophytes (e.g., *Anabaena*), living variously in association with plants (especially, legumes) or freely in soil, transform molecular nitrogen into forms (ammonia, nitrate) that can be taken into plants where nitrogen is incorporated into organic molecules. Such plant proteins and other molecules are consumed by herbivores that, in turn, are consumed by carnivores.

Putrefactive bacteria decompose some organic material and nitrogenous wastes and thereby return some nitrogen to the atmosphere. Nitrosifying bacteria oxidize ammonia to nitrites, and nitrifying bacteria oxidize nitrites to nitrates; this two-step process of nitrification yields molecules that are usable by plants.

Survey of the carbon and nitrogen cycles has revealed three important principles. What are these principles?

(3) Mineral cycling. Chapter 3 presents many of the minerals that are required for life. In part, these minerals (e.g., sulfur, phosphorus, etc.) cycle through organisms much as do carbon and nitrogen (Fig. 46.2). However, the reservoirs are not atmospheric. Rather, these minerals come from geological sediments. They are released by weathering of rock. The soluble salts of weathered minerals next enter the water cycle where they are available to organisms. Minerals can return to geological formations via sedimentation at the bottoms of bodies of water. Subsequent release from sediments involves natural processes that may require many hundreds of thousands or many millions of years. Humans short-circuit this cycle by mining minerals.

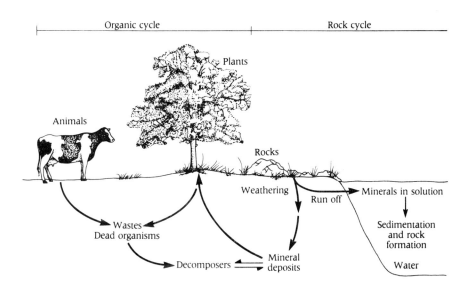

Fig. 46.2: Diagram of the mineral cycle

(4) Pollutant cycles. These are a relatively new concern. Whereas carbon, nitrogen, water, and minerals have been cycling for billions of years, pollutants are new, generally

synthetic chemicals with which the biosphere has never before had to cope. How these will cycle through the biogeosphere is <u>unknown</u>. We do not know the chemical forms in which many of these pollutants will cycle. Although some of these compounds are known to be toxic, we do not know the possible effects of most pollutants on organisms, including ourselves. Quite possibly, our introduction of novel compounds into the environment will upset the ecological balance of our planet. The results could be disastrous.

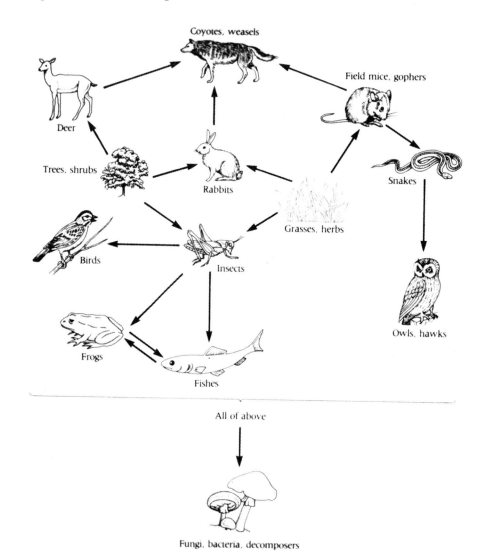

Fig. 46.3: A food web

We have learned some painful lessons; the pesticide DDT, for example, has provided an understanding of how pollutants pass through <u>trophic levels</u> of food chains and webs. DDT is persistent; it is not modified or detoxified over short intervals of time. It accumulates in tissues (paticularly in lipids, such as body fat and mother's milk). By this process of <u>biological magnification</u>, tissue concentrations of DDT can reach dangerous levels and cause damage to the nervous system. DDT also has the effect of weakening shells of bird eggs, with the result of greatly reduced reproductive output.

E. <u>Transfers of Energy</u>.--As presented in earlier chapters (8-11), biology can be viewed from the perspective of energetics. The movement of energy (from one compound to another, or from one organism to another) is unidirectional. Energy degrades from higher forms (light, electrical) to lower forms (heat). No transfer or transduction of energy is completely efficient; always some energy is lost in the translation.

In ecological situations, a "10% law" tends to apply: Transfer of energy between adjacent trophic levels (e.g., solar energy to producer; producer to first-level consumer) is only about 10% efficient; this means that about 90% of the energy is lost. [The actual efficiencies range from 0.05% to 20%.] The loss of energy from the biosphere tends to be balanced by new, incoming solar energy. But, eventually, the Sun will exhaust its energy reserves . . . what then for planet Earth?

F. Chains and Pyramids.--The dynamic flow of materials and of energy through interacting members of a community can be depicted by chains and pyramids. A chain shows the sequence of species through which materials and energy pass; each position in the chain represents a trophic level. Producers (autotrophs) occupy the first level where solar (or perhaps some other source of) energy is converted into biomass. Variable numbers of consumers (heterotrophs) follow. Decomposers (or reducers; also heterotrophs) act upon energy and materials contained in dead biomass at all levels. Chains depict trophic interactions more simply than they probably occur in reality (Fig. 46.3). The more realistic, complex interactions of species are better represented by a web.

Pyramids offer another way of showing relative amounts of energy and materials in trophic levels (Fig. 46.4). Pyramids may be based on energy (measured in joules or other units), mass of organisms (biomass), or numbers of individuals at each level. Generally, the earlier of any two adjacent trophic levels in a pyramid will be broader than the following level. [Your textbook notes some exceptions. What are they?]

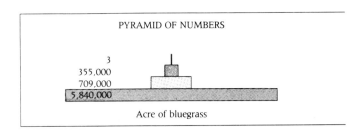

Fig. 46.4: An ecological pyramid of numbers

Three principles emerge from study of trophic dynamics: Complete food chains begin with photosynthesis and end with decay. Shorter food chains provide greater energy and materials to the top consumer; this is because energy is lost with every energy transfer. The number of individuals at any trophic level generally decreases with each energy transfer due to loss of energy at each step.

G. Budgeting.--Because the amount of solar energy reaching Earth is finite, and because there are a limited number of photosynthetic organisms, there are limits on the amount of available energy in the biosphere at any point in time. Likewise, the amounts of materials in the biosphere are limited. Hence, biological processes operate within the constraints of a budget.

TERMS TO UNDERSTAND: Define and identify the following terms:

ecosystem

homeostasis

abiotic environment

hydrologic cycle

climate

microclimate

niche

habitat

substratum

humus

biotic environment

tolerance range

euryecious

stenoecious

optimum

open system

closed system

biogeochemical cycles

carbon cycle

herbivorous animal

carnivorous animal

putrefaction

fossil fuels

greenhouse effect

nitrogen cycle

nitrogen fixation

legume plants

Rhizobium

Frankia

Anabaena

putrefactive bacteria

decomposers

nitrosifying bacteria

nitrifying bacteria

nitrification

denitrifying bacteria

sedimentation

pollutant cycles

biological magnification

10% rule of energy transfer

food chain

trophic level

producer

consumer

food web

pyramid

biomass

eutrophication

DISCUSSION QUESTIONS

1. Select any of the biogeochemical cycles and diagram it below. Indicate the biological and geological segments in the cycle. What species are involved? What is the amount of time required for material to move through the cycle? [This is an exercise that you should repeat for each mineral that cycles through the biosphere.]

2. What were the contributions of Rachel Carson and R. L. Lindeman to our understanding of ecosystems ecology?

3. How has human activity altered biogeochemical cycling? Provide examples. What might be the implications of human tampering with the delicate balance of ecology?

4. Compare and contrast food chains with food webs. Which of the two likely represents a more stable relationship among species? Why?

5. What is the role of microorganisms in cycling of materials and energy? Is the role of these microbes important enough that there would be problems if they somehow could not carry out their functions? Why, or why not?

TESTING YOUR UNDERSTANDING

For each of the test items below, three of the lettered alternatives are true and the other is false. Determine which alternative is false and write its letter in the blank to the left of the question number. On the blank line below alternative D, write the corrected version of the false statement.

    _____ 1. A. Ecosystem can be defined as a living community plus its physical environment.
                B. The concept of homeostasis applies at the ecosystem level as well as at the level of the individual organism.
                C. The greenhouse effect is a result of human intervention in the carbon cycle.
                D. The process of weathering is an important determinant of whether an organism can tolerate extreme conditions in newly colonized habitats.

            Correction: _____

___ 2. A. Chemosynthetic autotrophs depend on solar radiation as their energy source.
B. Effects of insolation include establishment of winds and distribution of precipitation.
C. Evaporation is the process by which water goes from liquid form in oceans to gaseous form in the atmosphere.
D. The hydrologic cycle is an important factor in the distribution of mineral salts that have been liberated from bedrock.

Correction: _____

___ 3. A. Climate is the average weather over a long period of time.
B. Microclimate is a more important determinant of species distribution than is climate.
C. Wavelengths of the electromagnetic spectrum that energize the photosynthetic apparatus of green plants are in the infrared part of the spectrum.
D. Habitat is the physical locale where an organism lives, whereas niche is the relationship between a species and the environment.

Correction: _____

___ 4. A. Characteristics of the substratum are important determinants of the distribution of organisms, especially for plants and burrowing animals.
B. Soils containing large amounts of humus usually are high in nitrogen content.
C. An aquatic species that could tolerate a broad range of salt concentrations would be called euryhaline.
D. A terrestrial species that could tolerate a narrow range of ambient temperature would be called eurythermic.

Correction: _____

___ 5. A. A species having a particular set of environmental requirements could be referred to as both hydrophilic and xerophobic.
B. Earth is a closed system with respect to energy, but is an open system with respect to minerals and other materials.
C. Phosphorus and sulfur are two of the many minerals that pass through biogeochemical cycles.
D. The amount of time required to cycle carbon is shorter than that required to cycle minerals.

Correction: _____

___ 6. A. Producer organisms feed upon decomposer organisms; herbivores feed on producer organsims.
B. Carnivores feed upon herbivores.
C. Decomposers obtain energy and materials from all other trophic levels.
D. Transfer of energy from one trophic level to the next involves oxidation reactions.

Correction: _____

_____ 7. A. Fungi and bacteria are the primary groups of organisms involved in putrefaction.
B. The temperature of Earth's atmosphere is increasing, with the expected result that global sea levels will rise and inundate coastal areas with water.
C. For carbon to be incorporated into biomass, a particular group of microorganisms must convert carbon dioxide into a form that can be used by plants.
D. All of the following are organisms whose chemical activities are crucial in providing nitrogen in forms that can be used by other organisms: *Azolla, Frankia, Rhizobium,* legumes.

Correction: _____

_____ 8. A. Nitrogen fixation is a process in which gaseous nitrogen is converted into ammonia or nitrates.
B. Nitrification is a two-step process in which nitrates are converted into ammonia.
C. Denitrification is a process by which nitrates, nitrites, and ammonia are reduced into gaseous nitrogen.
D. The nitrogen cycle does not involve the process of sedimentation.

Correction: _____

_____ 9. A. In the cycling of certain pollutants, animal tissues function as a reservoir for pollutant molecules.
B. Pollutants that are not harmful when released into the environment sometimes are converted into chemical forms that are harmful to organisms in the environment.
C. Among the various activities that can be considered as polluting are activities that speed up certain steps of an existing material cycle.
D. Applying fertilizer to lawns and crops is not a form of pollution.

Correction: _____

_____ 10. A. The "10% law" of energy transfer indicates that 10% percent of the energy in a lower trophic level is lost as energy is transferred to the next level.
B. Trophic relationships in most communities are better depicted by food webs than by food chains.
C. The lower layers of an energy pyramid are broader than are the upper layers.
D. Factors that limit life processes on Earth include limits to the available amounts of solar energy striking Earth and materials on Earth.

Correction: _____

CHAPTER 47--COMMUNITIES

CHAPTER OVERVIEW

Two chapters ago, we examined aspects of the biology of populations. We next moved to the level of ecosystems, units made of all populations and all environmental factors within a defined area. In this chapter, we will consider the biological portion of an ecosystem, the community.

A community consists of all populations in an ecosystem. Interestingly, communities have properties that their component populations do not have. These properties include interspecific competition, symbioses, predator-prey interactions, succession, and diversity. It is these properties on which we will focus here in Chapter 47.

TOPIC SUMMARIES

A. <u>What Is a Community</u>?--We can define <u>community</u> is several ways. A community is the biotic portion of an ecosystem (e.g., all organisms living in a pond). A community is all populations living in an ecosystem. A community is any subset of species occupying a particular area (e.g., the rodent community, the bird community). Just as each species has its own ecology, so does a community. Study of <u>community ecology</u> involves study of interspecific competition, symbioses, parasitism, and predator-prey relationships.

B. <u>Niches and Competition</u>.--As noted in our study of populations (Chapter 45), the intensity of competition between species increases with increasing similarity of the niches of the interacting species. Gause's <u>principle of competitive exclusion</u> recognizes that two or more species of identical niche cannot coexist indefinitely. When one or more components of the niche (e.g., food, nest sites, etc.) are available in limited supply, one of the species will "out-compete" the other(s), with the other(s) being excluded from the area . . . or, perhaps, becoming extinct. What organisms did Gause study to arrive at this understanding?

In many cases, the ability of ecologically-similar species to coexist (and, hence, their evolutionary success) relates to the ability of one or more of the competing species to <u>flex their requirements or activities</u>, at least temporarily. When <u>ecological bottlenecks</u> occur (e.g., times of limited food, etc.), a species might shift its diet enough for all involved species to survive through the period of limited resource. The secondary resource choice is called a <u>refugium resource</u>. Alternatively, a competing species might <u>partition the limited resource in time or space</u>. For example, one of two competing species might be active during twilight hours and the other be active at midday. Or, one species might feed high in the crowns of trees with the other species feeding in parts of the same trees nearer the ground.

Such shifting of resource usage in response to competition might, over many generations, lead to the phenomenon of <u>character displacement</u>. Indeed, ecologists offer the species of <u>finches in the Galapagos Islands</u> as an evolutionary product of character displacement. It is thought that these islands were colonized by one species of finch. Presumably, the islands harbored no predators of these finches. Hence, the finches flourished and reached high population densities. Intense intraspecific competition for limited food resources favored those individuals that could make use of foods different that which most individuals consumed. The ability to use different foods correlated with differences in <u>morphology of the bills</u>. Hence, selection favored enhanced reproduction of birds having divergent morphologies of bills. The ensuing <u>adaptive radiation</u> produced today's community of finches on the Galapagos Islands.

C. <u>Another Set of Interspecific Interactions: Symbioses</u>.--In addition to competition and predator-prey interactions, there is another set of interspecific interactions. <u>Symbiotic relationships</u> between species are described and named according to the degree of benefit or damage it acrues from the interaction.

<u>Commensalism</u> is a relationship wherein one species benefits and the other is neither harmed nor benefited. The barnacles attached to the skin of a whale neither benefit not harm the whale. The barnacles feed while being carried about; rather than relying on sea currents to bring food-laden water to them (as must barnacles attached to piers), the whale continuously carries them to waters where food is available.

All participants in a <u>mutualistic relationship</u> benefit. Lichens, discussed in Chapter 40, are an excellent example: The fungal participant provides the structural framework in which the photosynthetic participant (an alga) resides. The fungus receives food from the alga. The degree of dependency of one partner on the other allows recognition of a spectrum of mutualistic relationships; some may be <u>obligatory</u>.

<u>Parasitsm</u> is a situations where one species benefits and the other is harmed. A prudent <u>parasite</u>, however, does not (at least not quickly) kill its <u>host</u>. The unwilling host receives annoyance or perhaps disease as a result of the relationship. Would you classify the relationship of vampire bats feeding on blood of cattle as parasitism or as a predator-prey relationship?

D. <u>Predator-Prey Interactions</u>.--A broad definition of predator-prey relationship includes that between plants and the organisms that feed upon them (herbivores). Probably every predator-prey relationship represents an evolutionary "cat-and-mouse" game, with each participant evolving in ways to thwart the other's defensive efforts.

Responses of plants to herbivory include both <u>morphological</u> (spines, thorns, etc.) and <u>chemical defenses</u>. Among the many chemicals various plants have evolved to protect themselves against herbivores and parasitic fungi and bacteria are alkaloids, steroids, nicotine, digitoxin, caffeine, and many others. Humans have learned to adapt some such compounds for dietary and medicinal purposes.

Responses of animal prey to predators are of various types, including chemical. Examples include batrachotoxins produced by poison-arrow frogs, cardiac glycosides (procured from plants) in monarch butterflies, and stinging cells comandeered from hydroids by other marine animals.

E. <u>Ecological Succession</u>.--Just as individual organisms grow and change during their ontogeny, so do communities. The process of maturation of a community over decades and centuries is called <u>succession</u>. Fig. 47.1 shows successional stages on sand dunes of Lake Michigan. Initially, habitats (perhaps at this stage we should call them merely parcels of environment) are barren, devoid of life. The early stages of succession in which life becomes established in such a barren setting are called <u>primary succession</u>. Examples include colonization of bare rock by lichens, and initial growth of plants on a barren sand dune. Subsequent stages where species replace each other are part of <u>secondary succession</u>.

Succession progresses through stages because the environment changes and can no longer provide the resources required by species of earlier successional stages. The primary source of changes is the activities of the community members themselves. For example, the original colonizing lichens can live only on bare rock. The activities of the lichens breakdown the rock into soil, in which the lichen cannot live. Next, early stage (pioneer) plants grow in the primitive soil, their roots further degrading the rock, producing more soil, and producing a substrate suitable for another species of plant. Such self-caused (<u>autogenic</u>) changes occur throughout the course of succession.

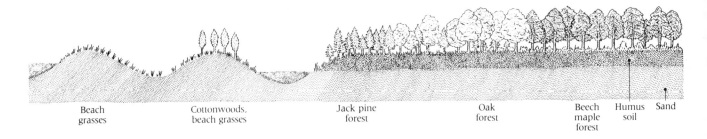

Fig. 47.1: Plant succession on sand dunes near Lake Michigan

External sources of change (<u>allogenic sources</u>) include fire, earthquake, flood, and many others. Such catastrophic changes set succession back to early stages, but usually not to pioneer or primary stages.

<u>Predictable changes</u> generally occur as succession progresses (Fig. 47.2). The number of species present (as measured by species diversity, richness, or abundance) in a community increases with increasing maturity of the community. <u>Diversity</u> tends to level near maturity. Total <u>biomass</u> in an ecosystem increases with increasing maturity; cycling of materials tends to produce an equilibrium at maturity. The <u>relations between species</u> tend to become more complex with increasing maturity. The proportion of <u>new biomass</u> contributed to the ecosystem by producers (autotrophs) is greater in early stages, with consumers contributing proportionately more biomass in later stages. Do these relationships mean that mature ecosystems are more stable and more efficient than early stage ecosystems . . . or, vice versa?

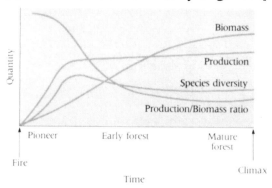

Fig. 47.2: Ecological trends in succession toward maturity

F. <u>Concept of Climax</u>.--The final (if there is one) product of succession is the <u>climax community</u>. A climax community tends to persist indefinitely with little change. In our study of biogeography in the next chapter, we will see that climax communities are useful in describing regional landscapes (<u>biomes</u>).

Much ecological debate and research has focussed on climax. Of specific interest is the question of whether succession in particular climatic conditions will always lead to the same result, an ecosystem with the same species. The <u>monoclimax theory</u> holds that succession in all situations within a climatic region will ultimately terminate in the same climax community (e.g., the oak-hickory forests of the Midwest). Persisting prairie communities in this region are regarded as <u>subclimaxes</u> that represent a temporarily arrested <u>sere</u> that eventually will achieve the climax.

Successional stages perceived by monoclimax adherants to be subclimaxes are viewed by supporters of the <u>polyclimax theory</u> as climaxes. Hence, both the oak-hickory forest and the prairie (and others) can be climaxes. Recognition of multiple climaxes within a climatic region recognizes that other factors (e.g., soils, fire, grazing, etc.) are important determinants of climax composition.

G. <u>Community Structure</u>.--We have noted already that the structure of communities changes over time--daily, seasonally, and over many years. Communities also are structured spatially--as described both horizontally and vertically.

In terrestrial communities, <u>vertical structuring</u> is dictated largely by plants. Different plant species (and animal species) occupy different vertical layers both above and below the ground. Hence, there are tall canopy trees, shorter understory trees, herbaceous ground cover, and underground roots. In aquatic environments, plants, animals, and other organisms also show vertical stratification, largely in response to depths of light penetration into water and to temperature changes with depth.

<u>Horizontal structuring</u> is apparent from a bird's-eye view. Such structuring is reflected in patchiness of vegetation or in water depths relative to distance to shore or to underwater topography.

H. <u>Species Abundance and Diversity</u>.--In many places, your textbook has spoken of <u>diversity</u> of communities and taxa. Generally, such references have referred to the number of species present in a community or in a taxon. Ecologists use <u>richness</u> to indicate the number of species present in an ecological situation. Yet, direct comparison of communities on the basis of richness is not very informative. Hence, ecologists use measures of <u>species diversity</u> that consider both richness and the <u>relative abundance</u> of each species present. Diversity increases as <u>equitibility</u> among species increases.

TERMS TO UNDERSTAND: Define and identify the following terms:

community

niche

habitat

principle of competitive exclusion

ecological bottleneck

refugium resource

character displacement

realized niche

symbiosis

commensalism

mutualism

parasitism

infectious diseases

plant defenses

secondary chemical compounds

batrachotoxins

elatol

primary succession

secondary succession

autogenic

allogenic

climax

monoclimax theory

subclimax

polyclimax theory

biomes

species diversity

species richness

abundance

equitability

island biogeography

MATCHING EXERCISE

Complete the following matching exercise to check your familiarity with various types of interspecific relationships. Match the descriptions and examples in the right column with the appropriate relationship in the left column. Some descriptions might apply to more than one interval.

RELATIONSHIP   DESCRIPTIONS AND EXAMPLES

____ 1. mutualism   A. deer feeding on acorns

____ 2. parasitism   B. two species of warblers feeding on insects in same tree

____ 3. predator-prey   C. lichen

____ 4. comensalism   D. high population densities exceeding carrying capacity of habitat

____ 5. intraspecific competition   E. cat infested with fleas

____ 6. interspecific competition   F. mistletoe growing on hackberry tree

____ 7. symbiosis   G. no examples provided for this relationship

PEOPLE TO KNOW: What is the significant biological contribution of each of these individuals as indicated in Chapter 47 ?

G. F. Gause

J. H. Connell

Jonathan Swift

H. C. Cowles

F. E. Clements

David Rhoades

DISCUSSION QUESTIONS

1. Define parasitism. Discuss diseases in humans with respect to your definition of parasitism.

2. Summarize the ecological trends that occur as an ecosystem matures.

3. Define herbivory, symbiosis, and parasitism. Give examples of each. Can you think of a particular example that simultaneously is herbivorous, symbiotic, and parasitic?

4. Distinguish between "niche" and "realized niche." What is the relevance of "ecological bottleneck" and "refugium resource" to the difference in types of niches? Provide an example.

5. Ecological succession occurs in response to changes that occur within an ecosystem. Provide examples of such changes. Discuss the source(s) of these changes.

TESTING YOUR UNDERSTANDING

For each of the test items below, three of the lettered alternatives are true and the other is false. Determine which alternative is false and write its letter in the blank to the left of the question number. On the blank line below alternative D, write the corrected version of the false statement.

_____ 1. A. The sum total of all populations in an ecosystem composes the community of that ecosystem.
B. Diversity is a property of populations but not of communities.
C. Succession is a property of an ecosystem but not of a population.
D. Change through time is a property of populations, communities, and ecosystems.

Correction: _____

_____ 2. A. A species' realized niche usually is narrower than that species' niche.
B. The intensity of competition encountered by a species determines the narrowness of its realized niche.
C. Character displacement can be an evolutionary result of intense competition.
D. The habitat and niche of a species are synonyms.

Correction: _____

_____ 3. A. The intensity of competition between two species increases as the amount of overlap of their niches increases.
B. Two ecologically-similar species will be in competition even if no resources are in limited supply.
C. Species involved in a commensal relationship do not compete for the same resources.
D. In commensalism, one species benefits and the other is neither harmed nor benefited.

Correction: _____

_____ 4. A. Some mutualistic relationships are obligatory.
B. A lichen is an example of both mutualism and symbiosis.
C. A hermit crab living in an abandoned mollusk shell is an example of commensalism.
D. The relationships of pollen- and nectar-feeding bats with certain plants are examples of commensalism.

Correction: _____

_____ 5. A. Termites can benefit from a diet of cellulose because of a mutualistic association with protists living within their digestive tracts.
B. Parasitism can be viewed as an example of a predator-prey relationship and a symbiosis.
C. The consumption of pecans or walnuts by humans can be viewed as examples of both herbivory and parasitism.
D. Tapeworms residing in the digestive tract of a human is an example of interspecific competition.

Correction: _____

_____ 6. A. The host species of a well-established parasite usually is not seriously damaged by their relationship.
B. Both plants and animals have evolved morphological and chemical means of defending themselves against herbivores and predators.
C. Chemical defenses in some animals involve use of chemicals that they obtained from plants.
D. Caffeine is a plant compound that thwarts herbivory by affecting the nervous systems of herbivores

Correction: _____

_____ 7. A. Alkaloids are a group of anti-herbivory compounds that appeared early in the evolution of flowering plants.
B. There is evidence that plants can communicate between individuals regarding the presence of herbivores.
C. Batrachotoxins produced by milkweed plants are used by monarch caterpillars to thwart predation by birds.
D. Elatol inhibits cell division.

Correction: _____

_____ 8. A. Primary succession occurs where previously existing plant communities have been disturbed or destroyed by agents such as fire.
B. During succession on land, grasses usually become established before woody plants become established.
C. Succession is driven by environmental changes caused by the plants themselves.
D. Flooding is an example of an allogenic force that prompts secondary succession.

Correction: _____

_____ 9. A. As an ecosystem matures, species diversity increases.
B. As an ecosystem matures, trophic relationships become more complex.
C. As an ecosystem matures, total biomass in the ecosystem decreases.
D. As an ecosystem matures, efficiency of energy transfer and of nutrient cycling increases.

Correction: _____

_____ 10. A. The monoclimax theory holds that all ecosystems in a climatic region will undergo succession towards the same climax.
B. For a supporter of the polyclimax theory, a prairie in the Midwest represents a subclimax of the oak-hickory forest.
C. Supporters of the polyclimax theory indicate that soils are an important determinant of the climax community.
D. The species-diversity index used by ecologists takes into account both the number of species present and the equitability of the occurrence of those species.

Correction: _____

CHAPTER 48--THE GEOGRAPHY OF LIFE

CHAPTER OVERVIEW

Whether you realize it or not, already you have a geographic perspective of life on Earth. For example, mention of elephants or of tigers instantly brings to mind thoughts of Africa and India. You know that not all kinds of organisms occur just anywhere, nor do any organisms occur everywhere.

In this chapter, we will examine the distributions of organisms in terrestrial and aquatic environments. We will identify the major biogeographic regions and will examine global patterns of physical factors that determine distributions of organisms. In addition to this ecological approach to explaining distributions, we also will take a historical perspective. We will see how large-scale changes in geography have affected distributions of organisms.

TOPIC SUMMARIES

A. <u>Factors Regulating Patterns of Life</u>.--Because no two species respond identically to all environmental parameters, probably each species on Earth has a unique distribution. Yet, many general patterns of distribution are evident. The species whose geographic distributions have been best studied are plants and animals. Because of their functional dependence on plants (e.g., for food, shelter, etc.), distribution patterns of animals often correspond to general patterns of plants with which the animals associate.

Distributions of plants and animals (either directly or indirectly) depend on many abiotic factors. Because of their <u>structure</u>, <u>moisture content</u>, and <u>nutrient content</u>, <u>soils</u> are important determinants of where plants can grow; fossorial (burrowing) animals can occur only where soils allow burrowing. Other critical factors include the amount and timing of <u>precipitation</u>, <u>humidity</u>, <u>temperature</u>, <u>solar radiation</u>, <u>air circulation</u>, and others. The timing and intensities of these factors depend on <u>latitude</u> (i.e., poleward distance from equator). Additionally, at any given latitude, <u>elevation</u> (i.e., height above sealevel) can significantly modify conditions suggested by latitude (Fig. 48.1). This relationship explains why temperatures in a desert might approach 40°C (>100°F), while in mountains only 25 kilometers away, but at an elevation 2,000 meters higher, snow might be falling!

B. <u>Terrestrial Communities: Biomes</u>.--Interestingly, on different continents where similar climatic, soils, and moisture conditions prevail, plant communities of similar structure have evolved. Perhaps as expected, the species in such <u>biomes</u> differ from one place to another. Yet, via <u>convergent evolution</u>, unrelated species have taken up similar functional roles in their communities.

We'll not repeat here the descriptions of major biomes. You should work with the textbook to become acquainted with characteristics and distributions of <u>tundra</u>, <u>taiga</u>, <u>temperate deciduous forests</u>, <u>tropical rain forests</u>, <u>grasslands</u>, and <u>deserts</u>.

C. <u>Aquatic Communities</u>.--Many kinds of aquatic settings exist, including oceanic, estuaries, bays, rivers, streams, ponds, lakes, etc. Such habitats have both horizontal and vertical zonation, as defined by factors such as light penetrance, motion of water, and temperature.

(1) Fresh water communities. Several zones can be recognized in many lakes and ponds (Fig 48.2). The <u>littoral zone</u> extends from water's edge to a point where depth is great enough that surface wave motion is not apparent. Open-water organisms (e.g., plankton) occupy the <u>limnetic zone</u>. In addition to the preceeding horizontal zones, vertical zones also can be

recognized: The profundal zone lies below the depth past which light does not penetrate. The thermocline is an intermediate layer of abrupt temperature change. Above it is the epilimnion and below is the hypolimnion. The depths of these zones vary with season.

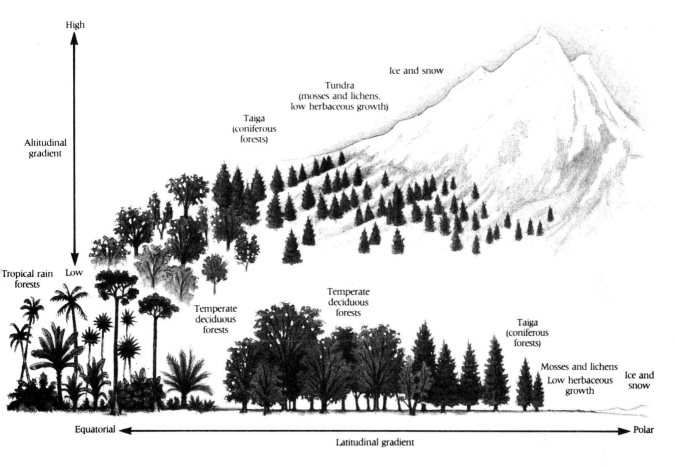

Fig. 48.1: Latitudinal and altitudinal zonation in habitats

(2) Ocean communities. Like ponds and lakes, oceans also possess statification (Fig. 48.3). The intertidal zone is that region between high and low tides. From low tide to a depth where waters are not stirred by wave action is the littoral zone. Extending from there to the edge of the continental shelf is the sublittoral zone. The preceeding zones together comprise the neritic province. The remainder is the oceanic province. Other ways of subdividing oceanic habitats are based on light penetrance and geomorphology of the ocean bottom.

Unlike most lakes, the oceans are salty. For physiological reasons, this feature effectively precludes most (but not all) fresh-water species from residing in the oceans; hence, oceanic and fresh-water floras and faunas are rather distinctive. Species of all kingdoms are present in diversity. Ocean waters circulate in a global system of warm and cold currents, flowing along eastern and western coastlines of the continents, respectively. These currents exert significant effects on climates of adjacent land masses; the Gulf Stream warms England and the California Current cools San Francisco.

As in terrestrial ecosystems, nutrients and minerals cycle. Materials that sink to the ocean bottom eventually are brought back into the biotic portions of the cycles by upwelling of cold waters, generally along western continental coasts. Areas of upwelling are extraordinarily productive.

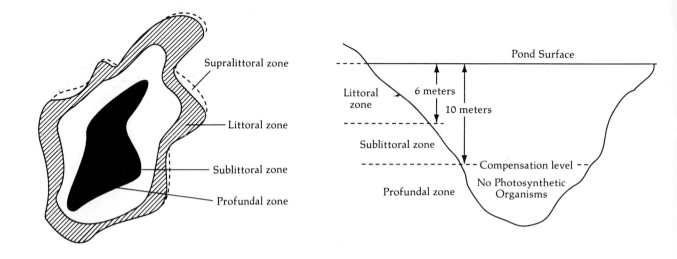

Fig. 48.2: Horizontal and vertical zonation in a lake

D. <u>Historical Biogeography</u>.--The preceeding portion of this chapter has examined distribution from an ecological perspective. Although ecological factors clearly are significant determinants of where organisms occur, events of past Earth history also help to explain modern patterns of distribution. Both ecological and historical information must be considered to understand biogeography. Historical biogeography builds explanations with the assistance of geological and palenotological evidence.

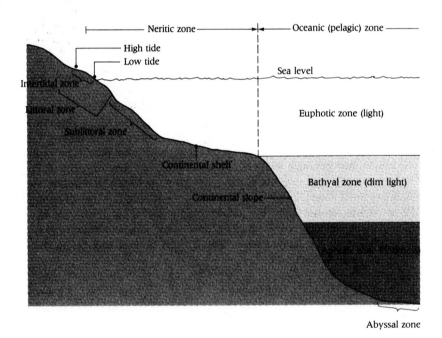

Fig. 48.3: Profile of oceanic zones

E. <u>Biogeographic Regions</u>.--Study of modern distributions of plants and of birds and mammals has facilitated subdivision of the globe into major <u>biogeographic regions</u>. Each region is characterized by at least a few <u>endemics</u>, species or groups of organisms that occur nowhere else. The seven <u>land faunal regions</u> are Nearctic (includes most of North America),

Neotropics, Palearctic, Holarctic, Oriental, Ethiopian, and Australian. Floral regions based on distribution patterns of angiosperms include the Boreal, Neotropical, Palaeotropical, Cape, Australian, and Antarctic regions.

F. Dispersal.--The entry of a species, or perhaps of entire regional faunas, into an area that it has not previously occupied is called dispersal. Dispersal involves movement of organisms, generally under their own natural powers, without requiring movements of land masses. Flying species are more vagile than are walking species than are burrowers. Hence, the extents of distributions might be expected to correspond with a species' dispersal ability. Many plants have seeds whose designs take advantage of winds and waters for dispersal. Happenchance (or sweepstakes) dispersal sometimes occurs as well, such as individuals on a floating mat of vegetation swept down a river, across an ocean, to a mid-ocean island. Mechanisms of dispersal are many!

Dispersal occurs along routes (tracks of habitat suitable to the dispersing organisms); it is inhibited by barriers (expanses of unsuitable habitat not suitable to dispersing organisms). Corridors are routes that do not ecologically resist entry and passage of the dispersers. A filter is a route that allows passage of some species, but not of others. Sweepstakes dispersal is a means of crossing expanses that otherwise are barriers.

G. Changes in Biotas as a Result of Changes in Geography.--An alternative approach to dispersal of individuals to new areas (a school referred to as dispersal biogeography) is to move new areas to the individuals such that barriers are removed (school of vicariance biogeography). Many seemingly anomalous distributions that could not be explained by dispersal have now been explained by vicariance.

The major process operating in vicariance is plate tectonics, a modernized version of continental drift (see textbook Box 48-1). Stated simply, the continents (or the plates of Earth's crust on which the continents ride) have not always been where they are now. In the interval since the Paleozoic (some 240 million years ago), Earth's land has transformed from a single supercontinent (Pangaea) to the current seven continents. The modern distributions of many organisms can now be interpreted in light of the manner in which Pangaea split apart and the routes followed by the smaller continents.

The Great American Interchange is the scenario that explains the earlier distinctness, and the later intermixing, of the mammalian faunas of North and South America. The fossil record before the Pliocene showed little similarity in faunas of these two continents; at this time the Americas were separated by an ocean trough near modern Panama. The few taxa represented on both continents during isolation must have resulted from sweepstakes dispersal. During the Plio-Pleistocene, geological events uplifted the ocean floor to establish land contact between North and South America. Using their own dispersal abilities, mammals (and other species) from each continent crossed into the other. It seems clear from this example that most biogeographic scenarios will include elements of both dispersal and vicariance.

Most continents or other smaller regions support diverse faunas and floras. 'We should note that the modern diversity of a region is the cumulative product of long periods of Earth history. For any region, probably many dispersal and vicariant events have occurred. Different events contributed subsets of species to the modern flora and fauna. Faunal and floral stratification embodies such recognition of subsets. Review the history of development of the mammalian fauna of the Americas as an example of stratificiaiton.

TERMS TO UNDERSTAND: Define and identify the following terms:

ecological biogeography

historical biogeography

plant formation

trade winds

westerlies

intertropical convergence

biome

tundra

permafrost

taiga

temperate deciduous forest

tropical rain forest

grassland

desert

estivation

hibernation

littoral zone

limnetic zone

profundal zone

epilimnion

hypolimnion

thermocline

eutrophic lake

oligotrophic lake

neritic province

oceanic province

sublittoral

intertidal zone

benthic

euphotic zone

aphotic zone

bathyal zone

abyssal zone

plankton

nekton

Nearctic region

Palearctic region

Holarctic region

Oriental region

Ethiopian region

Neotropical region

Australian region

disjunctive distribution

relicts

continental drift

Boreal region

Neotropical region

Palaeotropical region

Cape region

Australian region

Antarctic region

plate tectonics

seafloor spreading

subduction

dispersal

corridor

barrier

sweepstakes dispersal

Pangaea

Laurasia

Gondwanaland

MATCHING EXERCISE

Complete the following matching exercise to check your familiarity with various biomes. Match the descriptions in the right column with the appropriate biome in the left column. Some descriptions might apply to more than one biome.

BIOME   DESCRIPTIONS

____ 1. grassland   A. permafrost

____ 2. temperate deciduous forest   B. receives less precipitation than any other biome

____ 3. tundra   C. boreal forest

____ 4. tropic rain forest   D. generally lacks trees

____ 5. desert   E. found in areas of cold winters, warm summers and less than 40" annual precipitation

____ 6. taiga   F. has greatest species diversity of all biomes

PEOPLE TO KNOW: What is the significant biological contribution of each of these individuals as indicated in Chapter 48?

Ronald Good

F. B. Taylor

Alfred Wegener

DISCUSSION QUESTIONS

1. How is nutrient cycling in lakes and ponds functionally connected with thermal stratification in these bodies of water?

2. Compare and contrast how circulation of the atmosphere and circulation of ocean waters affect global climates.

3. Historical biogeography and ecological biogeography sometimes are presented as alternate ways of studying biogeography. Can you development the argument that these approaches actually cooperate to explain how modern distributions have come about?

4. Discuss the various means and routes by which organisms can enter new areas.

TESTING YOUR UNDERSTANDING

For each of the test items below, three of the lettered alternatives are true and the other is false. Determine which alternative is false and write its letter in the blank to the left of the question number. On the blank line below alternative D, write the corrected version of the false statement.

_____ 1. A. The distribution of life on Earth lacks uniformity.
B. Historical biogeography addresses the ecological forces underlying the distribution of plants and animals.
C. Although plants play a major role in distribution of animals, animals are not as important in affecting the distribution of plants.
D. Climate varies with latitude because solar energy is not equally distributed by latitudes.

Correction: _____

_____ 2. A. Climate varies with latitude because there are significant variations in the pattern of air currents at different latitudes.
B. Winds called the westerlies prevail between 30 and 60 degrees latitude.
C. Winds called the trade winds prevail between 30 degrees north and south; these winds blow towards the poles.
D. All of the great deserts of the world are located at about 30 degrees north and 30 degrees south.

Correction: _____

_____ 3. A. In the tropics, seasonal temperature variation is greater than daily variation in temperature.
B. Solar radiation is more intense in the tropics than in any other region of Earth.
C. At any particular latitude, temperature decreases with increasing elevation.
D. Variation in climatic factors in relation to latitude is an example of horizontal zonation.

Correction: _____

_____ 4. A. All of the grasslands on Earth form one type of biome.
B. Tundra, a vast treeless biome, occurs primarily in the southern hemisphere.
C. Sphagnum moss, lichens, and permafrost characterize tundra.
D. Taiga, known also as boreal forest, consists of vast coniderous forests.

Correction: _____

_____ 5. A. Rain forests occur in both the tropics and in the Pacific Northwest of North America.
B. Tropical rain forests support the greatest diversity of species of any terrestrial ecosystems.
C. Lianas and epiphytes occur abuddantly in temperate deciduous forests.
D. The presence of forests indicates a relatively abundant supply of water is available.

Correction: _____

_____ 6. A. Grasslands receive less precipitation than do temperate deciduous forests.
B. Soils of tropical rain forests contain relatively few nutrients.
C. In the North American Great Plains, tall-grass prairies occur to the west and short-grass prairies occur to the east.
D. C4-photosynthesis is characteristic of many desert plants.

Correction: _____

_____ 7. A. The sublittoral zone of a lake extends from the shoreline to a depth where wave action does not disturb underlying waters.
B. The thermocline is a stratum of water in which there is an abrupt change in water temperature.
C. Photosynthesic organisms generally do not occur in the profundal zone.
D. Seasonal changes in water temperature and water density are responsible for turnovers.

Correction: _____

_____ 8. A. A primary way in which nutrients are brought from the sediments to the ocean surface is by upwelling of cold waters.
B. Organisms living in sediments at the ocean bottom can be referred to as pelagic organisms.
C. Cold ocean currents flow southward along western margins of continents in the northern hemisphere.
D. Those species that swim freely in the oceans compose the nekton.

Correction: _____

_____ 9. A. Earth and its floras and faunas have changed greatly during Earth history.
B. During the Paleozoic, all major land masses on Earth were joined into a single supercontinent called Pangaea.
C. The process of continental drift is the currently accepted explanation for movements of Earth's land masses.
D. Most types of biomes can be found on most continents.

Correction: _____

_____ 10. A. Among the biogeographic regions of the southern hemisphere are the Ethiopian, Australian, Palearctic, and Neotropical regions.
B. A relict is a population of a species whose main distribution is elsewhere.
C. The botanist Ronald Good described modern floral regions on the basis of modern distribution of angiosperms.
D. Disjunctive distributions often can be explained by plate tectonics.

Correction: _____

_____ 11. A. Dispersal through a filter will allow successful invasion of new areas by some species but not by other species of the same fauna.
B. Dispersal can be the result of active effort by the dispersing organism, or can be passive.
C. A route of dispersal that allows virtually all members of a fauna to pass is called a corridor.
D. Filter dispersal is a common means by which dispersing species cross a barrier.

Correction: _____

_____ 12. A. Rafting is an example of sweepstakes dispersal.
B. The intermingling of the mammalian faunas of North and South America probably involved sweepstakes dispersal as well as movements along corridors and filter routes.
C. Australia, Africa, and India are modern continents that broke away from the supercontinent Gondwanaland.
D. The presence in Australia of marsupial versions of placental mammals found on other continents is an example of evolutionary divergence.

Correction: _____

CHAPTER 49--HUMAN POPULATIONS AND ENVIRONMENTS

CHAPTER OVERVIEW

In Chapter 1, we began our survey of biological diversity and processes by looking at life from a human perspective. In many places throughout the text, we have applied principles to human existence. We have studied examples of inheritance, diseases, behaviors, and structure and function in humans. This final chapter offers an opportunity for each of us to synthesize information and understandings gained from our earlier study.

This final chapter prompts reflection upon the role of humans in our biosphere. Here we summarize impacts that humans have had, and likely will have, on the biosphere. We address questions regarding the effects that medicine and hygiene have had on growth of the human population, regarding effects of agriculture, regarding effects of utilization of Earth's resources, and others. This chapter urges that we must be good stewards of our world.

TOPIC SUMMARIES

A. <u>Genetic Diversity Among Humans</u>.--Despite our differences, humans of different races are more alike than we are different. The processes that have produced differentiation between human populations are largely the same processes that yielded subspecies in mice or fruit flies or oak trees. Geographic isolation has been a primary determinant of genetic divergence of humans. Populations in different areas have evolved in response to different selection pressures. Differences in skin color, for example, tend to correspond to differences in the amount of solar radiation striking different areas of Earth. From previous chapters, you are aware that humans also vary in blood groups, size and shape of head, shape of nose, various biochemical traits, and others. With increased mobility, however, gene flow between formerly isolated populations has begun to decrease the degrees of differences between races.

B. <u>Growth of the Human Population</u>.--The factors that govern growth of populations of other organisms also govern growth of the human population. Study of Fig. 49.1 clearly reveals that the human population is in the <u>exponential phase</u> of growth. In 1986, Earth's population surpassed 5 billion! At what level our population growth will stop (where birth rate equals death rate; at <u>zero population growth</u>) is unknown. What is Earth's carrying capacity for humans?

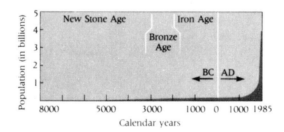

Fig. 49.1: Growth of the human population

Factors that have enabled at least some segments of Earth's population to flourish and live comfortably (i.e., in the developed contries) include human manipulation of food supplies and advances in medicine and hygiene.

C. <u>World Food Prospects</u>.--Our ability to grow crops and other foods in huge quantities has promoted rapid population growth. In countries experiencing the <u>Green Revolution</u> (where shifts from old agricultural techniques to new efficient methods occur), population booms follow. Production of foods at a greater than normal rate, of course, requires <u>subsidies</u> such as fertilizer, machinery, and automation. Some of these subsidies represent technological advances. Yet other subsidies represent "<u>loans</u>" from the world's limited supply of materials--energy (especially in form of fossil fuels), minerals (e.g., phosphorous, sulfur), soils, etc. We are using these materials more quickly than they can be replentished by natural processes. How long can we escape the consequences of this feeding strategy?

D. <u>Medicine and Hygiene</u>.--Until about the turn of the 20th century, <u>medicine</u> consisted of little more than comforting patients as illnesses ran their courses, oftentimes to death. Discovery of <u>antibiotics</u> this century has pushed infectious diseases down on the list of killers. Understanding of the causes of infectious diseases (generally, microbes) and of the biology of microorganisms has enabled us to avoid the causes of many diseases merely by practicing proper <u>hygiene</u>.

Hence, more adult humans are living longer, in part because of medications and hygiene, and in part due to our abilities to transplant many organs. Additionally, more infants survive now than in the past. As death rate decreases, population size increases even more rapidly. A relatively new concern, therefore, has become the active slowing of population growth by <u>contraceptives and abortions</u>. These and related issues stir considerable controversy, particularly in regard to religious and cultural values. Difficult decisions concerning human manipulation of birth and death rates await future generations of humans.

E. <u>Impact of Humans on the Environment</u>.--Perhaps no species has so greatly impacted its environment as have humans. We occur virtually everywhere on Earth. We leave few, if any, places untouched or unaltered. As biological entities, we might view human-caused effects as natural . . . perhaps.

Humans intensively use (in many cases, abuse) the <u>land</u>. Whether we eat the crops we grow, or whether we eat the animals that eat the crops we grow, we depend upon vegetation and, thereby, <u>we depend on the land</u>. We have destroyed many of the natural forests and prairies to develop land for agriculture. Such disturbances alter the hydrological cycle and they loosen soils that then are <u>eroded</u> by water and wind. Undoubtedly, removal of native species and replacing them with domesticated species <u>alters the intricate trophic relationships and balances</u> of natural ecosystems.

F. <u>Utilization of Resources</u>.--Utilization of natural resources is a natural activity of any species. Utilization often involves destruction. <u>Utilization and destruction</u> by humans of resources tends to exceed the rate at which the resources are capable of being renewed. <u>Wise strategies</u> would be to use resources at rates equal to (or slower than) the rate at which the resource is consumed. Our knowledge of <u>biogeochemical cycles</u> is inadequate develop such wise strategies. We use <u>soils</u> faster than they can be replentished. We burn <u>fossil fuels</u> much faster than they can be produced. Although our use of many <u>minerals</u> (e.g., copper, iron) is not destructive, we should think of these as <u>non-renewable</u> resources because their atoms are (because of human activity) so scattered as to be impractical to retrieve.

Clearly, however, <u>preservation</u> (abstaining from use of a resource) of resources would not allow the human population to grow or even to maintain status quo. The sensible middle ground is <u>judicious use and conservation</u> of resources.

TERMS TO UNDERSTAND: Define and identify the following terms:

Homo sapiens

zero population growth

Green Revolution

medicine

blood-group systems

fertility

hygiene

plant protein vs. animal protein

overfishing

antibiotics

acquired-immune-deficiency syndrome

prolongation of life

contraceptives

rehabilitation

organ transplants

productive working life

environmental impact

ecological balance

Industrial Revolution

land use

destruction of forests

draining of wetlands

irrigation of arid lands

disturbance of community organization

native species vs. introduced species

extinction

domestication

zoos

utilization of resources

preservation of resources

conservation of resources

renewable resources

nonrenewable resources

soil utilization

fertilizers

Dust Bowl

erosion

flood-control projects

solar technology

genetic engineering

preservation of biological diversity

DISCUSSION QUESTIONS

1. Beyond the esthetic reasons, why is it important to the well-being of the human population to preserve the biological diversity of Earth?

2. In many ways, all individuals of the global human population are "in the same boat" as regards the burgeoning growth of our population. However, some countries can be labelled as "haves" and others as "have nots" in terms of possessing agricultural and technological abilities to cope with the stresses of over-population. What types of social and political strains can you see developing between the "haves" and "have nots?" What might be the ultimate results of intense international strains? What might be the solutions for reducing such strains?

3. What are several ways to conserve our soils?

4. Analyze your (or your family's) lawn care and gardening program in terms of impact on rates of materials transfer and cycling.

5. Propose a way of determining the carrying capacity of Earth for *Homo sapiens*.

6. Humans have a habit of introducing non-native species (such exotic species include domestic dogs, rabbits, shrubs, trees, etc.) into areas where these species do not normally occur. What might be the ecological implications of such introductions of exotics?

7. Considering our advances in technology and medicine, do you consider humans to be subject to the pressures of natural selection? Explain why or why not.

TESTING YOUR UNDERSTANDING

For each of the test items below, three of the lettered alternatives are true and the other is false. Determine which alternative is false and write its letter in the blank to the left of the question number. On the blank line below alternative D, write the corrected version of the false statement.

_____ 1. A. The size ultimately achieved by Earth's human population will be determined by Earth's carrying capacity for this species.
  B. The human population is in the exponential phase of its growth curve.
  C. Since the beginning of this century, changes in birth rates and death rates have decreased the rate at which the human population is increasing.
  D. Successful organ transplants have the effect of decreasing the death rate of the human population.

  Correction: _____

_____ 2. A. Differentiation of the human races has been a result of geographic isolation.
B. Darkness of skin pigmentation in humans correlates with the latitudes in which evolution of the races has occurred.
C. The effect of technological advances has been to increase mobility of individuals and, thereby, to promote greater gene exchange among humans.
D. Cultural and social characteristics of various human subpopulations have not been an important barrier in genetic isolation.

Correction: _____

_____ 3. A. In the last 2,000 years of human history, the size of the global human population has increased from about 2 billion to over 5 billion.
B. Zero population growth is achieved when the rate at which new individuals are added to the population equals the rate at which individuals leave the population.
C. A current trend in human demographics is for an increasingly larger proportion of the population to be made up of elderly individuals.
D. "Green Revolutions" occur when a population or country shifts to more efficient agricultural practices that result in a significant increase in productivity.

Correction: _____

_____ 4. A. Intense use of land for agriculture usually requires subsidies in the form of energy and minerals.
B. Subsidy of agriculture with phosphorus fertilizer represents a loan from the "rock cycle" phase of a biogeochemical cycle.
C. Agricultural subsidies use resources as approximately the same rate that nature replentishes them.
D. The majority of the dietary protein in developing countries comes directly from plants, whereas the majority of dietary protein in developed countries is of animal origin.

Correction: _____

_____ 5. A. Biological diversity should be preserved because of its esthetic value.
B. Biological diversity should be preserved because further study of the biology of many species might reveal applications of those plants or of their products that can serve humankind.
C. Infectious diseases tend not to be the leading causes of death in humans.
D. Antibiotics have been used by western civilizations to fight infectious diseases for nearly four centuries.

Correction: _____

_____ 6. A. Contraceptives and abortions decrease the rate at which the human population is growing.
B. Human use of the land has decreased the rate at which soils are eroded.
C. Native forests are more effective in absorbing rainwater and at slowing runoff that are deforested areas.
D. Human activities are consuming soils at rates faster than the soils can be replaced.

Correction: _____

_____ 7. A. Rotation of crops (alternating of non-legumes with legumes) is an approach to agriculture that requires lesser subsidies of nitrogen than does continuous cultivation with non-legume crops.
   B. Agriculture interrupts natural ecological balances that exist before disturbance of natural communities.
   C. Many species of plants and animals now enjoy a greater geographic distribution because of modification of habitats by humans.
   D. All species of plants and animals now have a smaller geographic distribution because of modification of habitats by humans.

   Correction: _____

_____ 8. A. The rate at which humans are causing extinctions of species is expected to exceed the rate of extinctions during the Great Dying that ended the Cretaceous Period.
   B. The human population of Ethiopia has exceeded the carrying capacity of its immediate environment.
   C. Preservation refers to practices of sensible use of environmental resources.
   D. Excessive exploitation of natural resources by any species likely will lead to decreases in population sizes.

   Correction: _____

_____ 9. A. Metals, such as copper and iron, are renewable resources.
   B. Resources that operate in cycles are renewable.
   C. The wisest utilization of non-renewable resources is to use them as slowly as possible and with the least possible waste.
   D. Although an abundant long-term source of energy, the Sun is not an inexhaustible source of energy.

   Correction: _____

_____ 10. A. Fossil fuels and arable lands are finite resources.
   B. Preservation of other species probably will have no bearing on the future well-being of humans.
   C. Soils are a key factor in supplying food.
   D. Future well-being of humans probably will depend on many factors, such as education on environmental issues, solar technology, and genetic engineering.

   Correction: _____

ANSWERS TO MATCHING QUESTIONS

Chapter 1
1. A, E, F   2. C, E   3. A, D, E   4. C, E   5. E, G
6. A, B, E   7. A, B, E

Chapter 2
1. M   2. C   3. N   4. I   5. K
6. F   7. B   8. A   9. E   10. L
11. J   12. G   13. H

Chapter 5
Matching A
1. F   2. G   3. I   4. J   5. A
6. B   7. D   8. E   9. H   10. C
Matching B
1. A   2. E   3. B   4. D   5. C

Chapter 7
1. F   2. E   3. A   4. H   5. C
6. B   7. G   8. I   9. D

Chapter 11
1. C   2. D   3. E   4. B   5. A

Chapter 19
1. A   2. B   3. C   4. B   5. B
6. B   7. B   8. A   9. C   10. A

Chapter 24
Matching A
1. C   2. D   3. A   4. E   5. B
Matching B
1. B   2. D   3. E   4. C   5. A

Chapter 26
1. B   2. D   3. A   4. C

ANSWERS TO TESTING YOUR UNDERSTANDING

Chapter 1
1. B   2. D   3. A   4. B   5. D
6. D   7. B   8. C   9. B

Chapter 2
1. D   2. C   3. A   4. A   5. A
6. B   7. C

Chapter 3
1. C   2. A   3. B   4. D   5. B
6. D   7. A   8. C   9. D   10. C
11. A   12. C   13. B   14. B   15. B
16. D   17. A   18. A   19. A   20. D

Chapter 4
1. D   2. B   3. C   4. B   5. D
6. A   7. C   8. D   9. A   10. C
11. A   12. B   13. D   14. B

Chapter 5
1. D   2. B   3. A   4. D   5. B
6. A   7. C   8. D   9. D   10. B
11. A   12. A   13. C   14. D   15. B

Chapter 6
1. D   2. B   3. B   4. A   5. D
6. A   7. C   8. A   9. A   10. B
11. B   12. C

Chapter 7
1. B   2. C   3. A   4. C   5. B
6. A   7. D   8. D   9. B

Chapter 8
1. D   2. B   3. C   4. B   5. D
6. D   7. D   8. B   9. A   10. A
11. C   12. B   13. D   14. B   15. A

Chapter 9
1. D   2. B   3. A   4. C   5. A
6. D   7. C   8. C   9. B   10. A
11. A   12. D

Chapter 10
1. C   2. A   3. A   4. B   5. D
6. B   7. A   8. D   9. A   10. D
11. C   12. A   13. B   14. D   15. B

Chapter 11
1. D   2. B   3. A   4. D   5. B
6. C   7. D   8. C

Chapter 12
1. A   2. D   3. C   4. D   5. C
6. B   7. A   8. C   9. B   10. A
11. C   12. A   13. C   14. A   15. C

Chapter 13
1. B   2. A   3. B   4. A   5. D
6. A

Chapter 14
1. C   2. D   3. A   4. C   5. A
6. A   7. A   8. B   9. D   10. C

Chapter 15
1. B   2. D   3. A   4. A   5. C
6. D   7. A   8. D   9. B   10. B

Chapter 16
1. D   2. A   3. B   4. A   5. D
6. A   7. D   8. A   9. C   10. D
11. B   12. D

Chapter 17
1. A   2. B   3. A   4. D   5. B
6. B   7. A   8. D   9. D   10. B
11. A   12. B

Chapter 18
1. A   2. D   3. C   4. D   5. C
6. C   7. B   8. D   9. D   10. B
11. A

Chapter 19
1. A   2. D   3. C   4. B   5. A
6. C   7. B   8. D   9. A   10. D
11. C   12. B

Chapter 20
1. C   2. B   3. D   4. D   5. C
6. B   7. D   8. A   9. C   10. B

Chapter 21
1. B   2. D   3. A   4. C   5. D
6. D   7. A   8. A   9. B   10. C
11. D   12. A   13. C   14. D   15. B

Chapter 22
1. A   2. A   3. C   4. D   5. B
6. B   7. D   8. D   9. A   10. C
11. D   12. B   13. D   14. C   15. A

Chapter 23
1. C   2. D   3. D   4. D   5. B
6. C   7. D   8. B   9. C   10. A
11. C

Chapter 24
1. A   2. C   3. A   4. C   5. D
6. D   7. C   8. A   9. D   10. A
11. C   12. C   13. A   14. B

Chapter 25
1. B   2. C   3. A   4. D   5. B
6. C   7. A   8. A   9. A   10. D

Chapter 26
1. C   2. B   3. A   4. D   5. B
6. D   7. A   8. B   9. D   10. B
11. C   12. A

Chapter 27
1. D   2. B   3. C   4. D   5. B
6. A   7. D   8. C   9. A   10. B

Chapter 28
1. C   2. B   3. A   4. D   5. B
6. A   7. D   8. D   9. B   10. D
11. B   12. A   13. C   14. A   15. C

Chapter 29
1. B   2. B   3. A   4. D   5. C
6. A   7. A   8. D   9. C   10. D

Chapter 30
1. D   2. B   3. A   4. C   5. C
6. C   7. A   8. C   9. A   10. D
11. B

Chapter 31
1. B   2. A   3. B   4. D   5. B
6. D   7. C   8. D   9. B   10. A

Chapter 32
1. D   2. B   3. D   4. B   5. B
6. A   7. C   8. A   9. D   10. D

Chapter 33
1. D   2. C   3. D   4. B   5. A
6. D   7. A   8. B   9. B   10. D

Chapter 34
1. A   2. A   3. C   4. D   5. B
6. C   7. C   8. D   9. A   10. D

Chapter 35
1. C   2. B   3. A   4. D   5. C
6. C   7. A   8. D   9. D

Chapter 36
1. D   2. B   3. A   4. D   5. A
6. C   7. A   8. A   9. D   10. B

Chapter 37
1. B   2. C   3. A   4. C   5. D
6. D   7. A   8. A   9. B   10. A

Chapter 38
1. B   2. A   3. C   4. C   5. D
6. C   7. B   8. A   9. C   10. D

Chapter 39
1. A   2. D   3. D   4. C   5. A
6. B   7. B   8. C   9. A   10. C
11. D   12. A   13. B   14. D

Chapter 40
1. A   2. B   3. D   4. A   5. C
6. B   7. B   8. D   9. C   10. B

Chapter 41
1. A   2. D   3. D   4. A   5. C
6. A   7. C   8. C   9. D   10. D
11. A   12. A   13. B   14. D   15. C

Chapter 42
1. C   2. D   3. B   4. D   5. A
6. B   7. A   8. D   9. D   10. A
11. D   12. C   13. B   14. A

Chapter 43
1. A   2. B   3. A   4. B   5. D
6. C   7. C   8. B   9. C   10. B
11. D   12. C   13. C   14. B   15. A

Chapter 44
1. A   2. C   3. D   4. A   5. A
6. C   7. B   8. C   9. D   10. A
11. B   12. D

Chapter 45
1. B   2. A   3. B   4. C   5. C
6. C   7. D   8. B   9. A   10. B

Chapter 46
1. D   2. A   3. C   4. D   5. B
6. A   7. C   8. B   9. D   10. A

Chapter 47
1. B   2. D   3. B   4. D   5. B
6. D   7. C   8. A   9. C   10. B

Chapter 48
1. B   2. C   3. A   4. D   5. C
6. C   7. A   8. B   9. C   10. A
11. D 12. D

Chapter 49
1. C   2. D   3. A   4. C   5. D
6. B   7. D   8. C   9. A   10. B